135
Advances in Polymer Science

Editorial Board:
A. Abe · A.-C. Albertson · H.-J. Cantow · K. Dušek
S. Edwards · H. Höcker · J. F. Joanny · H.-H. Kausch
T. Kobayashi · K.-S. Lee · J. E. McGrath
L. Monnerie · S. I. Stupp · U. W. Suter
E. L. Thomas · G. Wegner · R. J. Young

Springer
Berlin
Heidelberg
New York
Barcelona
Budapest
Hong Kong
London
Milan
Paris
Santa Clara
Singapore
Tokyo

Blockcopolymers
Polyelectrolytes
Biodegradation

With contributions by
V. Bellon-Maurel, A. Calmon-Decriaud,
V. Chandrasekhar, N. Hadjichristidis, J. W. Mays,
S. Pispas, M. Pitsikalis, F. Silvestre

Springer

This series presents critical reviews of the present and future trends in polymer and biopolymer science including chemistry, physical chemistry, physics and materials science. It is addressed to all scientists at universities and in industry who wish to keep abreast of advances in the topics covered.

As a rule, contributions are specially commissioned. The editors and publishers will, however, always be pleased to receive suggestions and supplementary information. Papers are accepted for „Advances in Polymer Science" in English.

In references Advances in Polymer Science is abbreviated Adv. Polym. Sci. and is cited as a journal.

Springer WWW home page: http://www.springer.de

ISSN 0065-3195
ISBN 3-540-63156-9
Springer-Verlag Berlin Heidelberg New York

Library of Congress Catalog Card Number 61642

This work is subject to copyright. All rights are reserved, whether the whole or part of the material is concerned, specifically the rights of translation, reprinting, re-use of illustrations, recitation, broadcasting, reproduction on microfilms or in other ways, and storage in data banks. Duplication of this publication or parts thereof is only permitted under the provisions of the German Copyright Law of September 9, 1965, in its current version, and permission for use must always be obtained from Springer-Verlag. Violations are liable for prosecution under the German Copyright Law.

© Springer-Verlag Berlin Heidelberg 1998
Printed in Germany

The use of registered names, trademarks, etc. in this publication does not imply, even in the absence of a specific statement, that such names are exempt from the relevant protective laws and regulations and therefore free for general use.

Typesetting: Data conversion by MEDIO, Berlin
Cover: E. Kirchner, Heidelberg
SPIN: 10573649   02/3020 - 5 4 3 2 1 0 - Printed on acid-free paper

Editorial Board

Prof. Akihiro Abe
Department of Industrial Chemistry
Tokyo Institute of Polytechnics
1583 Iiyama, Atsugi-shi 243-02, Japan
E-mail: aabe@chem.t-kougei.ac.jp

Prof. Ann-Christine Albertson
Department of Polymer Technology
The Royal Institute of Technolgy
S-10044 Stockholm, Sweden
E-mail: aila@polymer.kth.se

Prof. Hans-Joachim Cantow
Freiburger Materialforschungszentrum
Stefan Meier-Str. 21
D-79104 Freiburg i. Br., FRG
E-mail: cantow@fmf.uni-freiburg.de

Prof. Karel Dušek
Institute of Macromolecular Chemistry, Czech Academy of Sciences of the Czech Republic
Heyrovský Sq. 2
16206 Prague 6, Czech Republic
E-mail: office@imc.cas.cz

Prof. Sam Edwards
Department of Physics
Cavendish Laboratory
University of Cambridge
Madingley Road
Cambridge CB3 OHE, UK
E-mail: sfe11@phy.cam.ac.uk

Prof. Dr. Hartwig Höcker
Lehrstuhl für Textilchemie
und Makromolekulare Chemie
RWTH Aachen
Veltmanplatz 8
D-52062 Aachen, FRG
E-mail: 100732.1557@compuserve.com

Prof. J. F. Joanny
Institute Charles Sadron
6, rue Boussingault
F-67083 Strasbourg Cedex, France
E-mail: joanny@europe.u-strasbg.fr

Prof. Hans-Henning Kausch
Laboratoire de Polymères
École Polytechnique Fédérale
de Lausanne, MX-C Ecublens
CH-1015 Lausanne, Switzerland
E-mail: hans-henning.kausch@lp.dmx.epfl.ch

Prof. T. Kobayashi
Institute for Chemical Research
Kyoto University
Uji, Kyoto 611, Japan
E-mail: kobayash@eels.kuicr.kyoto-u.ac.jp

Prof. Kwang-Sup Lee
Department of Macromolecular Science
Hannam University
Teajon 300-791, Korea
E-mail: kslee@eve.hannam.ac.kr

Prof. J. E. McGrath
Polymer Materials and Interfaces Laboratories
Virginia Polytechnic and State University
2111 Hahn Hall
Blacksbourg
Virginia 24061-0344, USA
E-mail: jmcgrath@chemserver.chem.vt.edu

Prof. Lucien Monnerie
École Supérieure de Physique et de Chimie Industrielles
Laboratoire de Physico-Chimie
Structurale et Macromoléculaire
10, rue Vauquelin
75231 Paris Cedex 05, France
E-mail: lucien.monnerie@espci.fr

Prof. Samuel I. Stupp
Department of Materials Science
and Engineering
University of Illinois at Urbana-Champaign
1304 West Green Street
Urbana, IL 61801, USA
E-mail: s-stupp@uiuc.edu

Prof. U. W. Suter
Department of Materials
Institute of Polymers
ETZ,CNB E92
CH-8092 Zürich, Switzerland
E-mail: suter@ifp.mat.ethz.ch

Prof. Edwin L. Thomas
Room 13-5094
Materials Science and Enginering
Massachusetts Institute of Technology
Cambridge, MA 02139, USA
E-mail. thomas@uzi.mit.edu

Prof. G. Wegner
Max-Planck-Institut für Polymerforschung
Ackermannweg 10
Postfach 3148
D-55128 Mainz, FRG
E-mail: wegner@mpip-mainz.mpg.de

Prof. R. J. Young
Manchester Materials Science Centre
University of Manchester and UMIST
Grosvenor Street
Manchester M1 7HS, UK
E-mail: r.young@fs2.mt.umist.ac.uk

Contents

Nonlinear Block Copolymer Architectures
M. Pitsikalis, S. Pispas, J. W. Mays, N. Hadjichristidis 1

Polymer Solid Electrolytes: Synthesis and Structure
V. Chandrasekhar 139

Standard Methods for Testing the Aerobic Biodegradation of Polymeric Materials. Review and Perspectives
A. Calmon-Decriaud, V. Bellon-Maurel, F. Silvestre 207

Author Index Volumes 101 – 135 227

Subject Index 235

Nonlinear Block Copolymer Architectures

Marinos Pitsikalis[1,2], Stergios Pispas[1,2], Jimmy W. Mays[1], Nikos Hadjichristidis[*,2]

[1] Department of Chemistry, University of Alabama at Birmingham, Birmingham, AL 35294
[2] Department of Chemistry, University of Athens, Panepistimiopolis, Zografou, 157 71 Athens, Greece
[*] Also affiliated with Institute of Structure and Laser, FORTH, Heraklion, Crete, Greece

The synthesis and bulk and solution properties of block copolymers having nonlinear architectures are reviewed. These materials include star-block copolymers, graft copolymers, miktoarm star copolymers, and complex architectures such as umbrella polymers and certain dendritic macromolecules. Emphasis is placed on the synthesis of well-defined, well-characterized materials. Such polymers serve as model materials for understanding the effects of architecture on block copolymer self-assembly, in bulk and in solution.

List of Symbols and Abbreviations		2
1	Introduction	4
2	Synthesis	4
2.1	Star-Block Copolymers	4
2.2	Graft Copolymers	16
2.2.1	Grafting "Onto" Methods	20
2.2.2	Grafting "From" Methods	25
2.2.3	Graft Copolymers via Macromonomers	35
2.3	Miktoarm Star Polymers	79
2.3.1	The Chlorosilane Method	79
2.3.2	The Divinylbenzene Method	85
2.3.3	1,1-Diphenylethylene Derivative Method	86
2.3.4	Other Methods Using Anionic Polymerization	88
2.3.5	Living Cationic Polymerization Method	93
2.4	Other Architectures	93
3	Properties	111
3.1	In Solution	111
3.1.1	Theory	111
3.1.2	Experiment	114
3.2	In Bulk	119
3.2.1	Theory	119
3.2.2	Experiment	123

4 Concluding Remarks 129

5 References 129

List of Symbols and Abbreviations

α	ratio of outer block molecular weight to that of inner block molecular weight
ε	asymmetry parameter
χ	Flory-Huggins interaction parameter
A_2	second virial coefficient
Aam	acrylamide
AHS	aluminum hydrogen sulfate
AIBN	N, N'-azobisisobutyronitrile
B	butadiene
D	diffusion coefficient
D_3	hexamethyltrisiloxane
DCC	dicyclohexylcarbodiimide
DDPE	1,3-bis(1-phenylethyl)benzene
DEF	diethylfumarate
DMS	dimethylsiloxane
dn/dc	specific refractive index increment
DPE	1,1-diphenylethylene
DSC	differential scanning calorimetry
DVB	divinylbenzene
ECH	epichlorohydrin
EHMA	ethylhexylmethacrylate
EO	ethylene oxide
f	volume fraction
f_A	number of arms
$<G_{AXBY}^2>$	distance between the centers of mass of the two homopolymer A and B of the star
$<G_i^2>$	distance between the star center and the center of mass of the i homopolymer branch in a symmetric star homopolymer with the same number of branches as the miktoarm star
HEMA	2-hydroxyethylmethacrylate
HOBT	1-hydroxybenzotriazole
I	isoprene
I	polydispersity index (M_w/M_n)
IBVE	isobutyl vinyl ether
IR	infrared spectroscopy
LDA	lithium diisopropylamide
MAC	methacryloyl chloride
MAN	maleic anhydride
MeOz	2-methyloxazoline
MMA	methyl methacrylate
M_n	number-average molecular weight

MO	membrane osmometry
MOM	-CH$_2$OCH$_3$
MPEG	poly(ethylene glycol) monomethylether
M$_w$	weight-average molecular weight
n	refractive index
n	number of arms in miktoarm starblock copolymers
N	degree of polymerization
n$_A$, n$_B$...	number of arms of a particular type
NMR	nuclear magnetic resonance
n$_s$	refractive index of solvent
OBDD	ordered bicontinuous double diamond
P2VP	poly(2-vinylpyridine)
P4VP	poly(4-vinylpyridine)
PB	polybutadiene
PBuMA	poly(butyl methacrylate)
PDMS	poly(dimethylsiloxane)
PEG	poly(ethylene glycol)
PEO	poly(ethylene oxide)
PI	polyisoprene
PIB	polyisobutylene
PMMA	poly(methyl methacrylate)
POX	polyoxazoline
PPO	poly(propylene oxide)
PS	polystyrene
PtBS	poly(*tert*-butylstyrene)
PtBuA	poly(*tert*-butylacrylate)
PtBuMA	poly(*tert*-butylmethacrylate)
PTHF	poly(tetrahydrofuran)
PVA	poly(vinyl alcohol)
PVCz	poly(*n*-vinyl carbazole)
PVPr	poly(vinyl propionate)
q	scattering vector
r$_1$, r$_2$	copolymerization reactivity ratios
Rg	radius of gyration
Rh	hydrodynamic radius
RI	refractive index
RV	viscometric radius
S	styrene
$<S^2>$	mean-square radius of gyration
SAXS	small-angle X-ray scattering
SEC	size exclusion chromatography
TEM	transmission electron microscopy
THF	tetrahydrofuran
TMEDA	*N,N,N',N'*-tetramethylethylene diamine
UV	ultraviolet
V$_i$	volume displaced by arms A or B
VS	4-vinylphenyldimethylsiloxane

Z-LYSNCA   Z-L-lysine-N-carboxyanhydride
ℓ_A, ℓ_B chain packing parameter
δ_G dimensionless ratio expressing quantitatively the effects of heterointeractions between unlike units on the conformational properties of copolymers
ϕ Flory hydrodynamic parameter relating intrinsic viscosity to the radius of gyration

1
Introduction

Intense commercial and academic interest in block copolymers developed during the 1960s and continues today. These materials attract the attention of industry because of their potential for application as thermoplastic elastomers, "tough" plastics, compatibilizing agents for polymer blends, agents for surface and interface modification, polymer micelles, etc. Academic interest arises, primarily, from the use of these materials as model copolymer systems where effects of thermodynamic incompatibility of the two (or more) components on properties in bulk and solution can be probed. The synthesis, characterization, and properties of classical *linear* block copolymers (AB diblocks, ABA triblocks, and segmented (AB)n systems) have been well documented in a number of books and reviews [1-7] and will not be discussed herein except for the sake of comparison.

The purpose of the present article is to review the synthesis and properties of nonlinear block copolymers with emphasis on synthetic aspects. These include graft and star-block copolymers, for which a number of excellent earlier reviews already exist [4, 5, 8], as well as extremely interesting novel structures made possible by recent advances in synthetic techniques. The latter include miktoarm stars, H-shaped and super-H polymers, umbrella polymers, dendrimers, and cyclic block copolymers. Our focus throughout is on synthesis and properties of well-defined, well-characterized systems. For this reason, most of our attention to graft copolymers deals with macromonomer methods, which allow the synthesis of reasonably well-defined graft copolymers incorporating an extremely wide range of monomers. The synthesis and study of such materials are essential for developing our understanding of how macromolecular architecture influences block copolymer properties. Theoretical and experimental results on solution and bulk properties of nonlinear block copolymers are also reviewed, and issues requiring additional study are identified.

2
Synthesis

2.1
Star-Block Copolymers

Star-block copolymers can be envisioned as star polymers where each arm is actually a diblock or a triblock copolymer. The presence of a central connecting point in this kind of macromolecule is expected to bring about differences in the properties of the material compared with the linear diblock and triblock copolymers, as has been proven over the last two or so decades.

Many different approaches have been used to synthesize star-block copolymers including anionic, cationic, radical, and condensation polymerization techniques, and even combinations of them [9]. The majority of the molecules produced thus far were prepared by anionic polymerization procedures. The dominant way of preparing star-block copolymers by anionic polymerization is the coupling of preformed diblock or triblock "living" copolymer chains with a suitable compound to produce the central linking point. In this way divinylbenzene (DVB) was first used in order for a central core to be created [10]. This was achieved by adding a predetermined amount of the divinyl compound to a solution of living diblock chains (Scheme 1).

Copolymerization of DVB results in formation of the core due to the existence of two polymerizable double bonds in the same molecule and the production of a network of small dimensions, which still bears active anionic sites. Deactivation with methanol produces the stable star-block material. By this method star molecules with a relatively broad distribution in the number of arms are prepared due to the statistical nature of the last step (DVB polymerization). In other words, different molecules will have different numbers of arms although the average functionality (f_A) of the stars can be controlled, primarily, by the ratio [DVB]/[chain end], the number of arms being increased by increasing this ratio. The average functionality is also determined by other factors, like the overall concentration of living ends, the molecular weight of the arm, and the reaction time and temperature. The molecular weight distributions of the star molecules are relatively narrow ($I \leq 1.2$, where I is the ratio of the weight-average, M_w, to number-average, M_n, molecular weights) and the amount of unlinked arms can be very small (sometimes less than 5%), although fractionation can be employed for the isolation of the star material. Star-block copolymers having as many as 29 arms per molecule have been prepared [11,12]. Since then a variety of star-block copolymers with different kinds of blocks have been synthesized with this method [13-15].

A better way to synthesize star molecules with a precisely defined number of arms is the deactivation of the living arms with an electrophilic linking agent (usually a chlorosilane compound) having the desired number of reactive bonds, equal to the target functionality of the star-block copolymer. Chlorosilane compounds with numbers of Si-Cl bonds ranging from 3 to 18 have been employed, giving star-block copolymers with controlled functionalities [16] (Scheme 2).

In this case, an excess of living diblock arm is used to ensure complete reaction and the star material is isolated by fractionation. Star-block copolymers with very narrow molecular weight distributions have been prepared in this way ($I < 1.1$). These materials also exhibit a well defined functionality, close to the func-

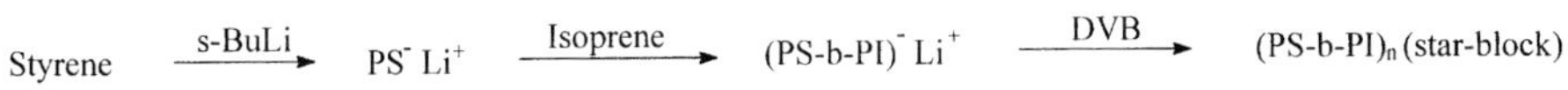

Styrene $\xrightarrow{\text{s-BuLi}}$ PS$^-$ Li$^+$ $\xrightarrow{\text{Isoprene}}$ (PS-b-PI)$^-$ Li$^+$ $\xrightarrow{\text{DVB}}$ (PS-b-PI)$_n$ (star-block)

Scheme 1

Scheme 2

Styrene $\xrightarrow{\text{s-BuLi}}$ PS$^-$ Li$^+$ $\xrightarrow{\text{Isoprene}}$ (PS-b-PI)$^-$ Li$^+$ $\xrightarrow{\text{CH}_3\text{SiCl}_3}$ CH$_3$Si(PS-b-PI)$_3$

s-BuLi + styrene $\longrightarrow$ PS$^-$ Li$^+$ $\xrightarrow[\text{benzene/TEA}]{\text{BD}}$ PS-PB$^-$ Li$^+$

$\xrightarrow{1/3\ \text{CH}_3\text{SiCl}_3}$ (PS-PB)$_3$SiCH$_3$ $\xrightarrow[\text{catalyst}]{\text{H}_2}$ (PS-PEB)$_3$SiCH$_3$

$\xrightarrow[\text{acetylsulfate}]{\text{CH}_2\text{Cl}_2}$ [(CH$_2$–CH)$_x$(PEB)$_y$]$_3$SiCH$_3$
(with pendant C$_6$H$_4$–SO$_3$H group)

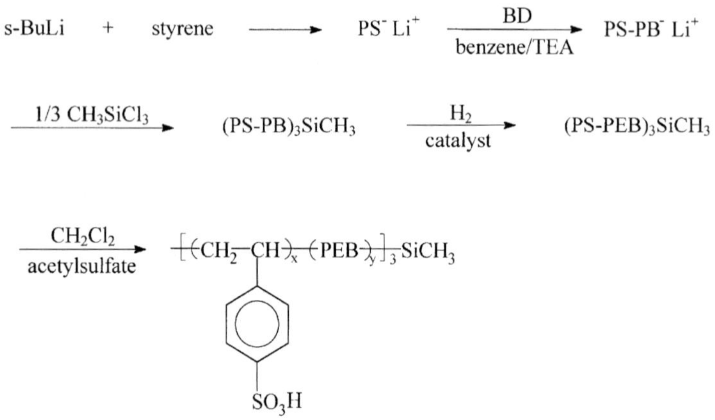

PEB = ethylene/1-butene copolymer

Scheme 3

tionality of the chlorosilane compound, as was determined from molecular weight characterization of the samples by size exclusion chromatography (SEC), ^{1}H- and ^{13}C-NMR, light scattering, and membrane osmometry.

Ionic star-block copolymers having three arms and short ionic outer blocks were synthesized by oligomerization of styrene followed by polymerization of butadiene and linking of the living polybutadiene (PB) ends with methyltrichlorosilane [17]. Hydrogenation of the polydiene blocks and sulfonation of the styrene segments gave the desired ionic star-block copolymers, with outer blocks bearing the sulfonate groups and hydrocarbon inner blocks (Scheme 3).

Four arm star-block copolymers with flexible inner blocks and rigid rod outer blocks have also been synthesized by a combination of anionic and condensation polymerization [18]. In the first step poly(dimethylsiloxane) (PDMS) flexible chains were produced by anionically polymerizing cyclic hexamethyltrisiloxane (D$_3$) with a blocked amine initiator derived from p-N,N-bis(trimethylsilyl)aminostyrene. The living polymers were reacted with tetrachlorosilane to produce 4-arm star PDMS of variable molecular weights and narrow molecular weight distributions. Deprotection of the amine end group by treatment with dilute aqueous hydrochloric acid solution gave aminotelechelic 4-arm star polymers. In the next step poly(p-benzamide) rigid blocks were grown from the aromatic amine end groups of the telechelic PDMS star polymers via the Yamazaki reaction, generating the flexible-rigid rod star-block copolymers [18]. Due to the synthetic procedure used, the characteristics of the end rigid blocks (molecular weight, polydispersity, etc.) are not well controlled and were not rigorously characterized.

Recently the synthesis and characterization of a new star-block architecture, the inverse star-block copolymer, was reported [19]. These 4-arm star molecules, with poly(styrene-b-isoprene) arms, have two of the arms connected to the silicon atom by the polystyrene (PS) end of the diblock arm while the other two are connected to the central point by the polyisoprene (PI) end (Scheme 4).

$$\text{Isoprene} + \text{s-BuLi} \xrightarrow{\text{benzene}} \text{PI}^-\text{Li}^+ \xrightarrow[\text{THF (trace)}]{\text{styrene}} \text{PI-PS}^-\text{Li}^+ \quad \text{(I)}$$

$$\text{styrene} + \text{s-BuLi} \xrightarrow{\text{benzene}} \text{PS}^-\text{Li}^+ \xrightarrow{\text{isoprene}} \text{PS-PI}^-\text{Li}^+ \quad \text{(II)}$$

$$2\text{(I)} + \text{SiCl}_4 \xrightarrow[\text{SEC Monitoring}]{\text{titration}} (\text{PI-PS})_2\text{SiCl}_2 + 2\text{LiCl}$$
$$\text{(III)}$$

$$\text{(III)} + \text{excess (II)} \longrightarrow (\text{PI-PS})_2\text{Si}(\text{PI-PS})_2 + 2\text{LiCl}$$

Scheme 4

The reaction route involves the slow addition of the two first arms to the SiCl$_4$ solution. These arms have a bulky polystyryllithium reactive end, and steric effects strongly favor the formation of coupled products rather than stars under the reaction conditions. Removal of samples and immediate analysis with SEC enabled the determination of the point where exactly two arms had reacted with the silicon linking agent. By subsequent introduction of the other kind of arms, bearing the polyisoprenyllithium reactive ends, complete substitution of the remaining Si-Cl bonds was achieved. In order to minimize steric hindrance effects the PI$^-$Li$^+$ ends were capped with a few units of butadiene. The inverse star-blocks were separated from the excess of the second kind of arms by fractionation. Molecular characterization of the isolated products by SEC with UV (ultraviolet) and RI (refractive index) detectors, ^{1}H- and ^{13}C-NMR spectroscopy, light scattering, differential refractometry and membrane osmometry showed high degrees of molecular weight and compositional homogeneity as well as close agreement between the expected (based on the synthetic procedure) and measured molecular characteristics.

Living cationic polymerization techniques are also capable of producing well defined star-block copolymers. An approach similar to the DVB method described above for the case of anionic polymerization was employed in order to prepare amphiphilic star-block copolymers [20]. In one case, living diblock copolymers of vinyl ethers and ester-containing vinyl ethers, prepared by the initiating system HI/ZnI$_2$ in toluene, were reacted with a small amount of a difunctional vinyl ether to produce star shaped block copolymers (Scheme 5).

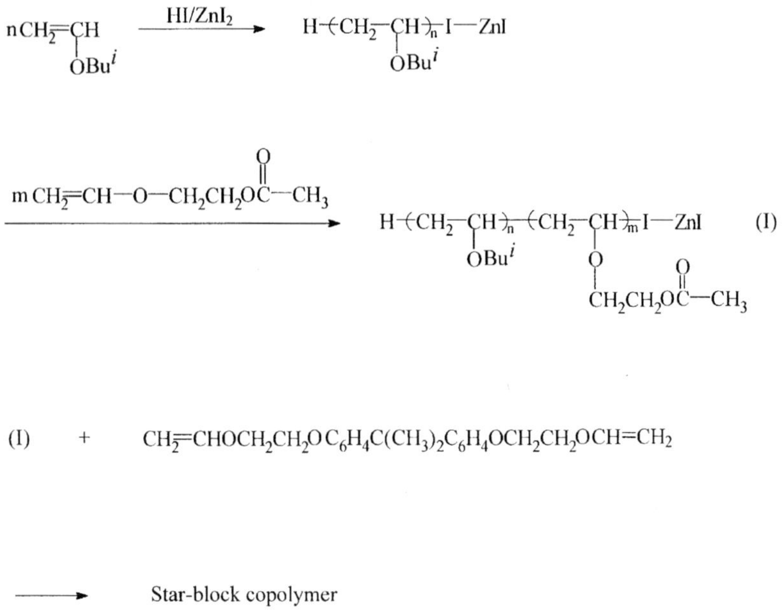

$$n\,CH_2{=}CH\text{-}OBu^i \xrightarrow{HI/ZnI_2} H{-}(CH_2{-}CH(OBu^i))_n{-}I{-}ZnI$$

$$m\,CH_2{=}CH\text{-}O\text{-}CH_2CH_2O\overset{O}{\underset{\|}{C}}\text{-}CH_3 \longrightarrow H{-}(CH_2{-}CH(OBu^i))_n{-}(CH_2{-}CH(OCH_2CH_2OC(O)CH_3))_m{-}I{-}ZnI \quad (I)$$

$$(I) + CH_2{=}CHOCH_2CH_2O\,C_6H_4C(CH_3)_2C_6H_4OCH_2CH_2OCH{=}CH_2$$

$$\longrightarrow \text{Star-block copolymer}$$

Scheme 5

Depending on the sequence of addition of the monomers during the preparation of the diblock arms, the ester-containing vinyl ether blocks were placed in the outer or in the core region of the star. Subsequent alkaline hydrolysis of these ester functions gave amphiphilic star-block copolymers with polyalcohol hydrophilic segments in the outer or the core region, respectively. Analysis by SEC showed that the molecular weight distributions of the star polymers were somehow broader than those of the diblock arms. The solubility behavior of these amphiphilic copolymers was very much dependent on the detailed arrangement of the two different kinds of blocks in the stars [20].

In a similar manner, star-block copolymers of trimethylcellulose-6-polyoxytetramethylene were obtained [21]. Partial cleavage of trimethylcellulose, in the presence of dry HCl, produced chloromethylene end groups in the polymer. Activation of these groups with AgSbF$_6$ initiated the living cationic polymerization of tetrahydrofuran (THF). The resulting active diblock arms were linked using low molecular weight poly(4-vinylpyridine) (P4VP) to form essentially star-shaped block copolymers (Scheme 6).

Obviously, due to the statistical nature of the cleavage reaction and the linking reaction, the polydispersity of the arm material with respect to molecular weight was high and the number of arms in the star-block not well controlled. However, this synthetic scheme is a nice example, demonstrating the production of thermoplastic elastomers from natural resources.

In another approach, Storey and coworkers used a trifunctional cationic polymerization initiator, derived from tricumylchloride, to prepare 3-arm star-block copolymers with polyisobutylene (PIB) inner blocks and PS outer blocks [22] (Scheme 7).

Titanium chlorides were used as a co-initiator and pyridine as an electron donor. Due to chain transfer reactions, which become significant in these sys-

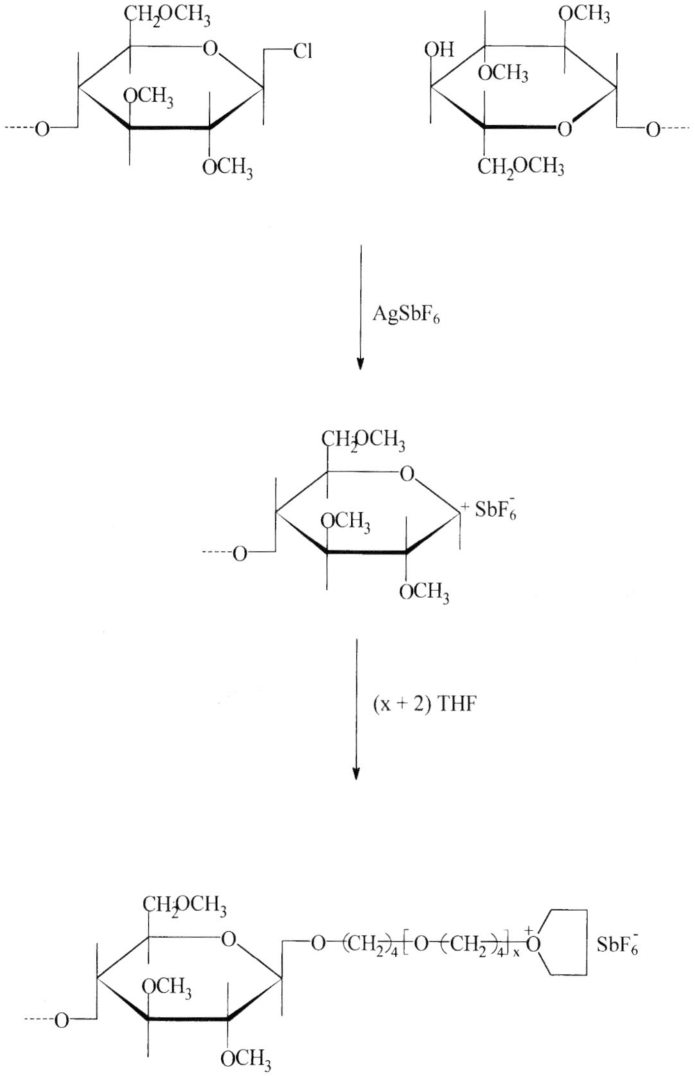

Scheme 6

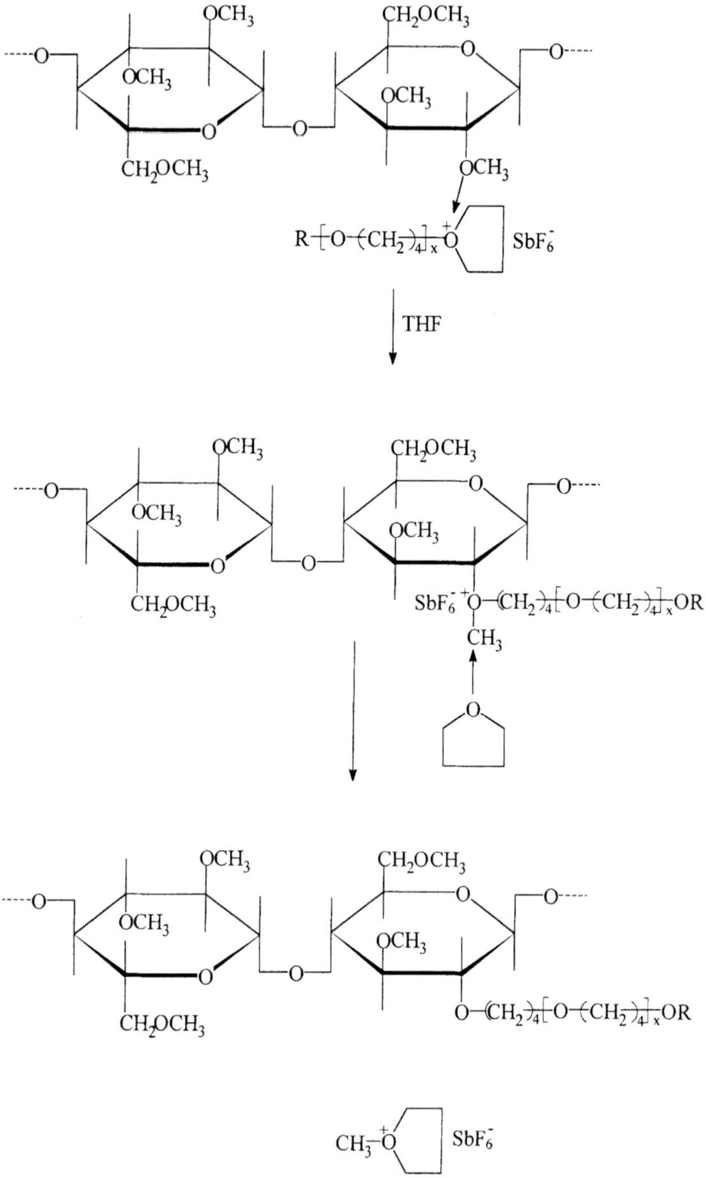

Scheme 6. (continued)

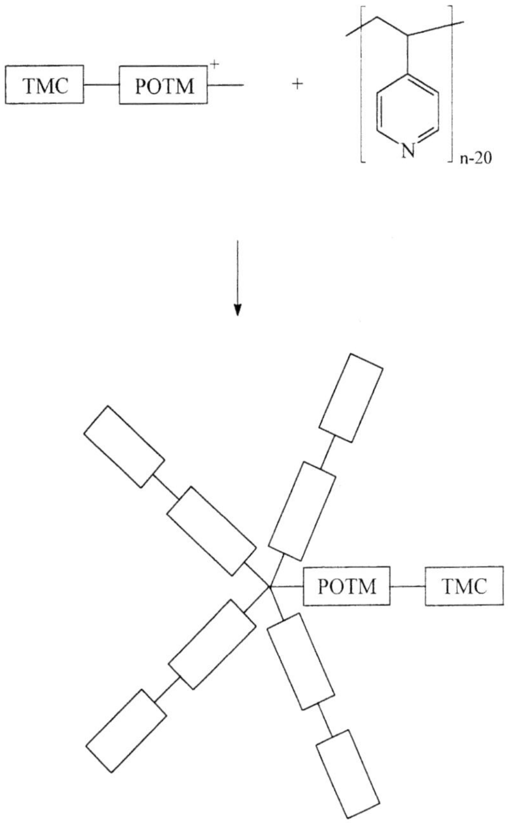

Scheme 6. (continued)

tems at high monomer conversion, the time of addition of the second monomer (styrene) to the reaction mixture containing living trifunctional PIB is of crucial importance. Even in the best polymerization runs there were homopolymers present (PIB and PS) which were removed from the final product by extraction. Detailed solution characterization data for the star material were not given.

Three arm amphiphilic star-block copolymers of isobutyl vinyl ether (IBVE) and 2-hydroxyethyl vinyl ethers were also prepared by Higashimura et al. [23]. The synthetic strategy involved the sequential living cationic polymerization of IBVE and 2-acetoxyethyl vinyl ether, initiated by the trifunctional initiator system composed of tris(trifluoroacetate) and ethylaluminum dichloride, with excess of 1,4-dioxane as a carbocation-stabilizing Lewis base (Scheme 8).

Hydrolysis of the ester groups gave the above-mentioned star-block amphiphilic copolymers with either segments as inner or outer blocks. The samples exhibited narrow molecular weight distributions and the experimentally determined apparent molecular weights, by SEC, were in good agreement with the corresponding stoichiometric ones. In a manner similar to the chlorosilane approach, amphiphilic tetra-armed star-block poly(vinyl ethers) were prepared [24] using

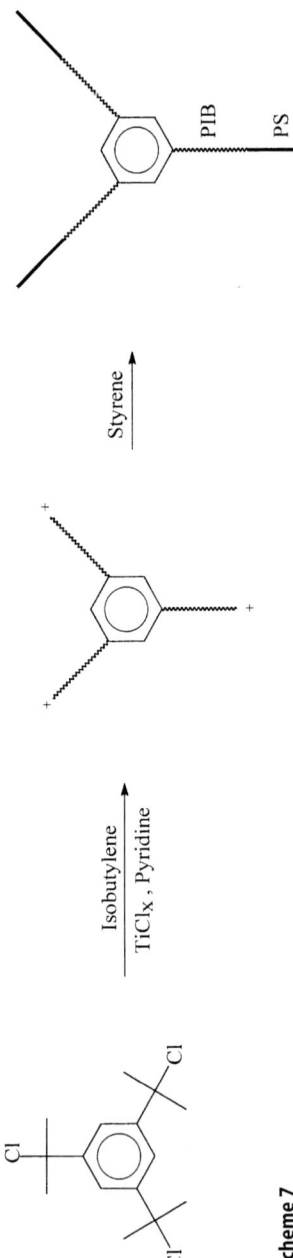

Scheme 7

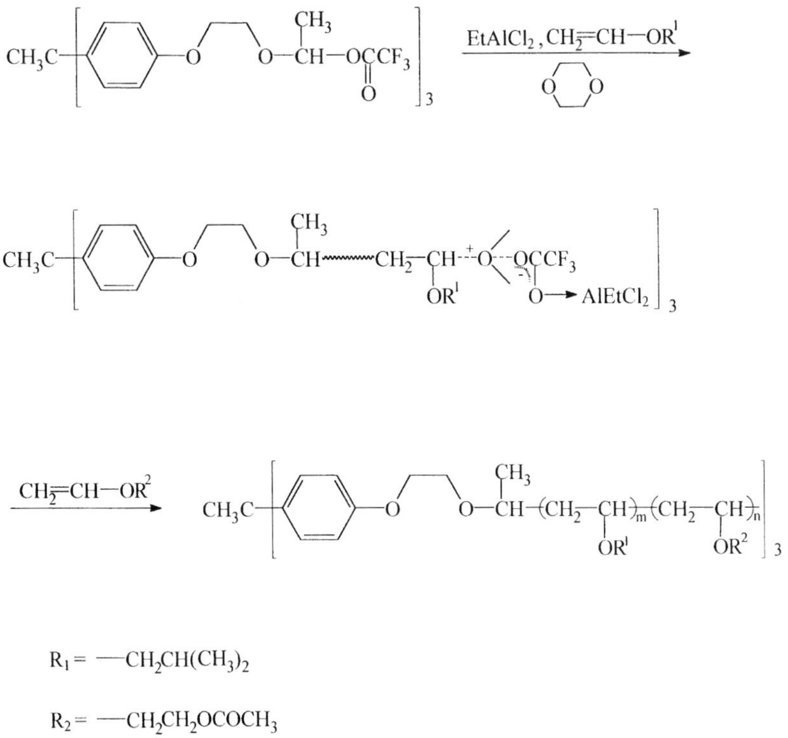

Scheme 8

a tetrafunctional silyl enol ether (C[CH$_2$OC$_6$H$_4$-p-C(OSiMe$_3$)=CH$_2$]$_4$). The products of the linking reaction had narrow molecular weight distributions as determined by SEC (Scheme 9).

Ring-opening metathesis polymerization was also used recently for the preparation of amphiphilic star-block copolymers [25]. Mo (CH-t-Bu) (NAr) (O-t-Bu)$_2$ was used as the initiator for sequential polymerization of norbornene-type, unfunctionalized and functionalized, monomers. The living diblocks were reacted with endo-cis-endo-hexacyclo-[10.2.1.1.^{3,4}1^{5,8}.0^{2,11}.0^{4,9}] heptadeca-6,13-diene, a difunctional monomer in a scheme analogous to the use of DVB in anionic polymerization, to form the central core of the star (Scheme 10).

The carboxylic moieties of the functional monomers, protected with trimethylsilyl groups during the polymerization reaction, were deprotected by hydrolysis, resulting in amphiphilic star-block copolymers. Molecular weight distributions of the stars were usually higher than 1.2, as determined by SEC, and only in one case was 1.06, although the polydispersities of the linear diblocks were lower than 1.05.

Attempts to prepare star-block copolymers via the macromonomer technique have also appeared in the literature [26-28]. The method involves the preparation of diblock arms and subsequent end functionalization with p-chloromethyl-

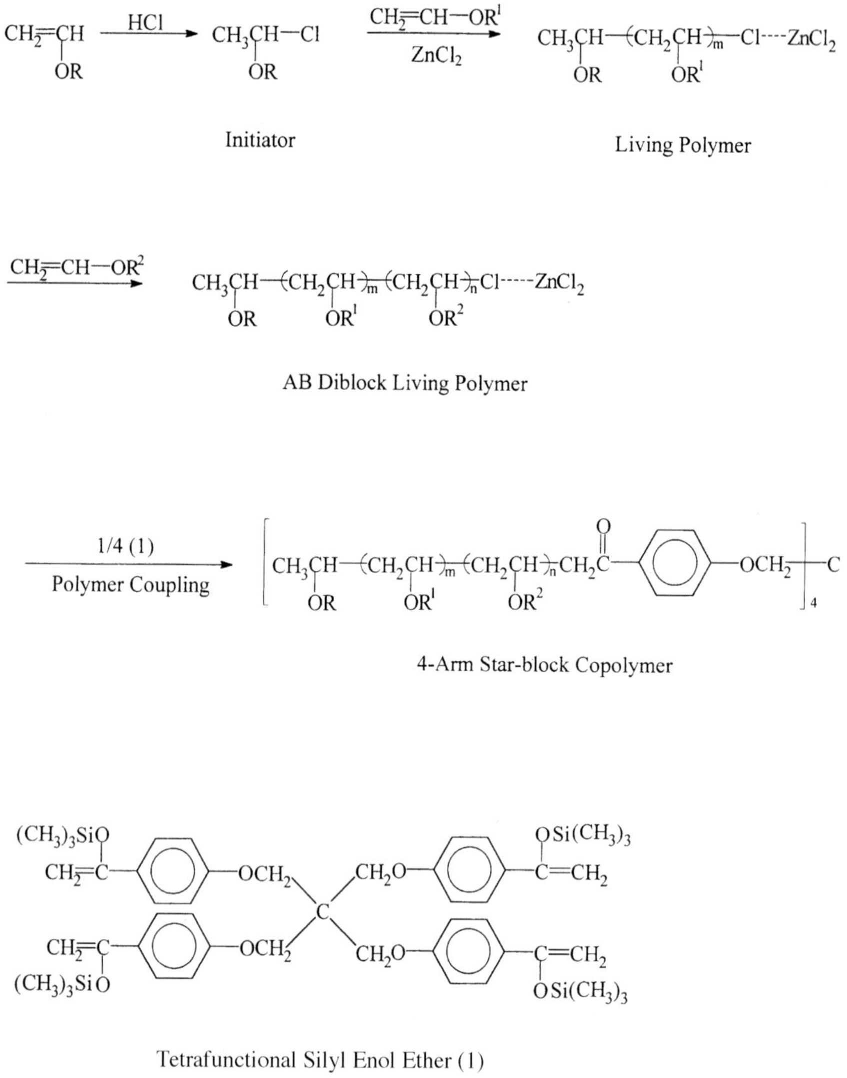

$$CH_2=CH \xrightarrow{HCl} CH_3CH-Cl \xrightarrow[ZnCl_2]{CH_2=CH-OR^1} CH_3CH-(CH_2CH)_m-Cl\text{----}ZnCl_2$$
$$\quad\;\, OR \qquad\qquad OR \qquad\qquad\qquad\qquad\quad OR \quad\;\; OR^1$$

Initiator   Living Polymer

$$\xrightarrow{CH_2=CH-OR^2} CH_3CH-(CH_2CH)_m-(CH_2CH)_n-Cl\text{-----}ZnCl_2$$
$$\qquad\qquad\qquad\quad OR \qquad OR^1 \qquad OR^2$$

AB Diblock Living Polymer

$$\xrightarrow[\text{Polymer Coupling}]{1/4\;(1)} \left[CH_3CH-(CH_2CH)_m-(CH_2CH)_n-CH_2\overset{O}{\underset{\|}{C}}-\bigcirc-OCH_2 \right]_4 -C$$
$$\qquad\qquad\qquad\qquad OR \qquad OR^1 \qquad OR^2$$

4-Arm Star-block Copolymer

Tetrafunctional Silyl Enol Ether (1)

Scheme 9

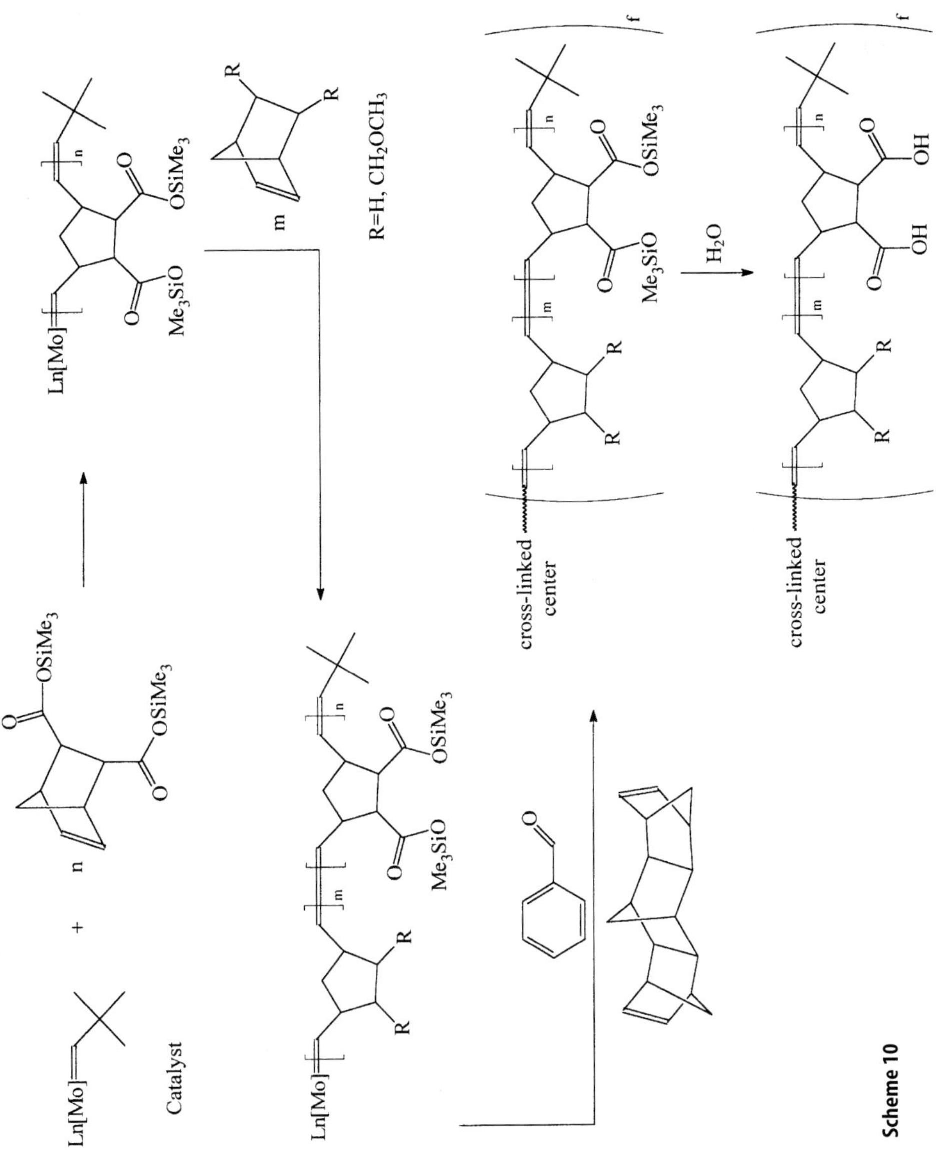

Scheme 10

styrene or DVB (anionic polymerization). In this way a polymerizable double bond is positioned at the end of the diblock chain. This bond can then be polymerized (ionically or radically) in solution or in the solid state to produce star-block copolymers, which however exhibit large variations in molecular weight and number of arms.

2.2
Graft Copolymers

Graft copolymers represent a valuable class of polymeric materials. They are composed of a main polymer chain to which one or more side chains are connected through covalent bonds. The branches are usually randomly distributed along the backbone. Synthetic methods initially developed for their preparation led to the formation of rather ill-defined polymers. These techniques were based mainly on free radical polymerization techniques because of their simplicity. More elaborate techniques were developed later to produce more homogeneous and well characterized graft copolymers.

There are three general methods for the preparation of graft copolymers [8]: a) grafting "onto", b) grafting "from", and c) via macromonomers. Grafting "onto" mechanisms involve the coupling reaction of pre-formed polymers having reactive chain ends with the backbone bearing functional groups. This method provides the advantage that both the backbone and the grafted chains can be characterized separately. If, in addition, the side chains are prepared by anionic polymerization methods, which provide the best control over molecular weight and molecular weight distribution, the resulting graft copolymers have controlled structures and are well defined. The most common case is the reaction of anionic living polymers with backbone electrophilic functionalities such as anhydrides, esters, nitrile, pyridine, or benzylic halide groups. A characteristic example is the functionalization of polybutadiene with chlorosilane groups, via the hydrosilylation reaction, followed by the coupling reaction with living polystyrene chains [29, 30] as shown in the following Scheme 11.

The grafting "from" procedure requires the generation of active sites on the main polymer chain which are capable of initiating the polymerization of a second monomer. Free radicals can be created by several methods such as irradiation of a polymer in the presence of oxygen [31, 32], chain transfer to the backbone [33,34] or redox reaction [35]. Several commercial products have been produced by these methods because they are simple and rather easy to perform. Nevertheless, significant amounts of homopolymers are produced, and, in combination with the poor control of radical polymerization, the final products are characterized by chemical heterogeneity.

Ionic grafting usually leads to well defined copolymers due to the limited (if any) termination reactions. Anionic sites can be created by metallation of the backbone. This can be accomplished by complexation of several types of C-H bonds (allylic, benzylic, aromatic) with organometallic compounds such as sec-BuLi. Usually chelating compounds, for example N,N,N',N'-tetra methyl ethylene diamine (TMEDA), act as solvating bases facilitating the reaction. By this method PB-g-PS [36-38] and PI-g-PS [39] graft copolymers have been prepared.

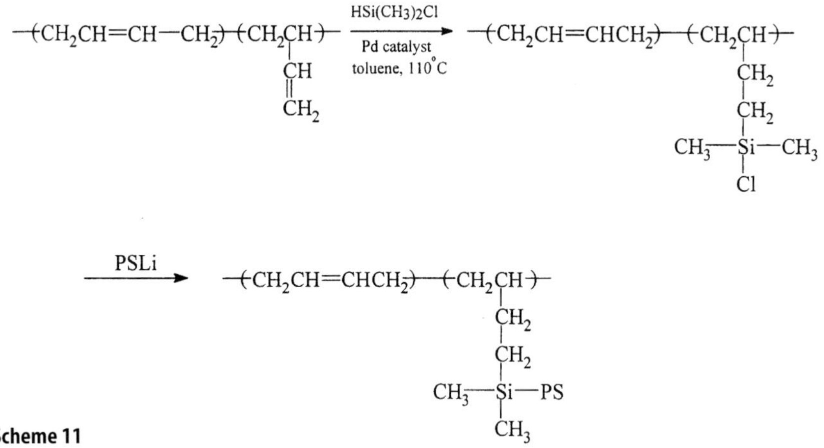

$$-(CH_2CH=CH-CH_2)-(CH_2CH)- \quad \xrightarrow[\text{toluene, 110°C}]{\text{HSi(CH}_3)_2\text{Cl}} \quad -(CH_2CH=CHCH_2)-(CH_2CH)-$$

(with CH=CH$_2$ side chain on left; and right product has side chain CH$_2$–CH$_2$–Si(CH$_3$)(CH$_3$)–Cl)

$$\xrightarrow{\text{PSLi}} \quad -(CH_2CH=CHCH_2)-(CH_2CH)-$$

(side chain: CH$_2$–CH$_2$–Si(CH$_3$)(CH$_3$)–PS)

Scheme 11

Several examples of cationic grafting have been reported in the literature [40, 41]. Polymer chains having labile halogen atoms in combination with various Lewis acids have been used. Examples are polychloroprene, poly(vinyl chloride), chlorinated styrene-butadiene rubber, chlorinated PB, etc. The monomers involved in the grafting reaction include isobutylene, styrenes, THF, alkyl vinyl ethers, etc.

The method most used for the preparation of graft copolymers is the macromonomer method [42–46]. A macromonomer is an oligomeric or polymeric chain bearing polymerizable end groups. Copolymerization with another monomer provides graft copolymers. The macromonomers can be monofunctional or difunctional depending on whether they have one or two end groups which can be polymerized.

Several variables must be considered during the synthesis of graft copolymers via the macromonomer procedure. One of the most important is the copolymerization behavior as expressed by the reactivity ratios, r_1 and r_2, of the macromonomer and the comonomer. These parameters determine how random the produced graft copolymer will be. A common problem is that the copolymerization is not homogeneous throughout the course of the reaction, since phase separation often occurs in these systems, leading to compositionally heterogeneous products. Several difficulties arising for the characterization of the graft copolymers, in addition to the possible chemical and compositional heterogeneity, make it necessary to use a combination of different characterization methods in order to prove whether the final products are well defined or not.

Macromonomers can be prepared by all the common polymerization techniques, e.g., free radical, anionic, cationic, condensation, or group transfer polymerization. Typical examples are given below:

a. Anionic polymerization [47] (Scheme 12)
b. Cationic polymerization [48] (Scheme 13)
c. Free radical polymerization [45] (Scheme 14)
d. Condensation polymerization [49] (Scheme 15)

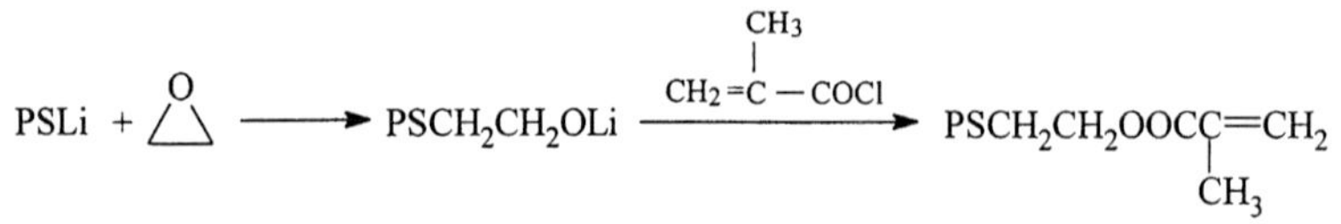

Scheme 12

PSLi + (epoxide) ⟶ PSCH$_2$CH$_2$OLi $\xrightarrow{CH_2=C(CH_3)-COCl}$ PSCH$_2$CH$_2$OOCC(CH$_3$)=CH$_2$

Scheme 13. (see next page) ⟶

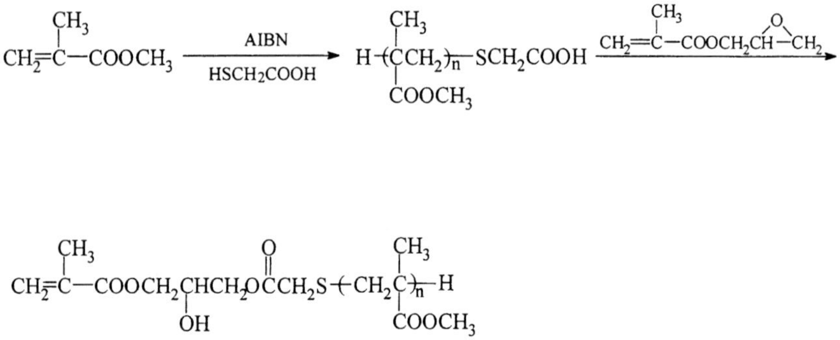

CH$_2$=C(CH$_3$)-COOCH$_3$ $\xrightarrow[\text{HSCH}_2\text{COOH}]{\text{AIBN}}$ H-(CH$_2$C(CH$_3$)(COOCH$_3$))$_n$-SCH$_2$COOH $\xrightarrow{}$ CH$_2$=C(CH$_3$)-COOCH$_2$CH(-O-)CH$_2$

CH$_2$=C(CH$_3$)-COOCH$_2$CHCH$_2$OCCH$_2$S-(CH$_2$C(CH$_3$)(COOCH$_3$))$_n$-H
    |
    OH

Scheme 14

nCH$_2$=CH-C$_6$H$_4$-CH=CH$_2$ + n C$_2$H$_5$NHCH$_2$CH$_2$NHC$_2$H$_5$ $\xrightarrow{C_4H_9Li}$

CH$_2$=CH-C$_6$H$_4$-CH$_2$CH$_2$N(C$_2$H$_5$)CH$_2$CH$_2$NH(C$_2$H$_5$) ⟶

CH$_2$=CH-C$_6$H$_4$-CH$_2$CH$_2$[N(C$_2$H$_5$)CH$_2$CH$_2$N(C$_2$H$_5$)CH$_2$CH$_2$-C$_6$H$_4$-CH$_2$CH$_2$]$_{n-1}$N(C$_2$H$_5$)CH$_2$CH$_2$NHC$_2$H$_5$

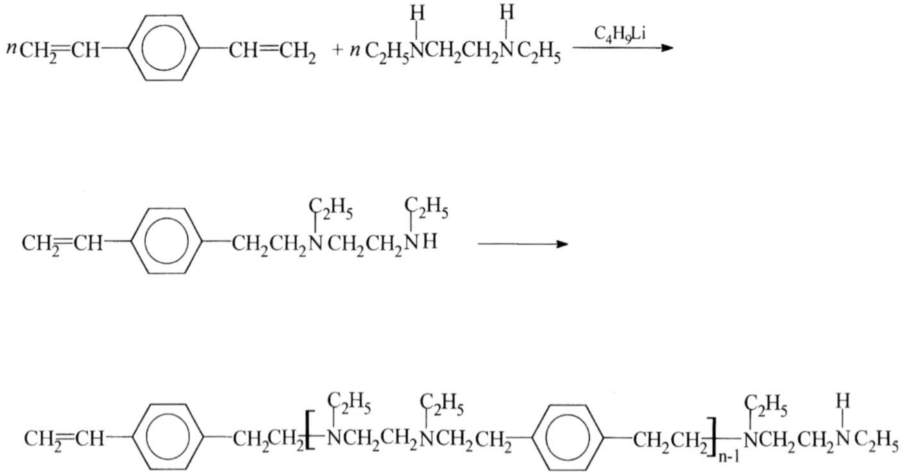

Scheme 15

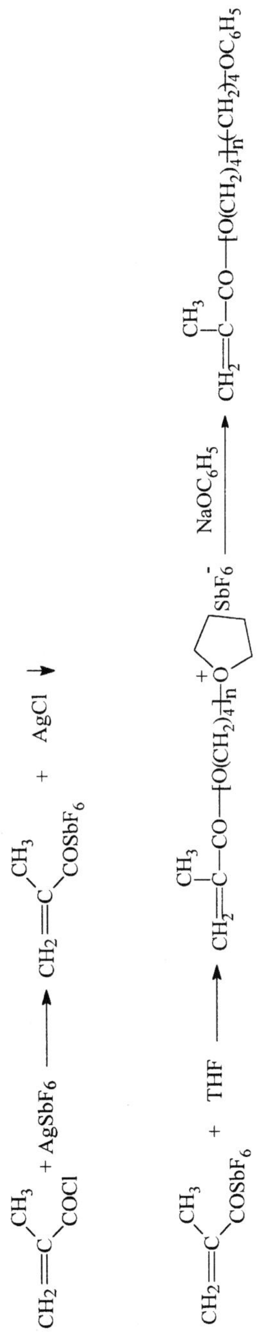
CH2=C(CH3)COCl + AgSbF6 → CH2=C(CH3)COSbF6 + AgCl↓
CH2=C(CH3)COSbF6 + THF → CH2=C(CH3)–CO–[O(CH2)4]n–O+(THF ring) SbF6-
NaOC6H5
CH2=C(CH3)–CO–[O(CH2)4]n–(CH2)4–OC6H5

Scheme 13

Anionic polymerization, although applicable to a rather small range of macromonomers, offers unique control of the molecular weight, molecular weight distribution, and chain end functionalization. Cationic polymerization is also limited to a few monomers and only systems without chain transfer or termination reactions provide well defined structures. Free radical or polycondensation methods are applied to a wide range of monomers but provide poor control for the preparation of macromonomers. Nevertheless, the products can be thoroughly characterized before copolymerization with another monomer. The macromonomers can be copolymerized via all the polymerization techniques depending on the type of the functional end group they have.

In the following sections recent developments (1985 and later) concerning the preparation of graft copolymers will be discussed. Several review articles cover the area for earlier years [4, 8, 45, 50-52]. Model graft systems recently developed, which allow rigid control over branch placement, will also be described.

2.2.1
Grafting "Onto" Methods

A method widely used for the preparation of graft copolymers is the chemical modification of the main chain with the introduction of functional groups capable of reacting with preformed polymers having active chain ends. A common procedure is the chloromethylation of PS [53-56] and the subsequent reaction with living polymers. Using this method Rempp et al. prepared PS-g-PEO graft copolymers [57] (PEO is poly(ethylene oxide)). In another study, Roovers et al. described the synthesis of PS-g-PI grafts [29]. In order to avoid the well-known side reaction involving Li-Cl exchange, Roovers transformed the -CH$_2$Cl group to the -SiMe$_2$Cl group before the reaction with the living polymers. Graft copolymers with polyvinylpyridine branches (PS-g-P2VP and PS-g-P4VP) were prepared by Selb and Gallot [58, 59]. All these polymers are well defined with regard to the molecular weights of the backbone and the branches, as well as the number of branches. They can have narrow molecular weight distributions, but the position of the branches is random.

Using the same method, George et al. prepared PS-g-PEO graft copolymers [60]. The chloromethylation of PS was done in the classical way using a CCl$_4$ solution of PS with chloromethyl methyl ether at 0 °C with SnCl$_4$ as a catalyst. The reaction conditions were controlled to provide low chlorine contents (<10% w/w). Ethylene oxide was polymerized in THF by cumylpotassium at 40 °C, and the living PEO$^-$K$^+$ solution was reacted with the chloromethylated PS at 50 °C for the preparation of the graft copolymer. The grafting efficiency, i.e., the ratio of the -CH$_2$Cl groups used for the grafting reaction over the total number of -CH$_2$Cl groups, was very low, in agreement with results reported by Selb and Gallot for the preparation of PS-g-P2VP and PS-g-P4VP graft copolymers. Characterization data by SEC and membrane osmometry (MO) confirmed the structure for the copolymer.

The synthesis of poly(N-vinyl carbazole)-g-PI graft copolymers by a similar method has been described [61]. Poly(N-vinyl carbazole), PVCz was synthesized by cationic polymerization under N_2 atmosphere. The chloromethylation of PVCz was performed in $CHCl_3$ solutions using $ZnCl_2$ and chloromethyl methyl ether at 0 °C. Living Pl^- Li^+ solutions prepared by anionic polymerization with n-BuLi were reacted with the chemically modified PVCz to produce the graft copolymers. Characterization data obtained by SEC, MO, NMR, and IR confirmed the graft structure. Also in this case the grafting efficiency was limited. For the two samples prepared, 28 and 70% of the chloromethylene groups remained unreacted, although the same PVCz backbone was used. This is an indication that the grafting reaction cannot be easily controlled. The author attributed the low grafting efficiency to steric hindrance effects.

Cellulose was modified so that living PS chains can react with it forming cellulose-g-PS graft copolymers [62, 63]. The performed tosylation reaction of the free -OH groups on cellulose acetate gives a better leaving group than the existing acetate group towards nucleophilic substitution and eliminates the possibility of homopolymer formation through the termination reaction of living PS with free -OH groups. The chemically modified cellulose reacted with living PS end-capped with 1,1-diphenyl ethylene (DPE), prepared by anionic polymerization. The final product was the graft copolymer with the grafting efficiency being very high (~90%). The unreacted acetate and tosyl groups were removed by mild hydrolysis with ammonia. The reaction series is given in the following scheme16.

UV and IR spectroscopy and elemental analysis were used to characterize the graft copolymers.

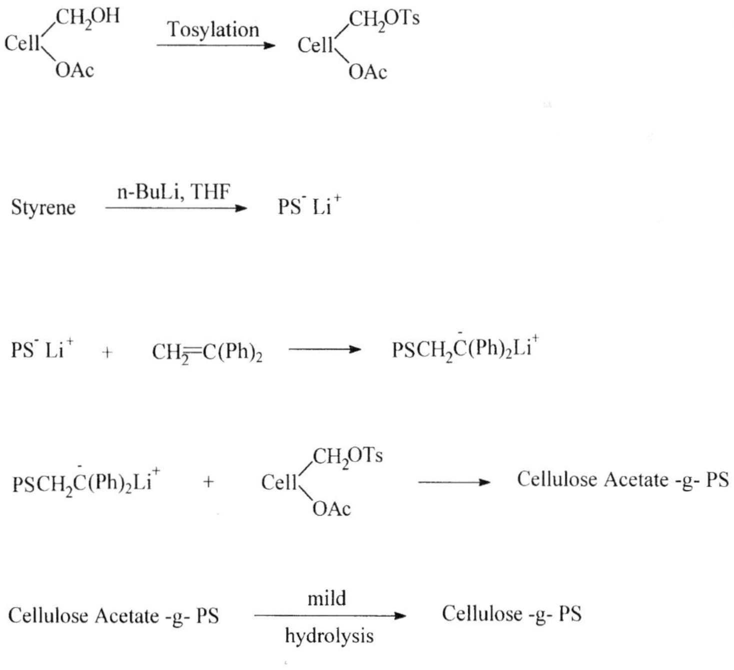

$$\text{Cell}\begin{matrix}CH_2OH\\OAc\end{matrix} \xrightarrow{\text{Tosylation}} \text{Cell}\begin{matrix}CH_2OTs\\OAc\end{matrix}$$

$$\text{Styrene} \xrightarrow{\text{n-BuLi, THF}} PS^-\ Li^+$$

$$PS^-\ Li^+ + CH_2{=}C(Ph)_2 \longrightarrow PSCH_2\bar{C}(Ph)_2Li^+$$

$$PSCH_2\bar{C}(Ph)_2Li^+ + \text{Cell}\begin{matrix}CH_2OTs\\OAc\end{matrix} \longrightarrow \text{Cellulose Acetate -g- PS}$$

$$\text{Cellulose Acetate -g- PS} \xrightarrow{\text{mild hydrolysis}} \text{Cellulose -g- PS}$$

Scheme 16

Watanabe et al. reacted living PS or PI chains with P2VP to produce P2VP-g-PS or P2VP-g-PI graft copolymers [64], according to the following scheme 17.

There was always an excess of the number of living polymers relative to the number of 2VP groups. The reaction was monitored by SEC indicating the incorporation of 6-65 grafted chains. Both the backbone and the grafted chains were prepared by anionic polymerization techniques possessing narrow molecular weight distributions. Nevertheless slightly higher distributions were observed for the copolymers, especially in the case of P2VP-g-PI samples.

Derand and Wesslen prepared graft copolymers having PEO branches [65]. The backbones were terpolymers of styrene, maleic anhydride, and methylmethacrylate (S/MAN/MMA), or styrene, maleic anhydride, and ethylhexyl methacrylate (S/MAN/EHMA), or styrene, maleic anhydride, and diethylfumarate (S/MAN/DEF), synthesized by radical copolymerization using AIBN as initiator. Poly(ethylene glycol) monomethyl ether (MPEG) was grafted to the backbones after reaction with the succinic anhydride groups. For example, Scheme 18 illustrates this reaction starting from an 5/MAN backbone.

Limited yields for the grafting reactions were observed (~60%), even though a large excess (~50%) of MPEG was used. When the reactions were carried out to high conversions, gelation occurred. A possible explanation for this behavior is that the commercial MPEG used in this study is known to contain some difunctional poly(ethyleneglycol) (PEG), in some cases as high as 25%. IR and NMR spectroscopy supported the formation of the above structures. Data concerning the polydispersity of the samples were not provided, although the nature of the radical polymerization suggests the existence of rather polydisperse copolymers.

Block-graft copolymers are copolymers having one linear block, whereas the other block is a graft copolymer (Scheme 19).

Using anionic polymerization techniques poly[styrene-b-(4-vinyl phenyldimethylsiloxane-g-isoprene)), P[S-b-(VS-g-I)] block-graft copolymer was prepared [66, 67] according to the following reaction sequence (Scheme 20).

Initially the block copolymer P(S-b-VS) was prepared by sequential addition of S and V. It was found that under specific conditions (THF as a solvent, cumyl Cs as initiator, 20 min. polymerization time at −78 °C and addition of N-

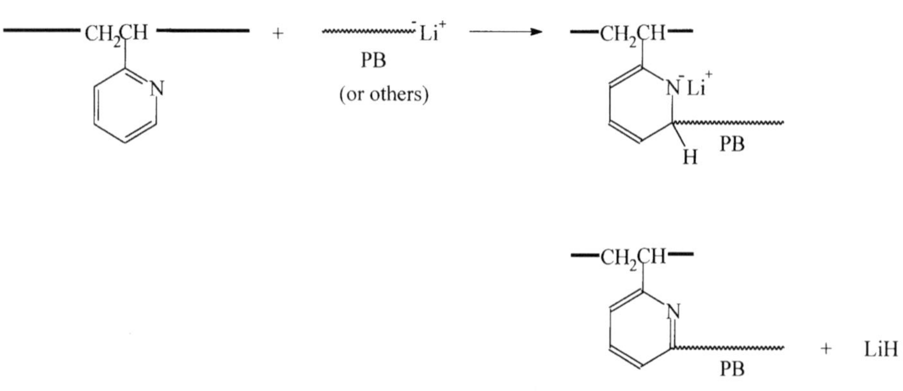

Scheme 17

Scheme 18

Scheme 19

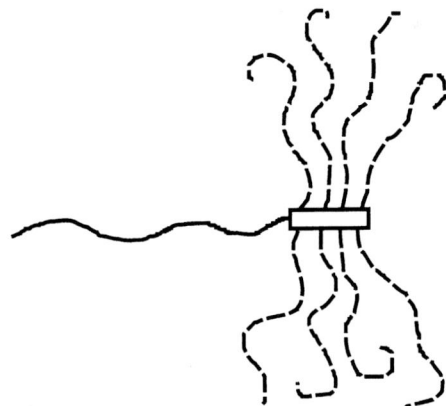

$$P(S\text{-}b\text{-}VS) + PI^-Cs^+ \longrightarrow P[S\text{-}b\text{-}(VS\text{-}g\text{-}I)]$$

or PI⁻Li⁺

Scheme 20

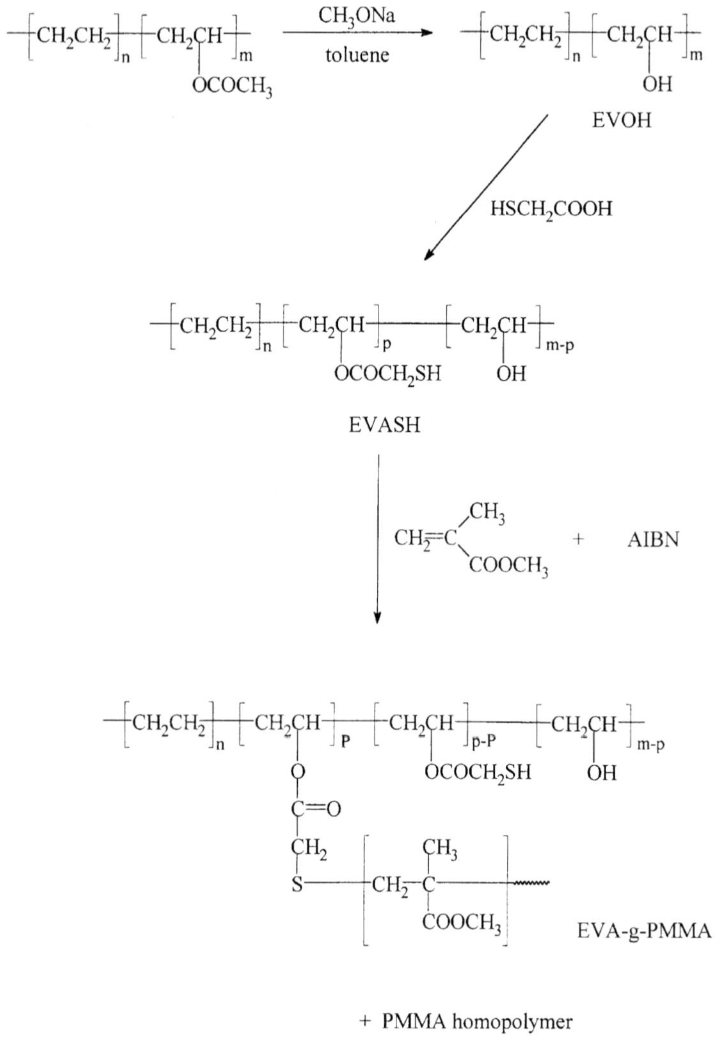

Scheme 21

methylpyrrolidone) the vinylsilane group remains unreacted during the block formation. The products were near monodisperse (I<1.15). The second step involved the preparation of living PI chains, which were reacted with the vinylsilane groups of the block copolymers to form the block-graft copolymers. SEC analysis showed that the products were near monodisperse (I<1.08). A small amount (~30%) of the V units was used for the grafting reaction due to steric effects.

Graft copolymers were prepared by grafting poly(ethyleneglycol monomethacrylate) (MPEG) onto polymethacrylate and polyacrylate copolymers via a transesterification reaction, using potassium methoxide as the transesterification catalyst [68]. The composition of the graft copolymers was chosen to lead to water soluble products. IR and ^{1}H-NMR spectroscopy and SEC were used to characterize the graft copolymers.

Random copolymers of PS and a small quantity of 4VP (~1% mol) were prepared by radical polymerization and used as the backbone for the grafting of the polytetrahydrofuran dication, prepared by living cationic polymerization [69]. The polymers were characterized by SEC. As an extension of this work the terminal hydroxyl groups of the graft copolymers were tosylated and 2-methyl oxazoline (MeOz) was added. Heating the mixture at 110 °C led to the polymerization of the MeOz.

Poly(S-*alt*-maleic anhydride)-g-PEO grafts were prepared by reacting monoamine terminated poly(ethylene oxide) with the styrene-maleic anhydride alternating copolymers [70]. The samples were characterized by SEC and UV-VIS and NMR spectroscopy.

Random copolymers of ethylene and vinyl acetate were treated with sodium methoxide, leading to the hydrolysis of the acetate groups. The hydroxyls thus generated were reacted with mercaptoacetic acid to introduce sulfidryl functional groups along the polymer chain. This copolymer acts as a macromolecular chain transfer agent for the radical polymerization of MMA using AIBN as initiator, leading to the formation of graft copolymers [71] according to the Scheme 21.

This method can be characterized as an in-situ grafting onto technique and has the disadvantage compared with the classical grafting onto method that the branches cannot be isolated and thus characterized independently. The uniformity of the molecular weight distribution can be adjusted by controlling the [AIBN]/[SH] molar ratio. The general behavior observed for these samples is that the grafting efficiency is increased by increasing the concentration of AIBN, although this usually leads to non-uniform branches.

2.2.2
Grafting "From" Methods

A linear polymer can be modified in several ways in order to form active sites capable of initiating the polymerization of a second monomer. Cationic grafting techniques were performed for the preparation of poly(ethyl vinyl ether-g-ethyloxazoline) graft copolymers [72]. A random copolymer of ethyl vinyl ether with a small quantity of 2-chloroethyl vinyl ether was synthesized by cationic polymerization techniques, using aluminum hydrogen sulfate as the initiator (AHS). The pendant alkyl chloride groups act as initiators for the polymerization of ethyloxazoline in the presence of sodium iodide (Scheme 22).

Based on the same idea, of using alkyl halides as initiators for the ring opening polymerization of 2-alkyl oxazolines, Schulz and Dworak partially hydrolyzed

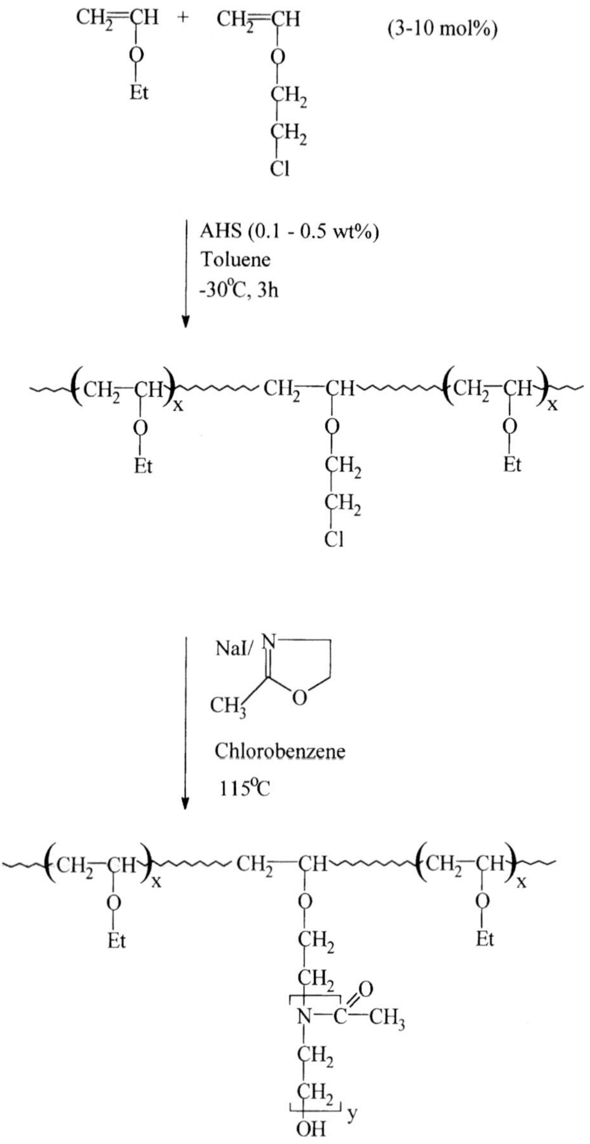

Scheme 22

poly(vinyl acetate) and reacted the produced hydroxyl groups with phosgene or diphosgene to incorporate acid chloride groups along the polymer chain [73]. These functional groups were used to initiate the polymerization of 2-phenyl or 2-methyl oxazoline in the presence of potassium iodide which facilitates the acceleration of the initiation reaction (Scheme 23).

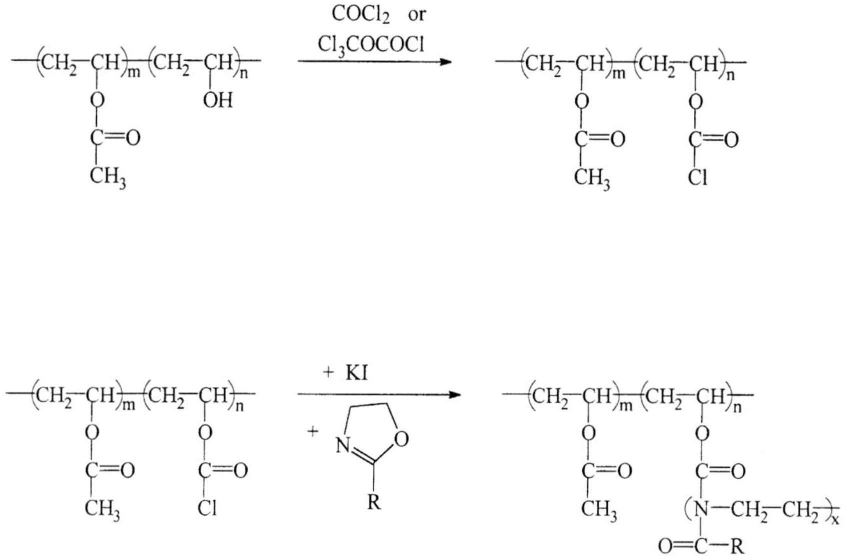

Scheme 23

The products were characterized by SEC and NMR and IR spectroscopy.

Jiang and Frechet performed a Friedel-Crafts acetylation on polystyrene chains followed by a Grignard reaction and acetylation of the resulting tertiary alcohol [74] as outlined in the following Scheme 24.

Subsequent addition of BCl$_3$ to the above product creates active initiation sites for the polymerization of isobutylene (Scheme 25).

Using an excess of BCl$_3$ a high grafting rate and high conversion of IB was observed. However, the excess of BCl$_3$ also leads to transfer reactions and the formation of PIB homopolymer. After purification, SEC analysis showed that the graft copolymer was monomodal, but with a rather high polydispersity index (I = 1.79).

Cationic polymerization techniques were also performed for the grafting of indene or styrene from poly(isobutylene-co-p-chloromethylstyrene) [75]. The backbone was prepared using TiCl$_4$ and t-BuCl as the initiator system. The p-chloromethylstyrene content was kept lower than 10% in all cases. These chloroalkyl groups, on addition of AlEt$_2$Cl, are capable of initiating the polymerization of indene. The grafting efficiency was as low as 40-50%, whereas the number of grafted polyindene chains per backbone was lower than five, and a rather high quantity of homopolymer was formed. When styrene was used for the grafting reaction the grafting efficiency was very high (>80%), the number of grafted chains per backbone was higher than eight in all cases and the homopolymer formed was on the order of 10%.

Polysilanes can be prepared by reaction of dichlorosilanes with sodium dispersion. Phenyl groups attached to polysilanes can be removed by strong protonic acids such as trifluoromethane-sulfonic acid. The resulting triflate polysi-

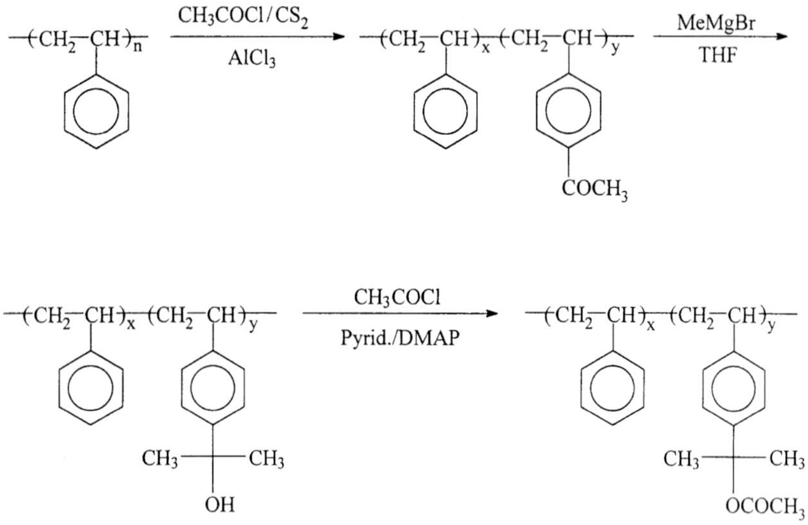

Scheme 24

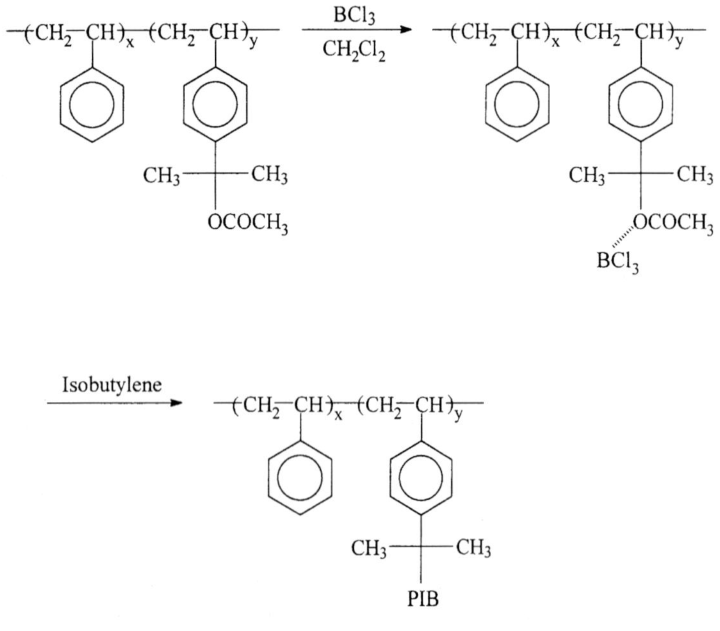

Scheme 25

lanes are capable of initiating the polymerization of other monomers. This method was used for the preparation of graft copolymers [76] as follows. Dearylation reactions were performed on poly(phenyl methyl silane). It was observed that not only the desired dearylation but also degradation reactions occur to a small extent. This side reaction was more pronounced for higher molecular weight polymers. The triflate groups introduced to the main chain act as initiator for the polymerization of THF leading to the formation of graft copolymers. Photodegradation studies showed that 2.4 polyTHF chains were, on average, incorporated into each polysilane chain.

The triflated polysilanes were also used to initiate the polymerization of methyl methacrylate by group transfer polymerization. The initiation sites were prepared in-situ using equimolar amounts of triflic acid and arylsilanes, methylpropionate and triethylamine. Tris(dimethylaminosulfonium) bifluoride was the catalyst. It was found by degradation studies that each polysilane chain contains on average 2.7 PMMA branches. No additional details were reported on the characterization of these graft copolymers.

It was observed that typical transition metal catalysts for the hydrosilylation reaction can be used for the ring-opening polymerization of heterocyclic monomers. Despite the actual mechanism taking place in the presence of a silicon cocatalyst, the initiating silicon atom remains at the end of the growing polymer chain. These observations led to the formation of poly(methyl hydrogen siloxane)-g-poly(cyclohexene oxide) graft copolymers [77] according to the reaction (Scheme 26).

Decomposition studies, by cleaving the Si-O-C bonds, were performed to confirm the structure of the copolymers. Molecular characterization data were not provided in this study.

It was already mentioned that anionic polymerization offers the best control for the synthesis of graft copolymers. A well-known method for generating anionic centers along a polymer chain is the lithiation reaction of polydienes using TMEDA-s-BuLi. From these sites the anionic polymerization of styrene may be initiated, resulting in the preparation of PI-g-PS copolymers. The synthesis was reported by Hadjichristidis and Roovers [39] and more recently by Al-Jarrah et al. [78]. It was found that the lithiation reaction was best controlled using high vacuum line conditions. In all cases the formation of homopolystyrene was observed either due to residual s-BuLi or to a side reaction which consumes part of the s-BuLi to form a lithium compound which can act as initiator for the polymerization of styrene. The grafting efficiency was increased by the addition of ~5% THF, which breaks the association of the growing PS chains. It was also observed that the grafting efficiency increases by increasing the cis 1,4 content in the PI backbones and by decreasing the PS content. In all cases the polydispersity of the copolymers is rather broad (I~1.4).

PS-g-PEO graft copolymers having alkyl groups on either the PS or the PEO chains were prepared using the grafting "from" method [79]. The main chain, consisting of a random copolymer of S and acrylamide (~4-14% AAm), was prepared by radical polymerization techniques using AIBN as initiator. The backbone was alkylated by ionizing the amide groups with potassium *tert*-butoxide and finally by reacting the anionic sites formed with 1-bromoalkanes. The rest

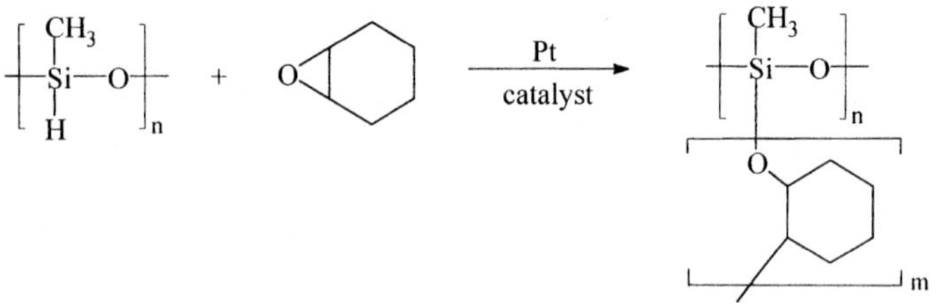

Scheme 26

of the amine anions were used for the polymerization of EO. A different route was utilized for the preparation of graft copolymers with alkylated side chains. Initially the PEO chains were grafted from the amide anions and then the alkoxide anions were terminated by reactions with 1-bromoalkanes. The alkylation yield was rather low (49-79%) and PEO homopolymer was observed. The molecular weight distributions of the main chains were rather broad (I~1.9), as expected for radical polymerization. The graft copolymers were characterized by elemental analysis, ^{1}H NMR and IR spectroscopy, and SEC (Scheme 27).

Almost the same method was used in another study for the synthesis of PS-g-PEO graft copolymers [80]. These results suggest the existence of narrow molecular weight distribution branches attached to a rather polydisperse backbone.

Inoki et al. reacted poly(vinyl propionate), PVPr, with lithium diisopropylamide (LDA) in order to create anionic sites along the polymer chain, since LDA can replace a hydrogen atom from a methylene or methine group adjacent to a carbonyl group. Subsequent addition of MMA leads to the formation of a graft copolymer PVPr-g-PMMA [81] (Scheme 28).

The backbone was prepared by esterification of poly(vinyl alcohol), PVA, and had a broad molecular weight distribution (I=2.4). It was observed that a side reaction occurs during the treatment of the main chain with LDA. It is possible for the carbanion to attack the ester group of the penultimate monomer unit to give a β-keto ester. It was estimated that this side reaction takes place with less than 10% of the monomer units. Hydrolysis of the graft copolymer with methanolic sodium hydroxide results in PVA and PMMA, so it was possible to characterize the copolymer. In combination with the quantitative analysis of the amount of lithium incorporated into the main chain, it was found that the number of PMMA branches varied between one and seven and that only 1.9-11.3% of the anionic sites acted as initiators. The authors attributed this behavior to aggregation phenomena involving the enolate groups. The lower reactivity of these aggregates compared to the free enolates also leads to a broadening of the molecular weight distributions. The I values varied for the graft copolymers between 2 and 6.7.

Block-graft copolymers were also prepared by the grafting "from" method [67]. The block copolymers P(S-b-VS) were reacted with n-BuLi in THF at −30 °C for 1 h leading to metallation of the vinyl groups of the V segments. Then D$_3$ was added for the formation of the P[S-b-(VS-g-DMS)] block-graft copolymer. The molecular characterization revealed that 170 and 325 PDMS branches were incorporated into the backbone for the two samples that were prepared.

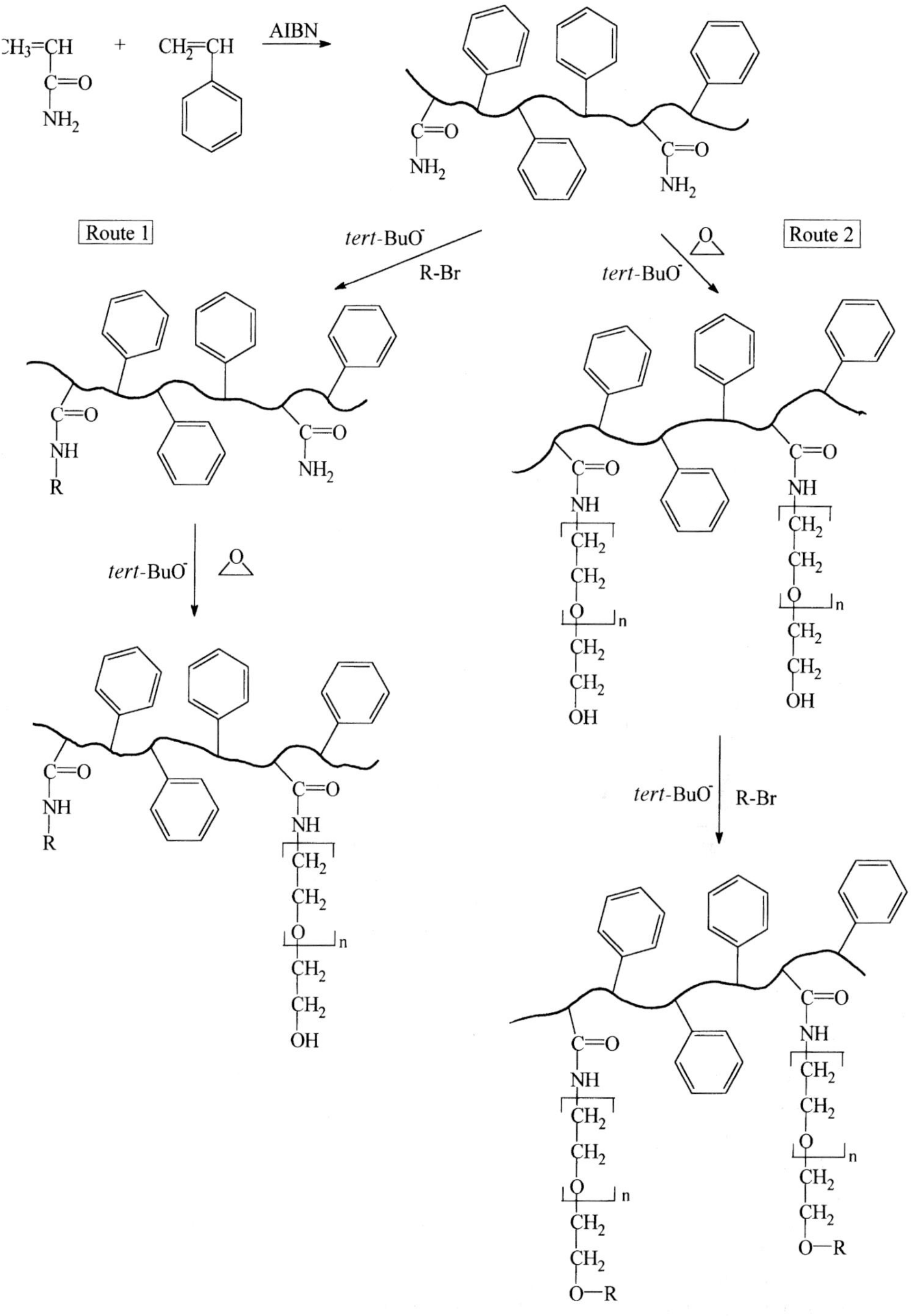

Scheme 27

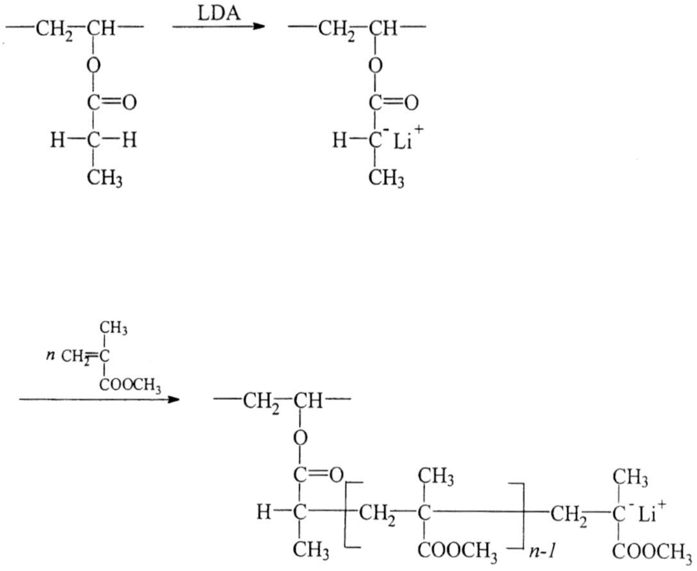

Scheme 28

The synthesis of the following block-graft copolymer: poly[styrene-b-(hydroxystyrene-g-ethylene oxide)-b-styrene] or P[S-b-(HS-g-EO)-b-PS] was reported in the literature [67] according to the Scheme 29.

The backbone was a triblock copolymer, poly(styrene-b-tert-butoxystyrene-b-styrene), prepared by anionic polymerization with sequential addition of monomers. It was found that the polymerization of tert-butoxystyrene in THF with n-BuLi at –78 °C for 15 min produces monodisperse macromolecules. The protective t-butyl group can be removed by treatment with HBr in an acetone-benzene solution leading to the formation of the P(S-b-HS-b-S) triblocks. The metallation of the hydroxyl groups was performed in THF for 1 h using either cumyl potassium or diphenylethylene potassium. A four-fold excess of the initiator over the molar concentration of the hydroxyl groups was used. Finally, the addition of EO generates the block-graft copolymer. The excess of the initiator also produced PEO homopolymer which was assumed to have the same molecular weight as the grafted PEO chains. Using this assumption it was found that 21 PEO chains were incorporated onto the main chain and that 54% of the -OH groups served as initiator sites for the polymerization of the EO.

Roha et al. incorporated photoinitiator fragments as pendant groups to linear chains and used them to initiate the polymerization of several monomers leading to the formation of graft copolymers [82]. Chlorobutyl rubber, neoprene, or chlorinated PVC were reacted with sodium isopropyl xanthogen, sodium mercapto-benzothiazole, or sodium diethyl-di-thiocarbamate to introduce the thio groups along the polymer chain. A variety of monomers like MMA, acrylonitrile, ethyl acrylate, and acrylic acid were polymerized with UV irradiation using the functionalized polymers as macroinitiators. SEC analysis revealed the existence

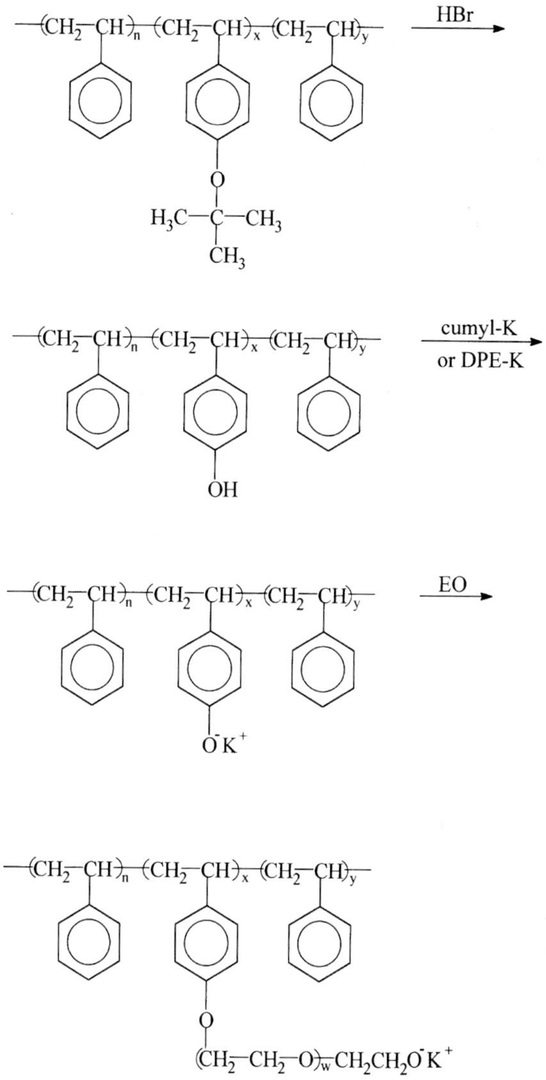

Scheme 29

of single peaks with only traces of homopolymers. A pseudo-living polymerization was observed, since the addition of a new quantity of monomer yielded higher molecular weight graft copolymers.

Copolymers containing PB (SBS and K-resin) were irradiated with UV light in the presence of anthracene, which acts as a photosensitizer, to create free radicals along the polymer chain. These radicals were used as initiation sites for the polymerization of methacrylic acid leading to the formation of graft copolymers [83]. Molecular characterization data were not provided in this study.

Graft copolymers with polyorganophosphagene backbones and PS branches were prepared by Gleria et al. [84]. Poly[bis(4-isopropylphenoxy)phosphagene] was peroxidized in the presence of benzoyl peroxide. The resulting hydroperoxide groups were thermally decomposed to produce free radicals. In the presence of St this procedure led to the formation of graft copolymers. PS homopolymer was observed during the grafting reaction and selective precipitation was used for the purification of the products. An inherent problem of the method is that the hydroperoxide groups are very reactive, leading to crosslinking reactions a

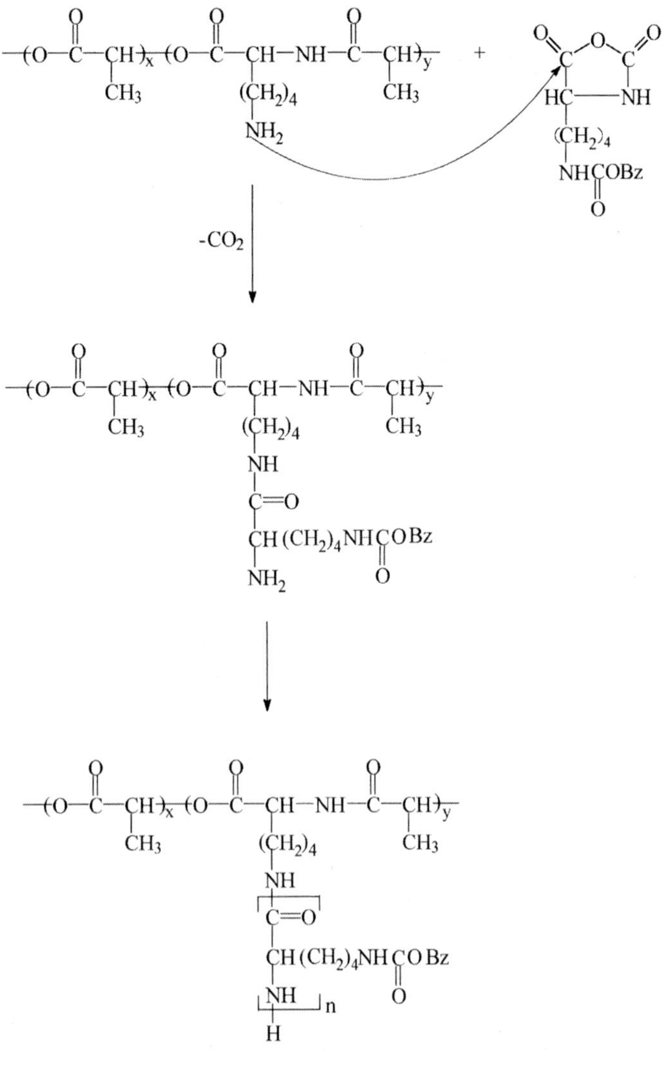

Scheme 30

few days after their formation even under inert atmosphere and in the dark. The samples were only characterized by SEC.

When lignin is treated with $CaCl_2$ and hydrogen peroxide in DMSO solutions and in the presence of 1-ethenylbenzene, poly(lignin-g-phenylethylene) is produced [85]. It was shown that free radicals are formed along the lignin chains, giving rise to radical polymerization of the 1-ethenylbenzene and thus leading to the formation of graft copolymers. The products were purified by fractionation but characterization data were not given by the authors.

Step-growth polymerization reactions can also be employed for the synthesis of graft copolymers. Using this technique poly(L-lactic acid-g-Z-L-lysine) graft copolymers were prepared [86]. The backbone was a random copolymer of L-lactic acid and ~2% N^ε-(benzyloxycarbonyl)-L-lysine (Z-L-lysine) prepared by the method of Barrera et al. Deprotection of the lysine groups was accomplished in an HBr/CH_3COOH solution to yield free ε-amine groups. These nucleophilic groups can attack the Z-L-lysine-N- carboxyanhydride (Z-LYSNCA), initiating its ring opening polymerization and leading to the formation of graft copolymers (Scheme 30).

Characterization studies by elemental analysis and IR and NMR spectroscopy showed that from one to seven lysine chains were incorporated per backbone and that the copolymers contained 7-72% of lysine groups.

2.2.3
Graft Copolymers via Macromonomers

2.2.3.1
PS-Based Macromonomers Prepared Anionically

The introduction of polymerizable end groups into a polymer chain can be rather easily achieved by anionic polymerization techniques. Functional initiation or electrophilic termination are the most common ways to incorporate the active groups. Due to their wide applicability, polystyrene macromonomers will be examined separately from the other anionically prepared macromonomers.

The most widely used method for the preparation of PS macromonomers is the one developed by Milkovich [46]. Living polystyrene solutions were reacted with ethylene oxide to form the less reactive alkoxide and then with methacryloyl chloride (MAC), according to the following reaction (Scheme 31).

Variations on this method were also reported in the literature. 5-Norbornene-2-carbonyl chloride can be used instead of methacryloyl chloride to prepare ω-norbornenyl-polystyrene [87] (Scheme 32).

Another approach involves the end-capping reaction of living polystyrene with propylene oxide and the subsequent treatment of the alkoxide with bicyclo [2.2.1] hept-5-ene-2,3-*trans*-dicarbonyl chloride to provide a macromonomer having a double bond in the middle of the polymer chain [88], as is shown in the following reaction (Scheme 33).

A common method for the preparation of PS macromonomers is the deactivation of the polystyryllithium anion with *p*-chlorovinyl benzene [89] as outlined in the reaction (Scheme 34).

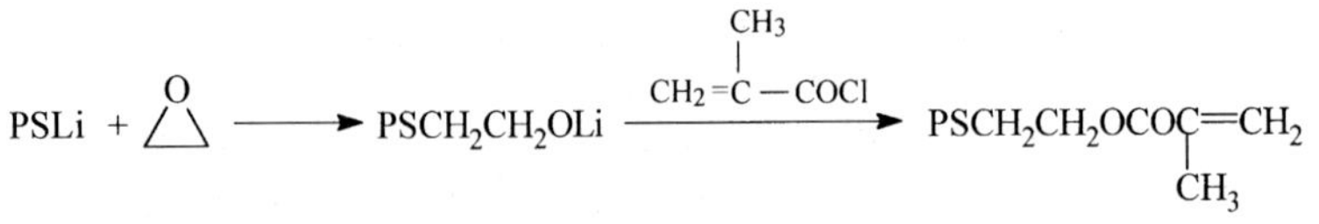

Scheme 31

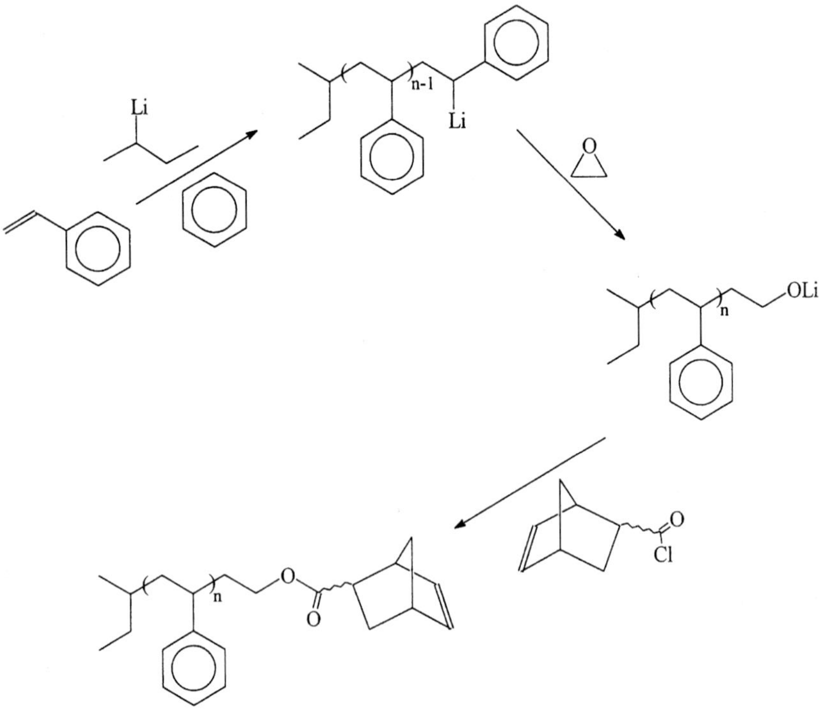

PSLi + (epoxide) → PSCH$_2$CH$_2$OLi $\xrightarrow{CH_2=C(CH_3)-COCl}$ PSCH$_2$CH$_2$OCOC(CH$_3$)=CH$_2$

Scheme 32

Post-functionalization reactions were employed for the synthesis of macromonomers. For example Tsukahara et al. used carboxyl-terminated PS and glycidyl methacrylate [90] for this purpose (Scheme 35).

The functional initiator 4-pentenyllithium was used to initiate the polymerization of styrene resulting in low polydispersity, vinyl terminated macromonomers [91].

Table 1 summarizes the literature data on the graft copolymers prepared by polystyrene macromonomers, which were synthesized by anionic polymerization.

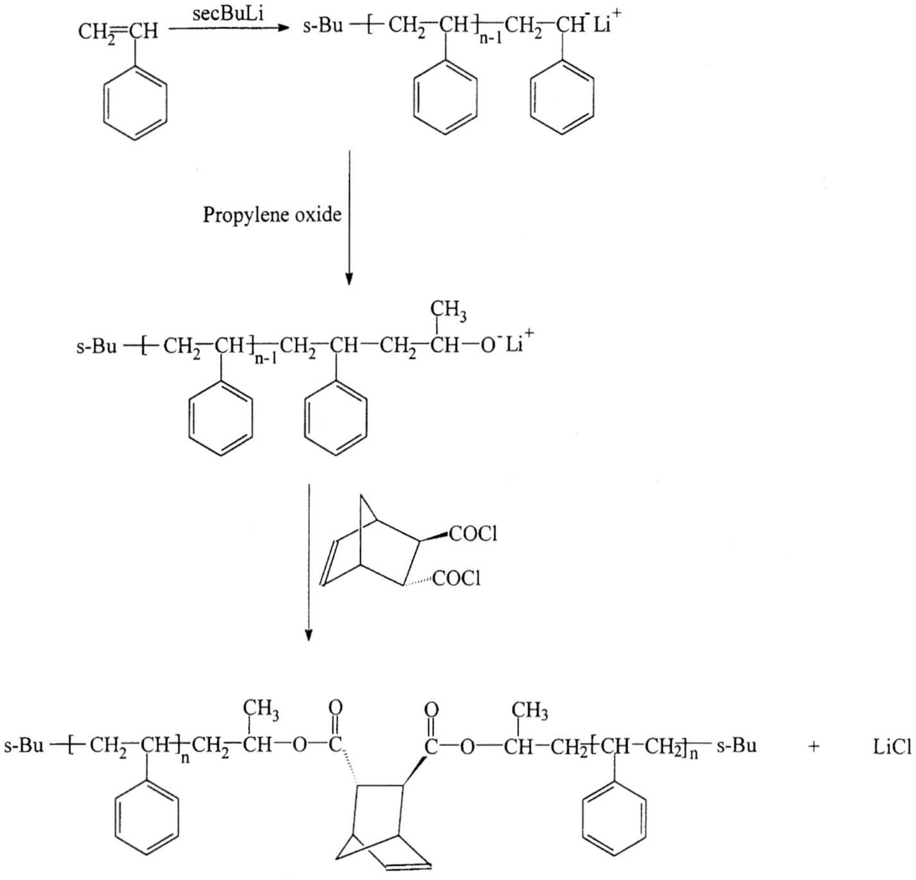

Scheme 33

Scheme 34

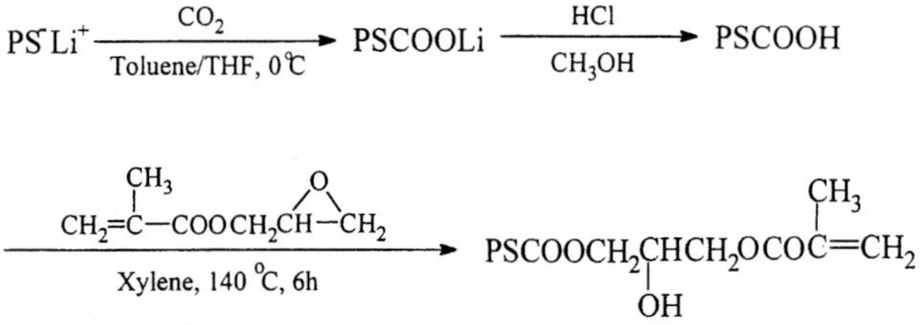

Scheme 35

Table 1. Graft copolymers prepared by polystyrene macromonomers synthesized anionically

End group	Comonomer	Copolymerization	Ref.
methacrylate	MMA	radical	92, 105
methacrylate	butylacrylate 2-ethylhexylacrylate acrylic acid	radical	93
p-vinylphenyl	MMA	radical	89
methacrylate	2,3-dihydroxypropyl methacrylate	radical	90, 94
norbornene	norbornene	olefin metathesis	87
norbornene	–	ring opening metathesis	88
norbornene	–	polymerization ring opening metathesis	95
hydroquinone or 4,4'-methylene diphenyl	terephthalic acid derivatives +t-butylhydroquinone	polycondensation	96
methacrylate	polyacrylamide	radical	97, 98
vinyl	–		91
p-vinylphenyl	butadiene	anionic	99
dicarboxyl	p-hydroxybenzoic acid + hydroquinone + t-butyl hydroquinone + isophthalic acid + biphenyl - 4,4'-diol + terephthalic acid	polycondensation	100
methacrylate	ethyl acrylate	radical	101
methacrylate	ethyl acrylate or acrylic acid	radical	102
dihydroxy	polycarbonate	polycondensation	103
methacrylate	HEMA/PHPMA/MAA/AMPS/FA	radical	104
methacrylate	butylacrylate + acrylic acid	radical	106
methacrylate	methacrylic acid	radical	107
vinyl	cycloheptene or hexene-2 or methyl pentene-1+SO$_2$		108

2.2.3.2
Macromonomers Prepared Anionically

Although styrene is the most common monomer used for the preparation of macromonomers anionically, other monomers capable of anionic polymerization have been used. Several papers have appeared in the literature concerning the anionic ring opening polymerization of hexamethyl cyclotrisiloxane (D_3) followed by suitable termination for the preparation of macromonomers. One of the methods that widely used functionalization reactions, capable of producing monodisperse macromonomers with controlled functionality, is illustrated in the following Scheme 36.

sec-BuLi initiated the anionic polymerization of D_3 in cyclohexane followed by the addition of THF to accelerate the propagation. The living siloxanolate was terminated with 3-methacryloxypropyl dimethylchlorosilane to produce the macromonomers [109].

Hexenyldimethyl chlorosilane was also used as a terminating agent to prepare the 5-hexenyl terminated polydimethylsiloxane (PDMS) [110]. (Scheme 37)

Si-H terminated PDMS was prepared by termination of the living polymer with dimethylchlorosilane. A hydrosilylation reaction was performed between the Si-H terminated PDMS and the product of the reaction between 4-bromo butene-1 and 4,4'-dinitro-2-hydroxybiphenyl to produce the dinitrobiphenyl-terminated PDMS. The subsequent reduction gave the macromonomer [111]. These macromonomers were used for the preparation of graft copolymers via polycondensation reaction with other comonomers (Scheme 38).

Anionic initiation of D_3 with allyllithium leads to the formation of macromonomers [112] (Scheme 39).

This method for the preparation of allyl-terminated PDMS is preferred over the method concerning the termination of living PDMS with allylchloride because the weak -Si-O-C-bonds in -Si-O-CH_2-CH=CH_2 are sensitive to moisture, alkali and acid.

Another monomer frequently used is methyl methacrylate. Functional termination is the most common method for the preparation of macromonomers. MMA macromonomers with methacryloyl end groups were prepared [113] according to the following reaction series (Scheme 40).

Styryl terminated PMMA macromonomer was prepared, reacting the living polymer with p-bromomethyl styrene [114] (Scheme 41).

Vinylmagnesiumchloride was also used to produce PMMA macromonomers [115]. Using similar techniques, several macromonomers have been prepared by the anionic polymerization of ethylene oxide, butadiene, isoprene, 2-vinylpyridine, α-methylstyrene, vinylpyrrolidone etc. Table 2 summarizes the data reported in the literature concerning the synthesis of graft copolymers with anionically prepared macromonomers.

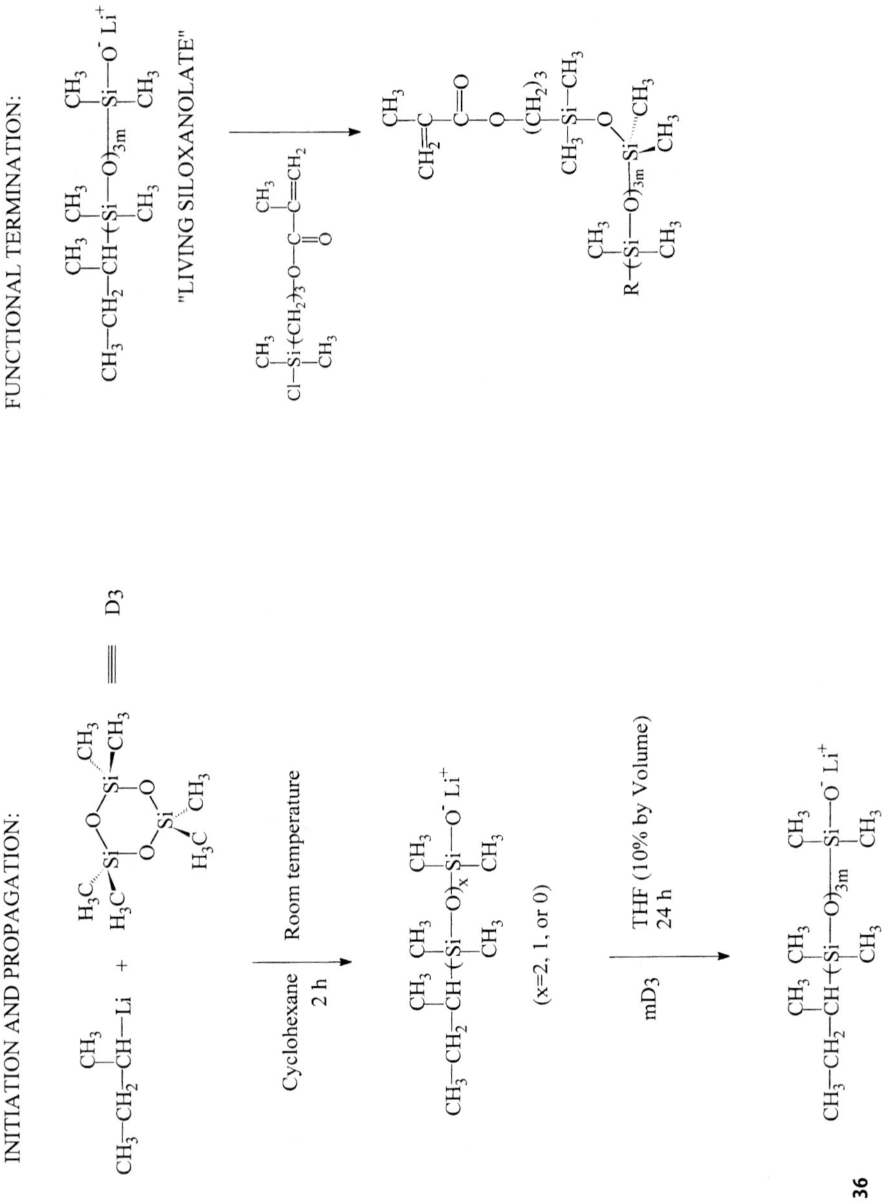
INITIATION AND PROPAGATION:
D3
Cyclohexane
2 h
Room temperature
(x=2, 1, or 0)
mD3
THF (10% by Volume)
24 h
FUNCTIONAL TERMINATION:
"LIVING SILOXANOLATE"

Scheme 36

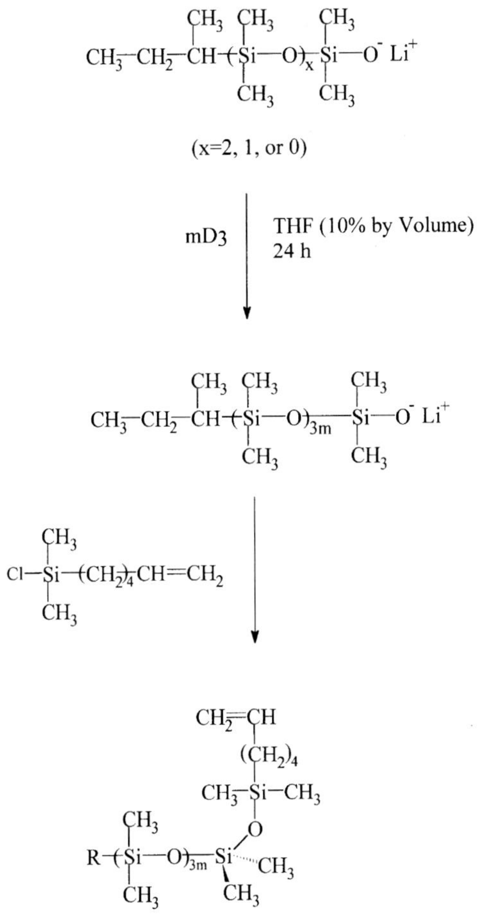

Nonlinear Block Copolymer Architectures

$$CH_3-CH_2-CH \underset{CH_3}{\overset{CH_3}{-}} Si \underset{CH_3}{\overset{CH_3}{-}} O)_x Si \underset{CH_3}{\overset{CH_3}{-}} O^- Li^+$$

(x=2, 1, or 0)

mD3 | THF (10% by Volume)
24 h

$$CH_3-CH_2-CH \underset{CH_3}{\overset{CH_3}{-}} Si \underset{CH_3}{\overset{CH_3}{-}} O)_{3m} Si \underset{CH_3}{\overset{CH_3}{-}} O^- Li^+$$

$$Cl-\underset{CH_3}{\overset{CH_3}{Si}}-(CH_2)_4 CH=CH_2$$

$$R \underset{CH_3}{\overset{CH_3}{-}} Si-O)_{3m} Si \overset{CH_3-Si-CH_3}{\underset{CH_3}{\overset{O}{|}}} CH_3$$
$$\quad\quad\quad\quad\quad CH_2=CH$$
$$\quad\quad\quad\quad\quad (CH_2)_4$$

Scheme 37

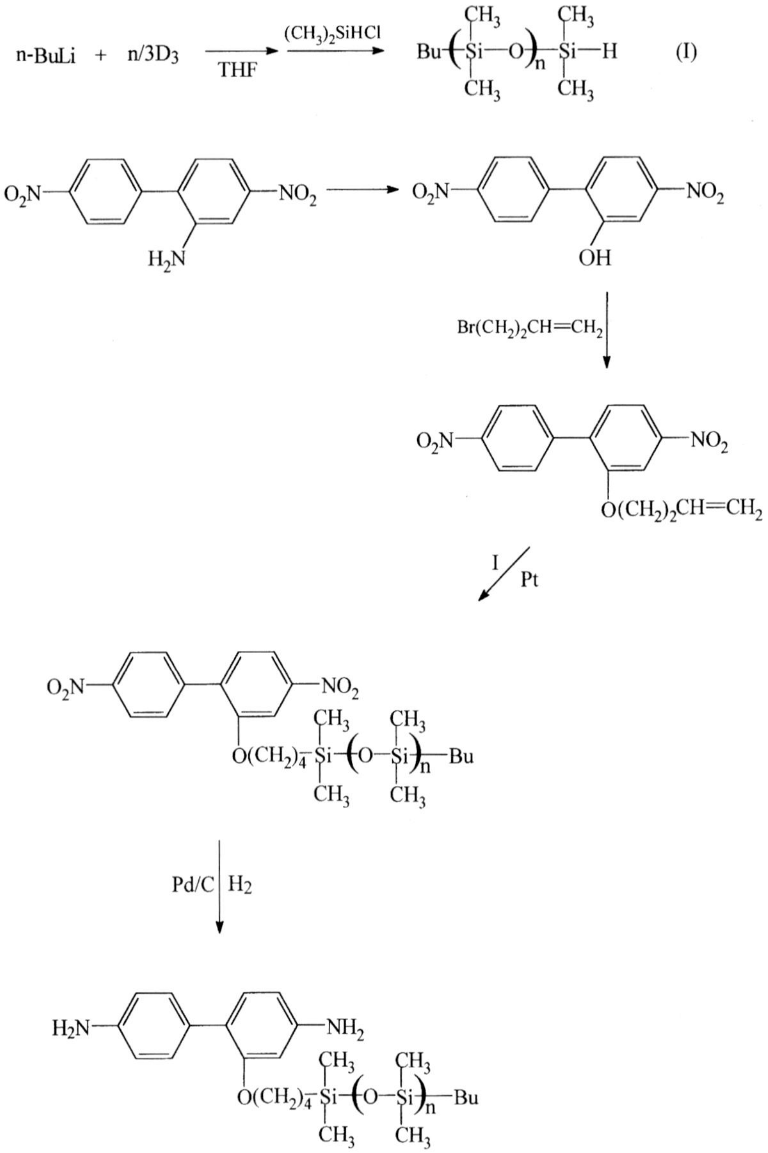

Scheme 38

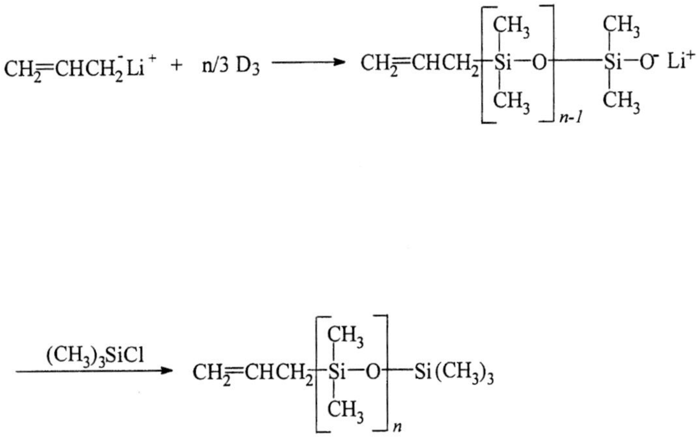

$$CH_2=CHCH_2^-Li^+ + n/3\ D_3 \longrightarrow CH_2=CHCH_2\left[\underset{\underset{CH_3}{|}}{\overset{\overset{CH_3}{|}}{Si}}-O\right]_{n-1}\underset{\underset{CH_3}{|}}{\overset{\overset{CH_3}{|}}{Si}}-O^-\ Li^+$$

$$\xrightarrow{(CH_3)_3SiCl} CH_2=CHCH_2\left[\underset{\underset{CH_3}{|}}{\overset{\overset{CH_3}{|}}{Si}}-O\right]_n Si(CH_3)_3$$

Scheme 39

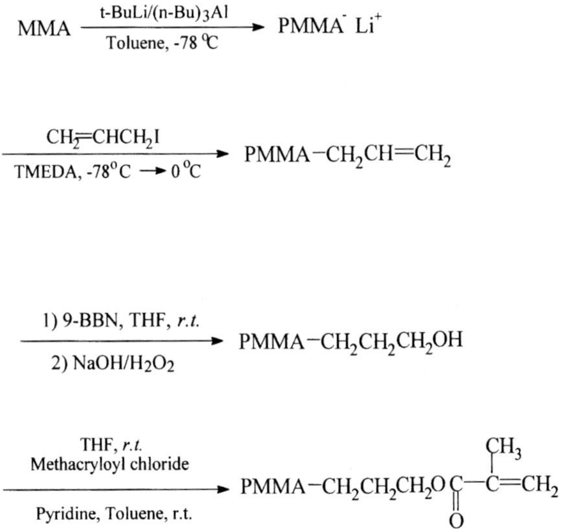

$$MMA \xrightarrow[\text{Toluene, -78 °C}]{\text{t-BuLi/(n-Bu)}_3Al} PMMA^-\ Li^+$$

$$\xrightarrow[\text{TMEDA, -78°C} \rightarrow 0°C]{CH_2=CHCH_2I} PMMA-CH_2CH=CH_2$$

$$\xrightarrow[\text{2) NaOH/H}_2O_2]{\text{1) 9-BBN, THF, r.t.}} PMMA-CH_2CH_2CH_2OH$$

$$\xrightarrow[\text{Pyridine, Toluene, r.t.}]{\substack{\text{THF, r.t.}\\ \text{Methacryloyl chloride}}} PMMA-CH_2CH_2CH_2O\underset{\underset{O}{\|}}{C}-\overset{\overset{CH_3}{|}}{C}=CH_2$$

Scheme 40

Table 2. Graft copolymers prepared by macromonomers synthesized anionically

Macromonomer	End group	Comonomer	Copolymerization technique	Ref.
PDMS	methacryloyl	MMA	GTP	116
PDMS	5-hexenyl	SO_2+1-butene	radical	110, 117
PS	1-butenyl	SO_2+1-butene	UV irradiation	118
polyamide	vinyl benzene	2VP	radical	119
PMMA	vinyl benzene	styrene	radical	120
PDMS	methacryloyl	MMA	radical	109, 121
PDMS	methacryloyl	MMA	anionic	122, 123
PDMS	methacryloyl	MMA	GTP	124, 125
polyamide	vinyl benzene	styrene	radical	126, 127
PEO	acryloyl	1,1-dihydroperfluorooctylacrylate	radical	128
poly(2-isoprenylnaphthalene)	vinylbenzene	n-BuMA	radical	129
PMMA	styryl	styrene	radical	114, 130, 131
PMMA	methacryloyl	butylacrylate	radical	132
PMMA	vinyl	styrene	radical	133
PB, hydrogenated PI	methacryloyl	butylacrylate, MMA	radical	134
PMMA	styryl	liquid crystalline monomer	radical	135
PMMA	methacryloyl	ethyl methacrylate	radical	113
PMMA	vinyl benzyl	styrene	radical	115
PI	methacryloyl	MMA	GTP	116
PDMS	methacryloyl	MMA	radical	136, 137
P2VP	-methylstyryl	styrene	radical	138
P-aMeSt	vinyl benzene	4VP, acrylic acid	radical	139
PS-b-PIs	methacryloyl	styrene	radical	140
PDMS	2-(3,5 diaminophenyl) ethyl	polyimides	condensation	141
PEO	methacryloyl	MMA butylacrylate MMA	radical	142
poly(methyl, phenyl siloxane)	2-methyl butenyl	hexene-2+SO_2, 2-methylpentene-1+SO_2	radical	108
PDMS	methacryloyl	acrylonitrile	radical	143
PDMS	4,4'-diaminobiphenyl	polyimide	condensation	111

Table 2. (continued)

Macromonomer	End group	Comonomer	Copolymerization technique	Ref.
PDMS	allyl	ethylene	Ziegler-Natta	112
PDMS	thiophene	thiophene	radical	144
PDMS	methacrylate	styrene	radical	145
PDMS	vinyl	–	–	146
vinyl pyrrolidone	vinyl benzene	7-(2-methacryloyloxyethyl theophylline N-(2-methacryloyloxyethyl)-4-fluorouracil	radical	147
PS-b-PDMS	diol	diphenylmethyl diisocyanate + butanediol	polycondensation	148
PS-b-PDMS	vinyl	vinyl acetate	radical	148
polyethylene glycol	methacryloyl	MMA	radical	149
PEO	vinyl benzene	2VP	radical	150
Pt-BuMA	dimethylvinylsilyl	4VP	radical	151
PDMS	3,5-diaminiophenyl	vinyl acetate	radical	152
PDMS	methacrylate	pyromellitic anhydride	condensation	141
poly(2-alkyl-oxazolines)		–	–	153
PEO	norbornene	homopolymerization	ROMP	154

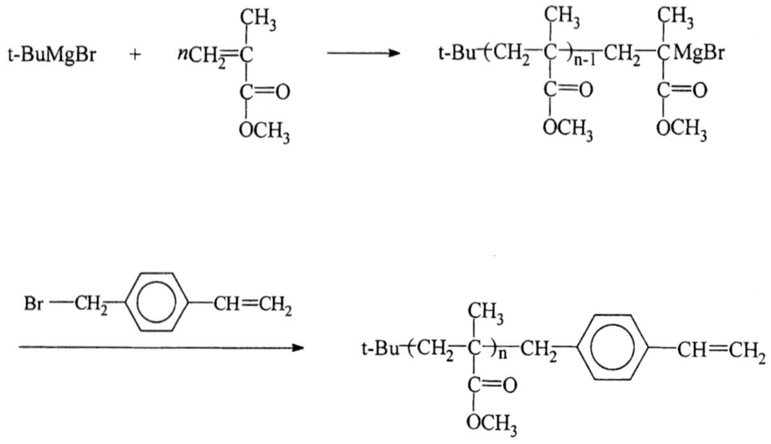

Scheme 41

2.2.3.3
Macromonomers Prepared by Radical Polymerization

The synthesis of macromonomers by free radical methods has been repeatedly discussed in the literature because of the advantages offered by this technique: a) the polymerization procedures are rather simple; b) there is no need for special purification of the reagents; c) many monomers can be subjected to radical polymerization offering unique opportunities for the preparation of graft copolymers. On the other hand, there are also several disadvantages associated with the nature of radical polymerization. The products have broad molecular weight distributions and there is poor control over the molecular weight. The basic method for the introduction of the polymerizable group is by using chain transfer agents. If these reagents are used in large excess, the termination usually leads to monofunctional polymers. However chain coupling by combination can produce difunctional samples, whereas disproportionation leads to non-functionalized materials. So ill-defined polymers are typically be prepared by radical polymerization techniques.

PMMA macromonomers have been prepared by several methods. The most common involves the use of thioglycolic acid, as a chain transfer agent for the preparation of polymers with carboxyl end groups, followed by the reaction with glycidyl methacrylate (GMA) to introduce a methacrylate terminal group [8] (Scheme 42).

Dicarboxyl terminated PMMA was prepared using thiomalic acid as a chain transfer agent [155] (Scheme 43).

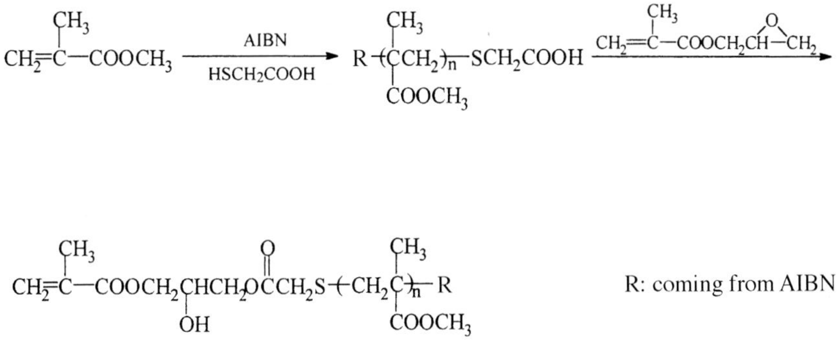

Scheme 42

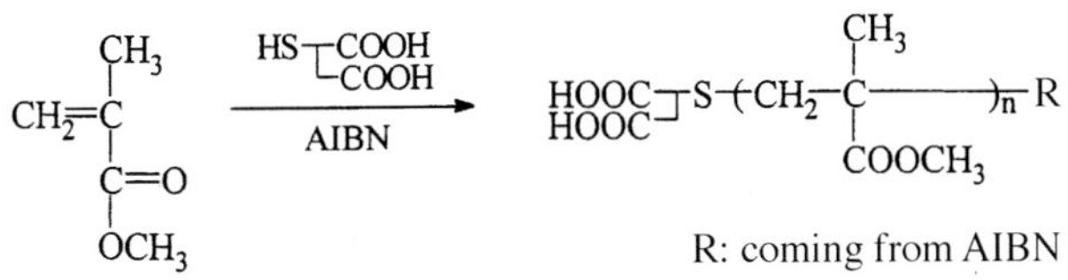

Scheme 43

α -(Bromomethyl) acrylate was also used as a chain transfer agent to produce PMMA with acrylic chain ends [156] (Scheme 44).

Aromatic dicarboxyl-terminated PMMA was prepared in two different ways [155]. Using 2-mercaptoethanol or 2-aminoethanethiol hydrochloride, -OH or -NH_2 end-groups were introduced, respectively. Finally, reaction with trimetallic anhydride provides the desired macromonomers. The following scheme describes these procedures (Scheme 45).

Dihydroxyl terminated PMMA was prepared using α-thioglycerol as a transfer agent [157] according to the following reaction (Scheme 46).

Similar methods have been employed for the preparation of polyvinylpyrrolidone, polystyrene, or polyacetoxyethyl methacrylate macromonomers.

Scheme 44. (see next page) ───────────────────────────────▶

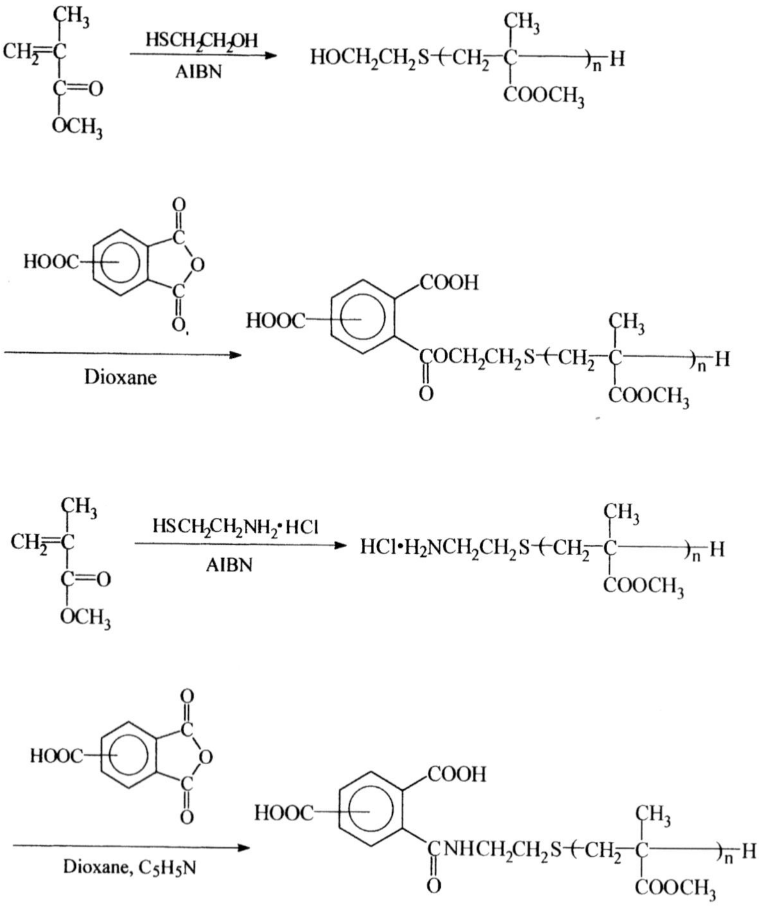

Scheme 45

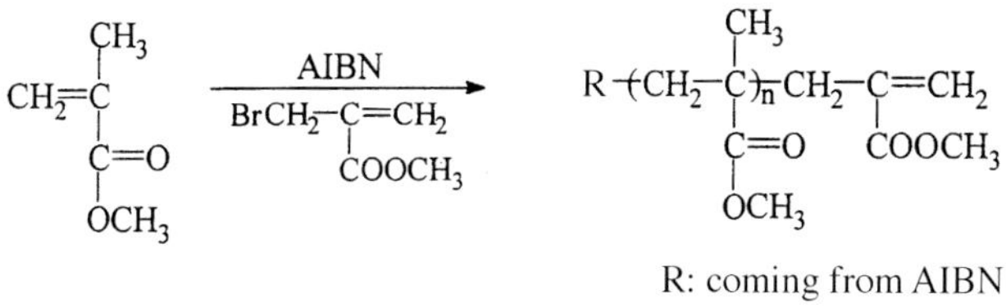

Scheme 44

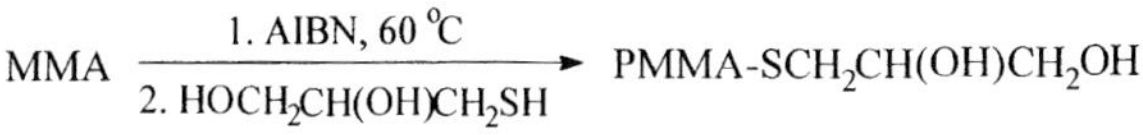

$CH_2=C(CH_3)COOCH_3$ →(AIBN, $BrCH_2-C(COOCH_3)=CH_2$)→ $R-(CH_2-C(CH_3)(COOCH_3))_n-CH_2-C(COOCH_3)=CH_2$

R: coming from AIBN

Scheme 46

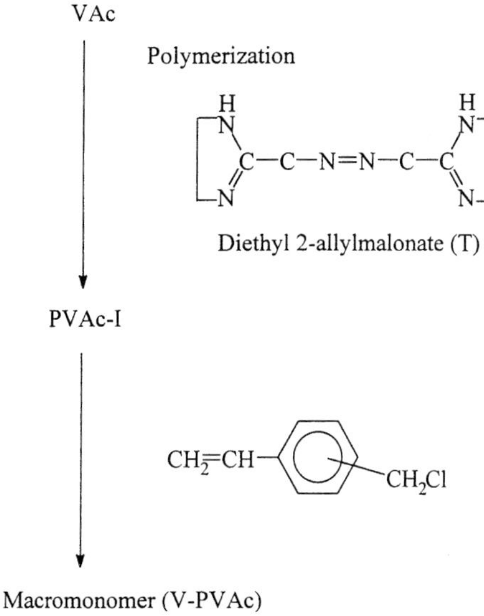

MMA →(1. AIBN, 60 °C; 2. $HOCH_2CH(OH)CH_2SH$)→ $PMMA-SCH_2CH(OH)CH_2OH$

Scheme 47

VAc
↓ Polymerization with Diethyl 2-allylmalonate (T) [imidazole-C-N=N-C-imidazole]
PVAc-I
↓ $CH_2=CH-C_6H_4-CH_2Cl$
Macromonomer (V-PVAc)

Functional initiators have also been used for the synthesis of macromonomers by free radical processes. 2-2'-Azobis(*N*-*N'*-dimethylene isobutyramidine) was used to prepare imidazol terminated PS followed by an end-capping reaction with chloromethylstyrene [158]. A similar initiator was used for the radical polymerization of vinyl acetate (VAc) followed by reaction with chloromethylstyrene or methacryloyl chloride [159] (Scheme 47).

Table 3 summarizes the literature data concerning the preparation of graft copolymers by the use of free radically prepared macromonomers (Table 3).

Table 3. Graft copolymers prepared by macromonomers synthesized by radical polymerization

Macromonomer	End group	Comonomer	Copolymerization technique	Ref.
PMMA	methacryloyl	hydroxyethyl methacrylate FA	radical	104
PMMA	methacrylate	acrylic acid	radical	160
PMMA	methacrylate	vinyl acetate or ethyl acrylate or St or acrylonitrile	radical	161
PMMA	dicarboxyl	m-phenylenediamine + p-aminobenzoic acid	condensation	162
PMMA	acrylate	styrene, vinyl chloride, vinyl acetate	radical	156
PMMA	2-phenylallyl	styrene	radical	163
PMMA	carboxyl	butyl acrylate + acrylic acid	radical	164
polyvinylpyrrolidone	styryl	styrene	radical	165
P4VP	styryl	styrene	radical	165
PMMA	aromatic dicarboxyl	terephthalic acid + bisphenol	condensation	155
PMMA	aromatic dicarboxyl	sebasic acid + m-phenylene diamine	condensation	155
poly(MMA + fluoroalkyl ethyl acrylate)	dicarboxyl		condensation	166
PMMA	methacrylate	2-dimethyl ammonium methyl methacrylate	radical	167
poly(2-dimethyl ammonium methyl methacrylate)	methacrylate	MMA	radical	167
poly(dodecyl acrylate)	acrylate	acrylamide	radical	168
PVAC		St, 4VP, p-chloromethyl styrene	radical	159
PS	carboxyl	ethyl cellulose	UV irradiation	169
PMMA	dihydroxyl	caprolactone	condensation	157
PMMA	methacryloyl	butyl acrylate	radical	170
oligovinyl pyrrolidone	methacrylate or styryl	MMA/styrene	radical	171
oligovinyl pyrrolidone	styryl	styrene	radical	172
oligoacrylamide	styryl	styrene	radical	158
PMMA	methacryloyl	methacrylic acid	radical	173

Table 3. (continued)

Macromonomer	End group	Comonomer	Copolymerization technique	Ref.
poly(ethyl-methacrylate)	methacryloyl	methacrylic acid	radical	173
poly(n-butyl methacrylate)	methacryloyl	methacrylic acid	radical	173
PVAc		4VP	radical	151
poly(2-methacryloyloxyethyl phosphoryl choline)	methacryloyl	butyl methacrylate	radical	174
MMA	phosphatidyl choline and methacrylate azobenzene moieties	–	–	175
vinylidene fluoride	methacrylate	ethyl acrylate	radical	176
PMMA	2-methyl-2-propenoate ethylbenzene	styrene	radical	177
poly(acetoxyethyl methacrylate)	2-methyl-2-propenoate	MMA	radical	177
PS	dihydroxyl	6;'-(4,4'-bi-phenylene dioxy) di-1-hexanol + di-phenyl N,N'-hexamethylenedicarbamate	condensation	178
PS	methacrylate	styrene + acrylonitrile	radical	179
poly(phenylene ether)	methacrylate	styrene + acrylonitrile + butyl acrylate	radical	179

2.2.3.4
Macromonomers Prepared by Cationic Polymerization

Only a few examples of macromonomers prepared cationically have appeared in the literature despite the development of many systems susceptible to living cationic polymerization. Similarly, with all the living procedures the polymerizable end group can be introduced by either functional termination or functional initiation.

Termination of poly(tetrahydrofuran) by 3-sodio-propyloxydimethylvinylsilane produces a macromonomer with vinyl silane end groups [180], as shown in the following reaction (Scheme 48).

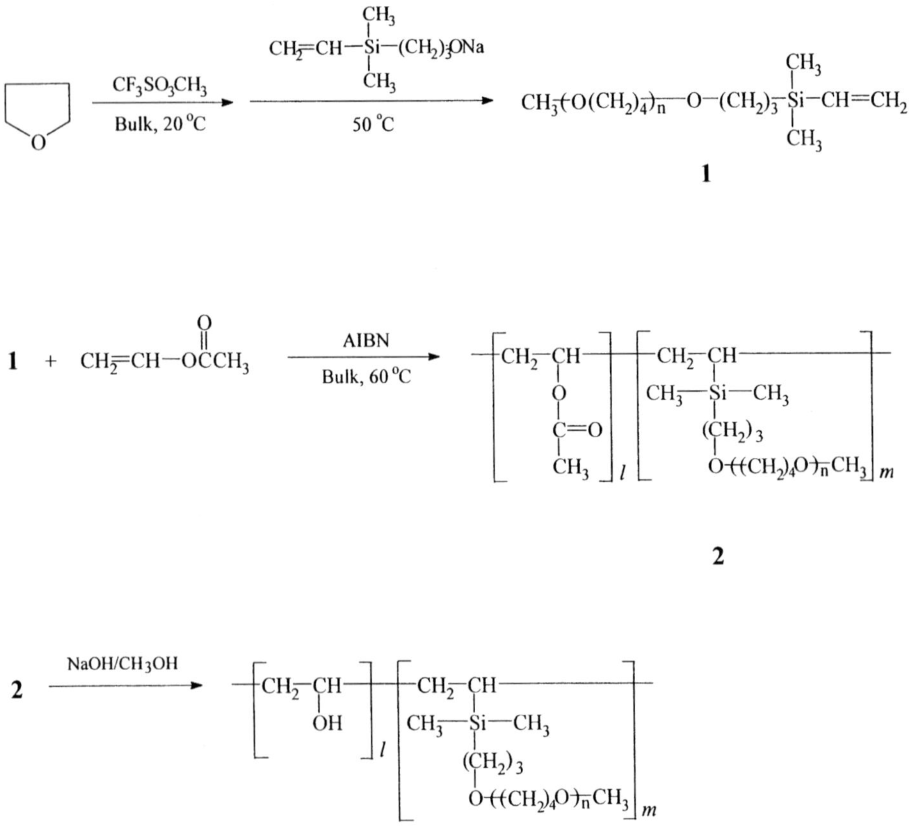

Scheme 48

These macromonomers were copolymerized with vinyl acetate, using AIBN as initiator to produce PVAc-g-PTHF graft copolymers. Subsequent saponification with NaOH resulted in the formation of poly(vinyl alcohol)-g-PTHF graft copolymers.

2-Alkyl-2-oxazoline macromonomers were prepared by termination of the living polymers with diethanolamine [181] as illustrated in the following reactions (Scheme 49).

These macromonomers were used for the preparation of polyurethane-g-poly(2-alkyl-2-oxazoline) graft copolymers. Initially a prepolymer was formed by the reaction of poly(ε-caprolactone) having 2-OH end groups with 4-4′methylenedi(phenylisocyanate). Copolyaddition of this product with the macromonomers provided the graft copolymers. The reaction is outlined in the following Scheme 50.

Poly(2-alkyl oxazoline)s having methacrylate or acrylate end groups were prepared by two methods [182]. a) Living polyoxazoline chains, prepared using methyl p-toluene sulphonate as initiator, were end-capped by reaction with metal salts or tetraalkylammonium salts of acrylic or methacrylic acid or a trialkylammonium salt or trimethylsilyl ester of methacrylic acid (functional termination). b) The living polymers were terminated with water in the presence of Na_2CO_3 to provide hydroxyl-terminated chains. Subsequent acylation with acryloyl or methacryloyl chloride in the presence of triethylamine led to the formation of the macromonomers. The procedures are outlined in the following Scheme 51.

Functional initiators were also used for the synthesis of macromonomers. Initiation of the polymerization of propylene oxide or epichlorohydrin (ECH) with hydroxyethylacrylate (or hydroxymethylmethacrylate) and BF_3 OEt_2 provided macromonomers with acrylate (or methacrylate) end-groups [183]. Subsequent radical copolymerization with S or MMA produced the graft copolymers: PS-g-PPO, PS-g-PECH and PMMA-g-PECH.

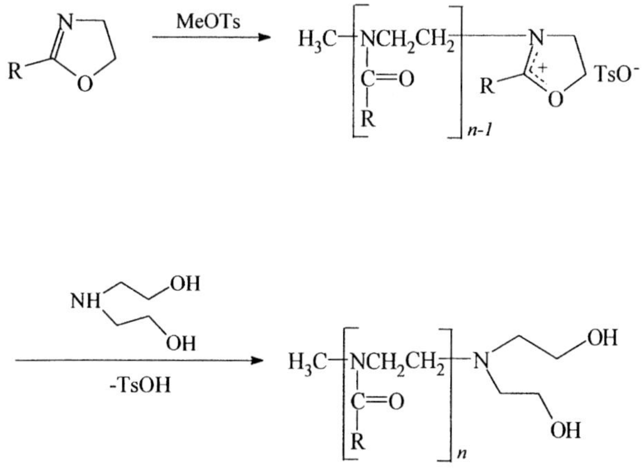

Scheme 49

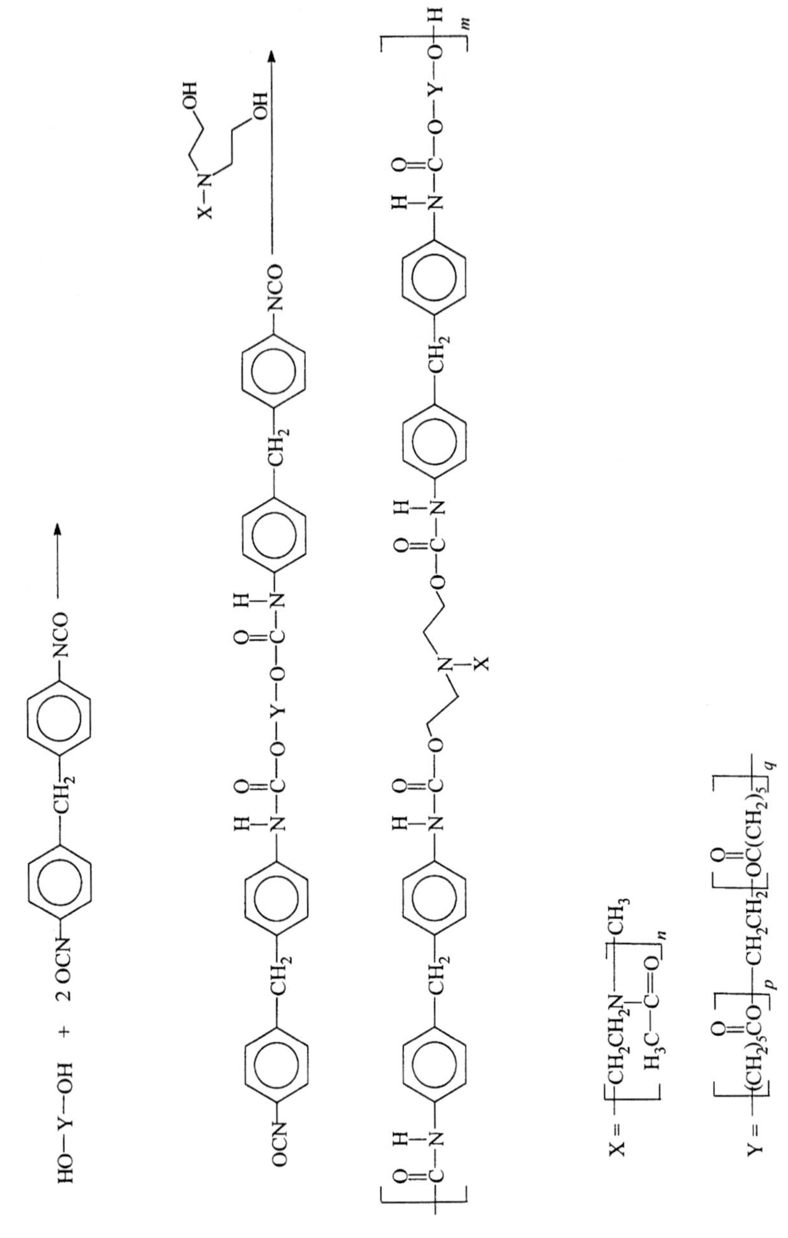

Scheme 50

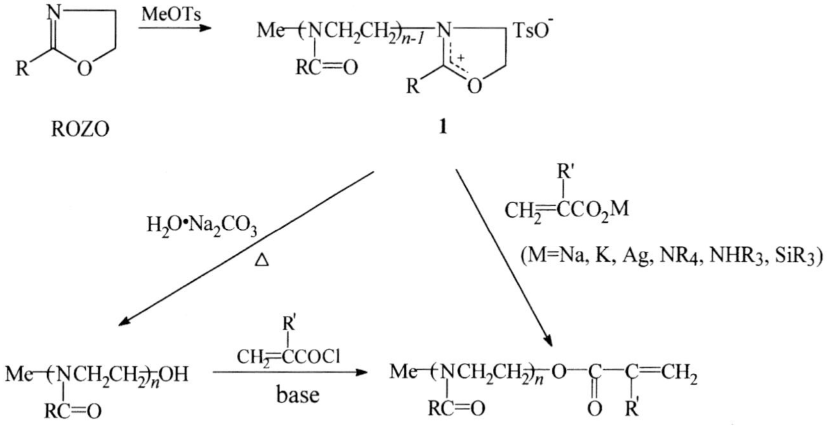

Scheme 51

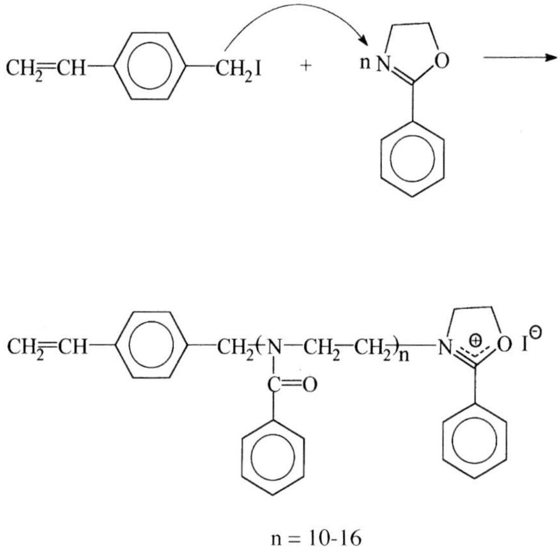

Scheme 52

p-Iodomethylstyrene proved to be an efficient initiator for the polymerization of 2-phenyl-2-oxazoline, resulting in macromonomers with styryl end groups [184], according to the following reaction (Scheme 52).

Radical copolymerization with styrene produced PS-g-poly(2-phenyl-2-oxazoline) graft copolymers.

3,3,5-Trimethyl-5-chloro-1-hexyl methacrylate was used as initiator for the polymerization of isobutylene, leading to macromonomers with methacrylate end-groups [185]. Group transfer polymerization of these macromonomers with MMA provided P(MMA-g-isobutylene) graft copolymers (Scheme 53).

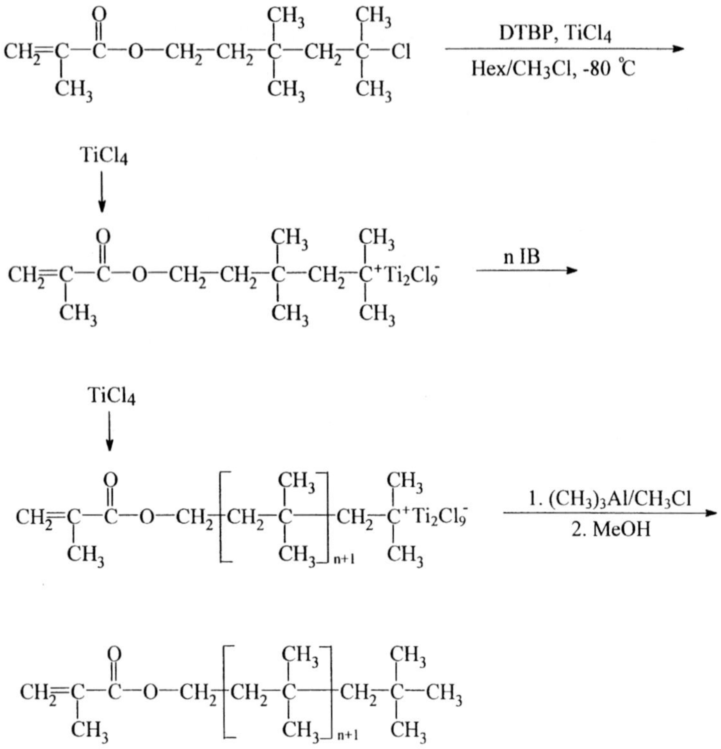

Scheme 53

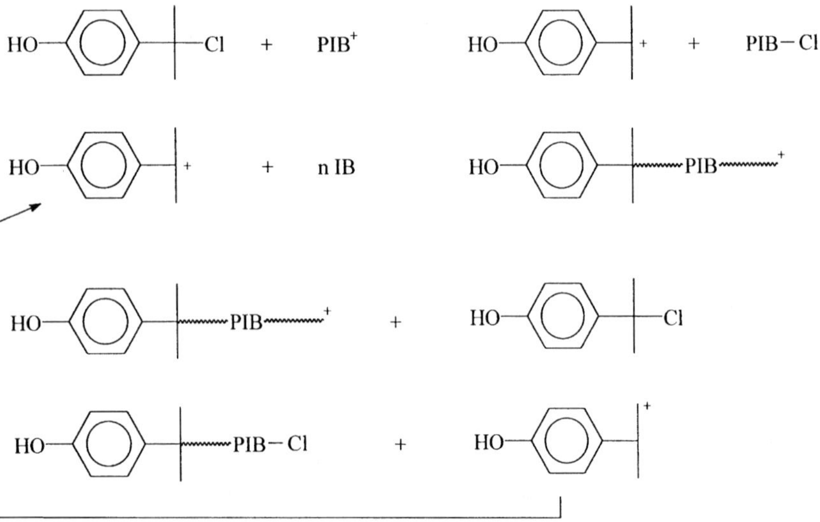

Scheme 54

Kennedy and Carter used the "inifer" method to prepare polyisobutylene (PIB) macromonomers with phenol end-groups [186]. The method involves the polymerization of IB with BCl_3 and p-hydroxycumylchloride as the initiation system. This system gives rise to transfer reactions but both these reactions and the ion splitting of the BCl_4^- anion produce the same species, as shown in the following Scheme 54.

These macromonomers were reacted with a halohydrin to yield phenyl glycidyl ether end-groups. Subsequent copolymerization with ethylene oxide or epichlorohydrin, using the Vanderberg catalyst (triethylamine/water mixture) produced graft copolymers [187].

2.2.3.5
Macromonomers Prepared by Condensation Polymerization

The polymers prepared by condensation techniques have broad molecular weight distributions but the method is applicable to monomers which cannot be polymerized otherwise. Therefore the synthesis of macromonomers and subsequently graft copolymers with polycondensation methods present special interest.

2-Amino-4-(m-amino-N-alkylanilino)-6-isopropenyl-1,3,5-triazines with ethyl and octadecyl alkyl groups were prepared [188]. Self-polyaddition of these monomers provides polyguanamine macromonomers (Scheme 55).

Poly(oxy-1,4-phenylene) macromonomers were prepared in a one step Ullmann type condensation of potassium 4-bromophenolate and 1-bromo-2,5-dimethoxybenzene [189] (Scheme 56).

The protective methyl groups can be removed with treatment with iodotrimethylsilane and subsequent acidification according to the following Scheme 57.

Poly(γ-benzyl-L-glutamate) macromonomers were synthesized starting with the polymerization of benzyl (S)-3-(2,5-dioxo-1,3-oxazolidin-4-yl) propionate (or γ-benzyl-L- glutamate-N-carboxyanhydride [189]) with the primary amino group of the N-methyl-N-(4-vinylphenethyl) ethylene diamine (1) [190] (Scheme 58).

Reaction of this product with aminoalcohols produces another type of macromonomer with side -OH groups (Scheme 59).

Polyamine macromonomers with *tert*-amino groups were prepared by self-polyaddition of the 1:1 adduct of 1,4-divinylbenzene and N,N-diethylethylene diamine using lithium acrylamide as a catalyst [191] according to the reaction (Scheme 60).

Table 4 summarizes the data reported on the synthesis of graft copolymers through macromonomers prepared by polycondensation methods.

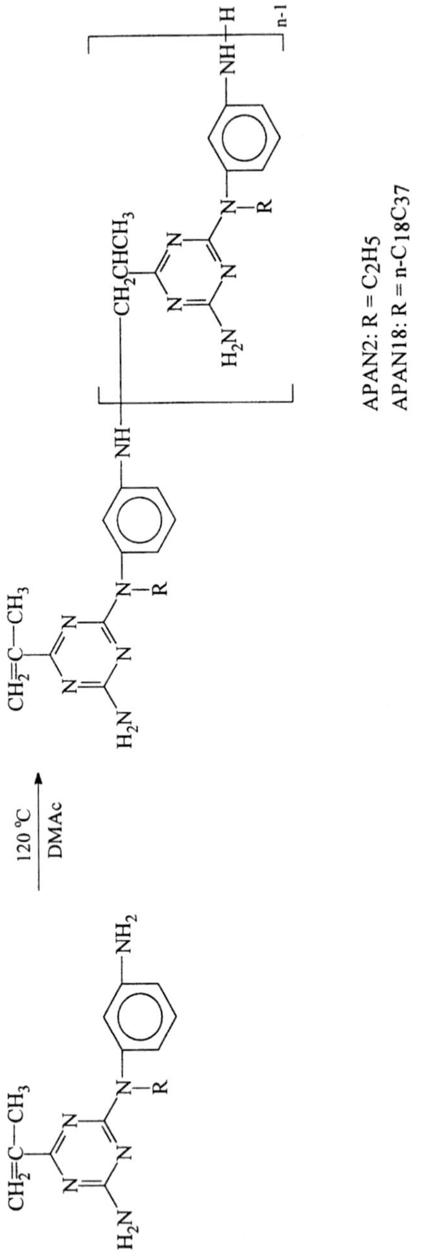

Scheme 55

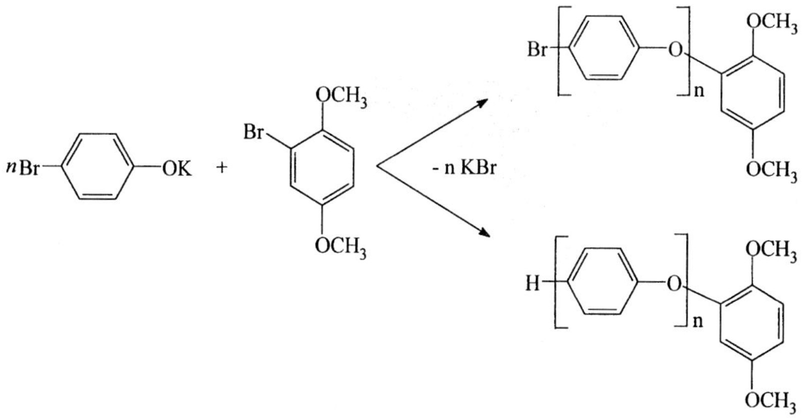

Scheme 56

Scheme 57. (see next page)

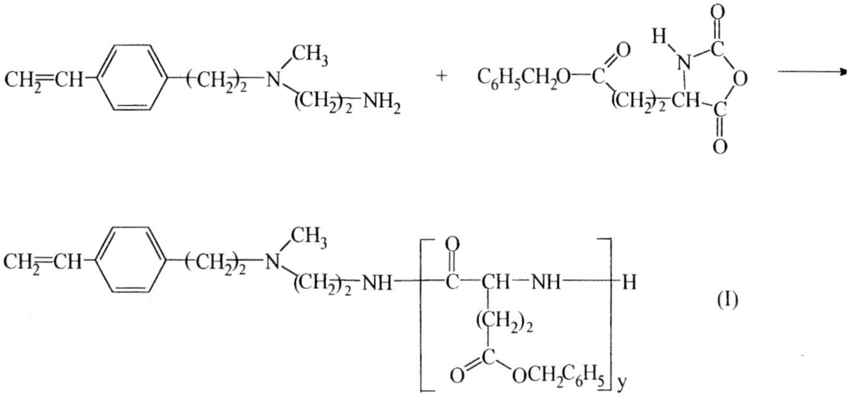

Scheme 58

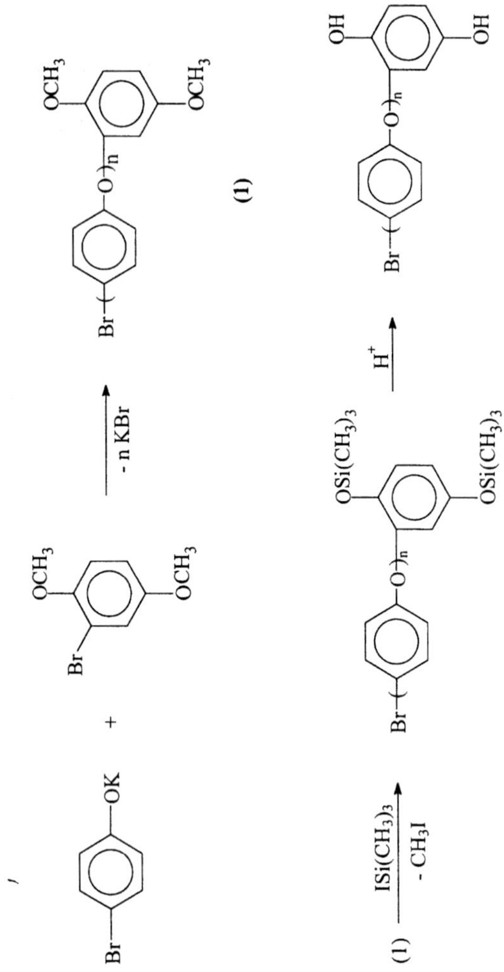

Scheme 57

Nonlinear Block Copolymer Architectures

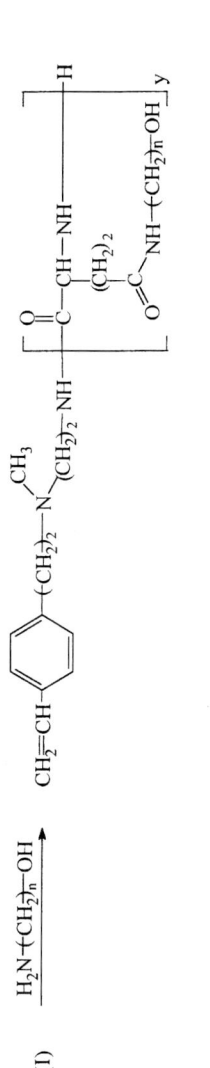
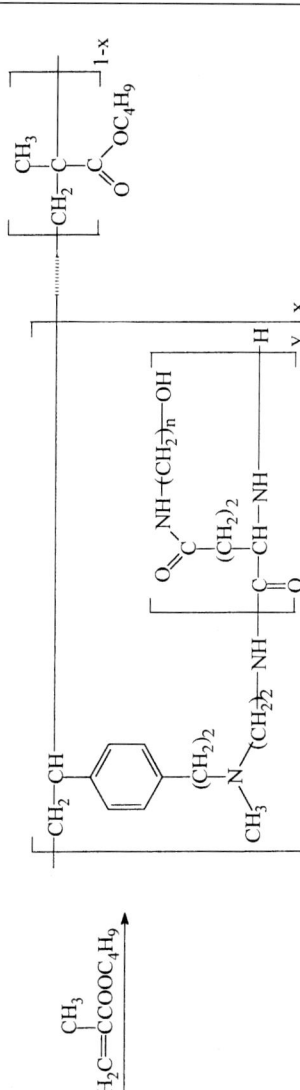

Scheme 59

Scheme 60

Table 4. Graft copolymers prepared by macromonomers synthesized by condensation methods

Macromonomer	End group	Comonomer	Copolymerization technique	Ref.
polyguanamine	Isopropenyl substituted triazine	styrene, MMA	radical	188
acrylamide-diisocyanodiamide		acrylamide	radical	192
poly(oxy-1,4-phenylene)	p-hydroxy-phenol	t-butylhydroquinone + trifluoro-methylterephthaloyl dichloride	polycondensation	189
2-(N-methyl-N-4-vinylphenethylamino) ethylaminopoly[N5-(2-hydroxyethyl) or (3-hydroxypropyl) glutamine	styryl	n-butyl-methacrylate	radical	190
polyurethane	methacrylate	MMA	radical	193
polyamine	styryl	2-hydroxylethyl methacrylate	radical	191
polyurethane	methacrylate	MMA	radical	194

2.2.3.6
Macromonomers Prepared by Post Polymerization Reactions

The most common example of macromonomers in this category are polyethylene oxide and propylene oxide. From their method of preparation they have one or two end -OH groups which can be used for post-polymerization reactions. Acryloylchloride, methacryloylchloride, *p*-vinyl benzyl chloride, and isocyanatomethylmethacrylate are some of the reagents reacted with PEO or poly(propylene oxide) (PPO) to prepare macromonomers. A few of these reactions [195, 196] are presented in the following schemes (Scheme 61).

Macromonomers were also prepared from poly(2,6-dimethyl-1,4-phenylene oxide) by reacting the terminal -OH group with *p*-vinyl benzylchloride, norbornylchloride, or 4-vinyl benzoylchloride [197], according to the following reactions (Scheme 62).

Another method reported in the literature involves the hydrosilylation reaction of compounds bearing double bonds with hydrosilyl-terminated polydimethylsiloxane under the appropriate conditions [198, 199] (Scheme 63, 64).

Oligourethane macromonomers were synthesized via the reaction of amino-terminated oligourethanes with dibenzyl-2-[(chloroformyl)oxy]methyl-N,N'-piperazine dicarboxylate [200], as illustrated in the following Scheme 65.

The protective groups were removed by catalytic hydrogenation to provide piperazine end groups.

The PS half ester of maleic acid was esterified with poly(ethylene glycol) monoether or PS-*b*-PEO in the presence of dicyclohexylcarbodiimide (DCC) and 1-hydroxybenzotriazole (HOBT) to give macromonomers having internal double bonds [201] (Scheme 66).

Table 5 summarizes the data reported in the literature on the synthesis of graft copolymers prepared from macromonomers synthesized by post polymerization reactions.

2.2.3.7
Macromonomers Prepared by Group Transfer Polymerization (GTP)

The development of group transfer polymerization was followed shortly by the formation of macromonomers prepared by this technique. The method involves a nucleophilic polyaddition of trialkylsilyl vinyl ether monomers using an aldehyde as initiator in presence of a Lewis acid which acts as a catalyst [217]. The most common method for the preparation of macromonomers by this technique is the use of aldehydes having functional groups (functional initiation).

Poly (vinyl alcohol) macromonomers having the hydroxyl groups protected as silyl ethers were prepared using *p*-formylstyrene as initiator [218] (Scheme 67).

Trimethylsilyloxy ethyltrimethylsilyl dimethyl ketene was the initiator used for the preparation of PMMA macromonomers. Tetrabutylammonium benzoate was the catalyst. Treatment with dilute HCl provided the -OH functional macromonomers, whereas the subsequent reaction with acryloylchloride formed the final structure of the macromonomer [219] (Scheme 68).

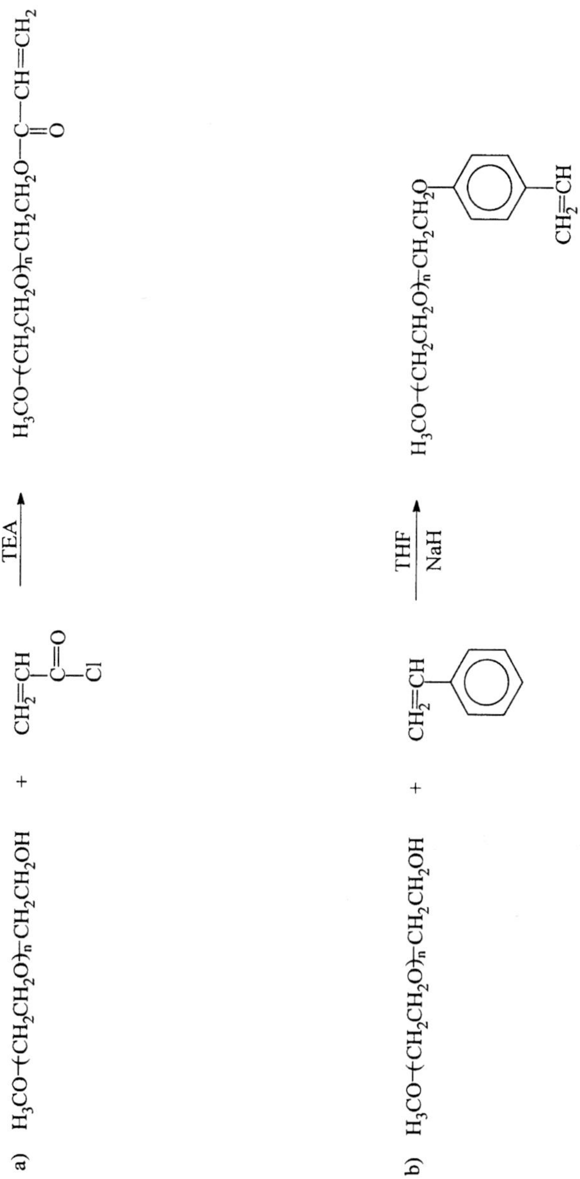

a) $H_3CO\text{-}(CH_2CH_2O)_n\text{-}CH_2CH_2OH$ + $CH_2\text{=}CH\text{-}C(\text{=}O)\text{-}Cl$ $\xrightarrow{\text{TEA}}$ $H_3CO\text{-}(CH_2CH_2O)_n\text{-}CH_2CH_2O\text{-}C(\text{=}O)\text{-}CH\text{=}CH_2$

b) $H_3CO\text{-}(CH_2CH_2O)_n\text{-}CH_2CH_2OH$ + $CH_2\text{=}CH\text{-}C_6H_5$ $\xrightarrow[\text{NaH}]{\text{THF}}$ $H_3CO\text{-}(CH_2CH_2O)_n\text{-}CH_2CH_2O\text{-}C_6H_4\text{-}CH\text{=}CH_2$

Scheme 61

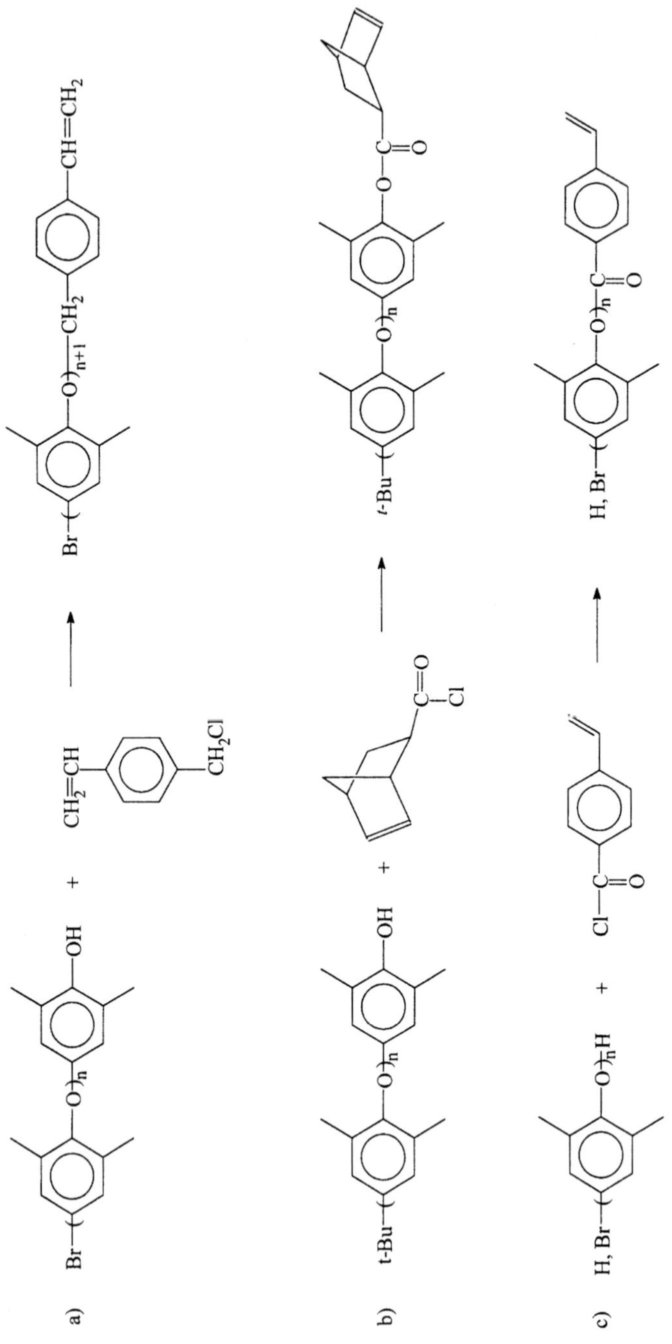

Scheme 62

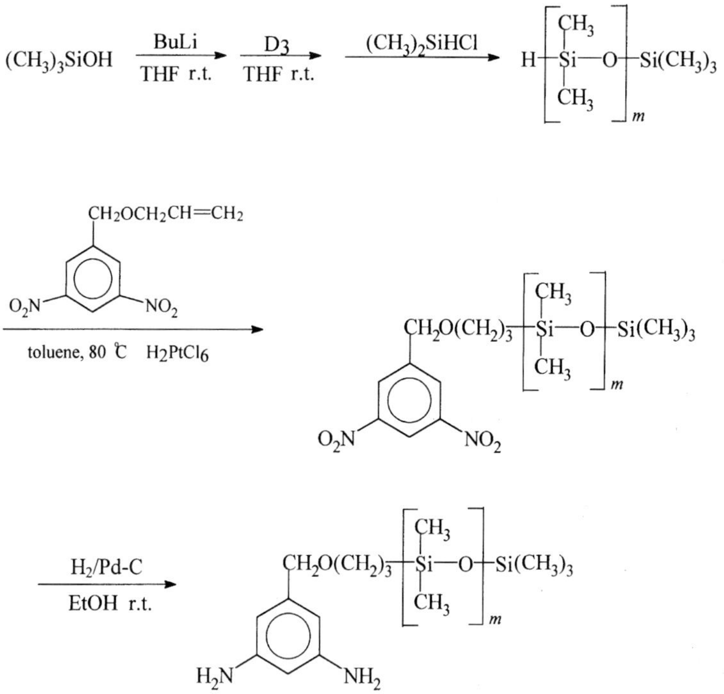

Scheme 63

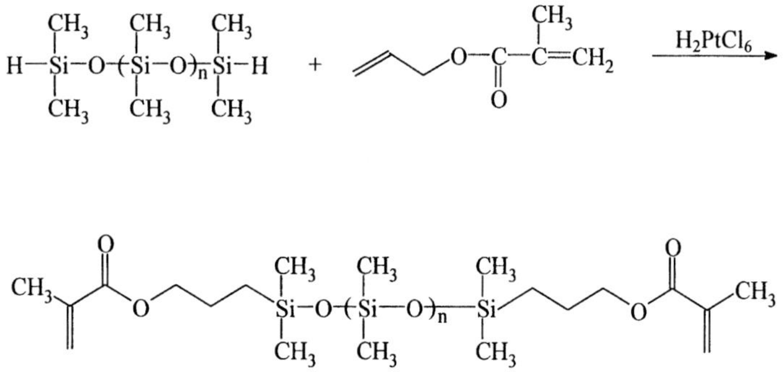

Scheme 64

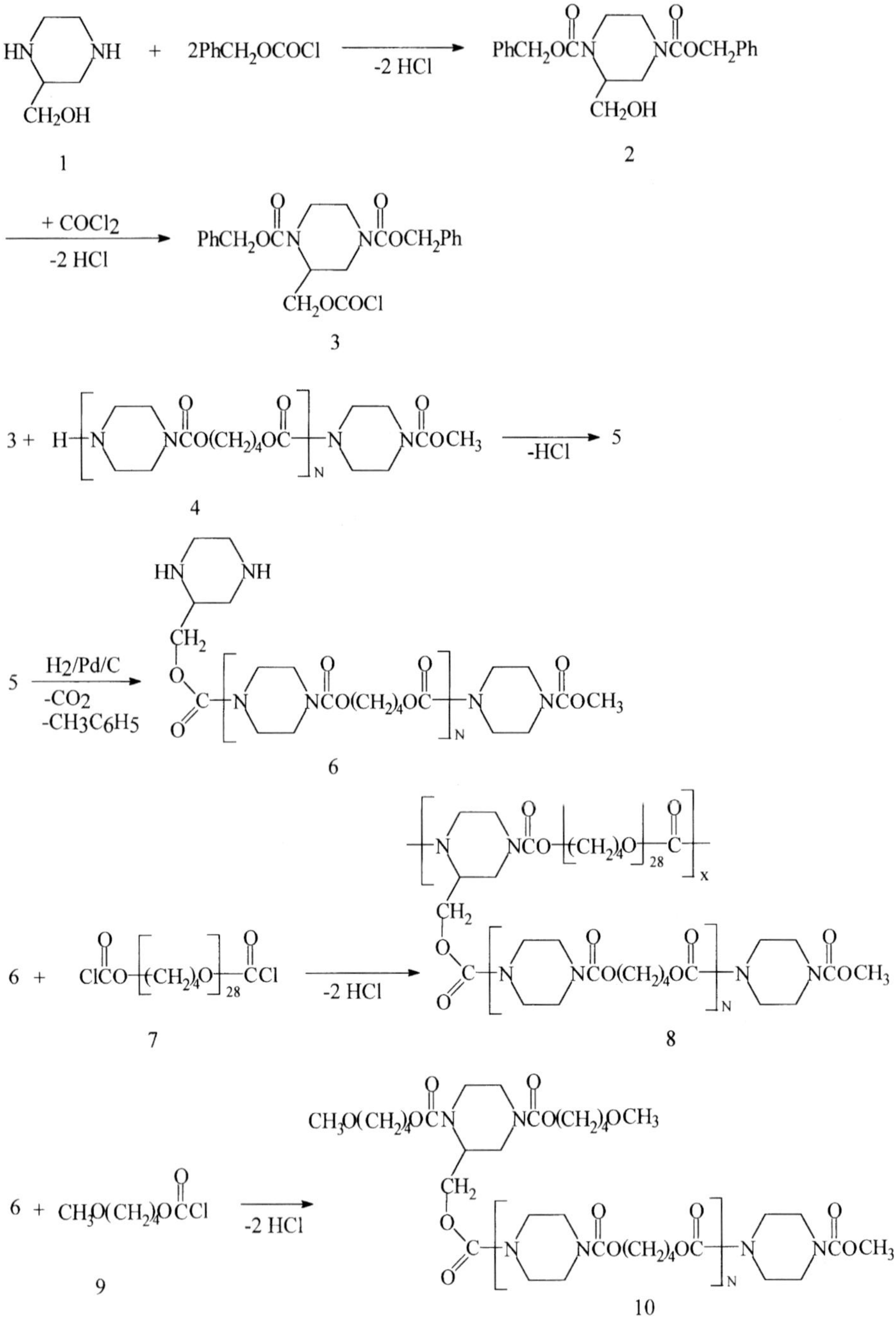

Scheme 65

Nonlinear Block Copolymer Architectures

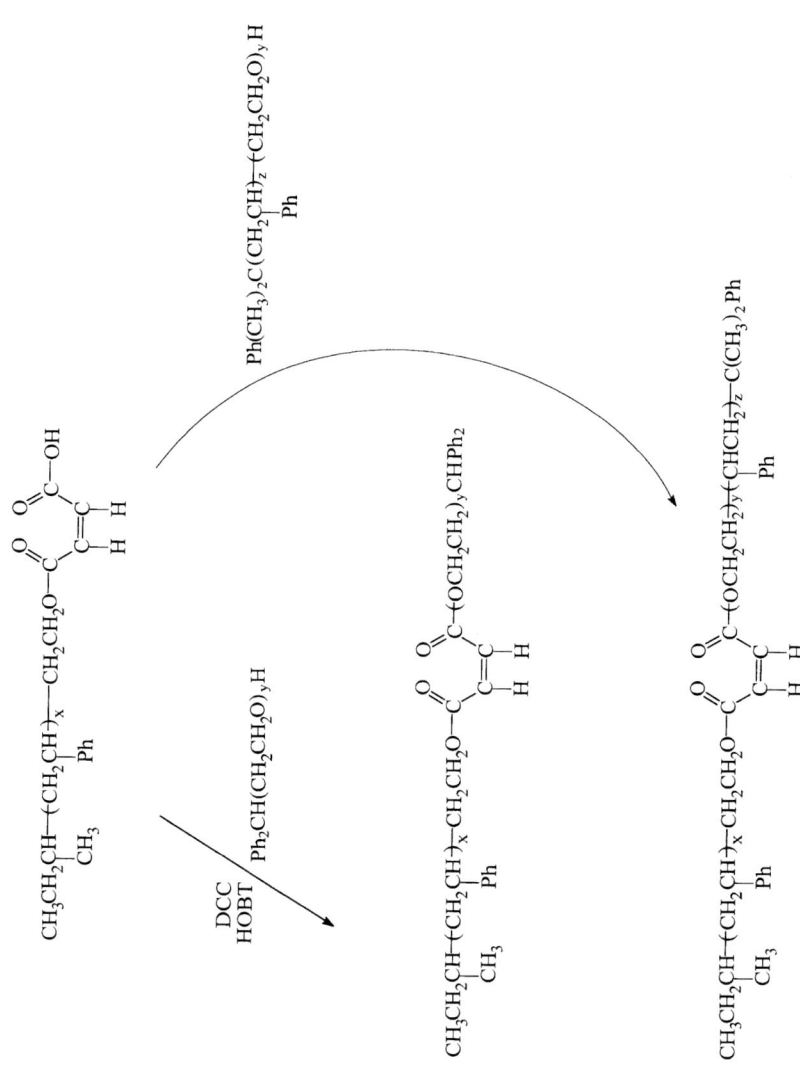

Scheme 66

Table 5. Graft copolymers prepared by macromonomers synthesized by post polymerization reactions

Macromonomer	End group	Comonomer	Copolymerization technique	Ref.
PEO	acrylate	styrene	radical	195
poly(2,6-dimethyl-1,4-phenylene oxide)-PPO	vinyl benzyl	acrylonitrile	radical	202
PDMS	methacrylate	styrene	radical	199
PPO	norbornene	norbornene derivative	ring opening metathesis	203
PEO	vinyl benzyl	n-butylmethacrylate	radical	196
w-stearyl PEO	acrylate	styrene	radical	204
PPO	4-vinyl benzyl	isoprene, butadiene, styrene, ethyl and butylacrylate	radical	197
PEO	methacrylate	styrene, MMA	radical	205
PEO	methacrylate	acrylonitrile	radical	206
PEO	methacrylate	acrylamide	radical	207
PEO	methacrylate	MMA	radical	208
PEO	methacrylate or acrylate	–	radical	209
PEO	methacrylate	MMA or 2-ethyl hexyl acrylate	radical	210
PIB	methacrylate	MMA	anionic	211
propylene oxide PPO	methacrylate	styrene	radical	212
PEO	methacrylate	MMA	radical	213
PS-b-PEO	central double bond	styrene	radical	201
PDMS	3,5-diaminobenzyl	polyimide	polycondensation	198
polyurethane	piperazine	poly(oxytetramethylene)	–	200
polyurethane	α-methyl styrene	MMA	radical	214
poly(β-methyl-δ-valerolactone)	vinyl or isopropenyl	MMA or perfluororo-octyl ethyl methacrylate or N-vinyl pyrollidone + 2HEMA	radical	215
hydroxy-functional oligosiloxane	methacrylate	BuMA + hydroxypropyl methacrylate	radical	216

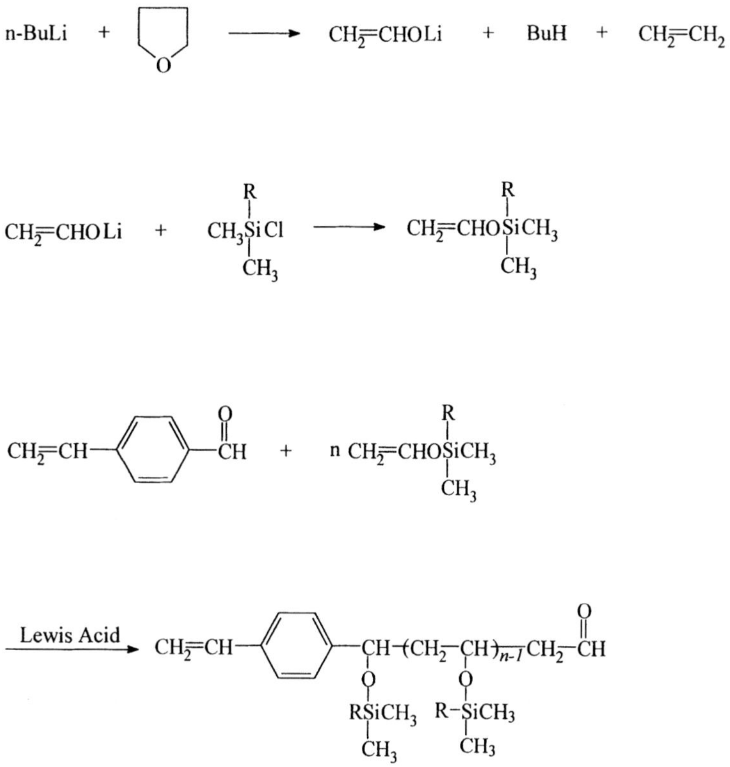

$$\text{n-BuLi} + \underset{\text{O}}{\square} \longrightarrow CH_2{=}CHOLi + BuH + CH_2{=}CH_2$$

$$CH_2{=}CHOLi + \underset{CH_3}{\overset{R}{\underset{|}{CH_3{-}Si{-}Cl}}} \longrightarrow \underset{CH_3}{\overset{R}{\underset{|}{CH_2{=}CHO{-}Si{-}CH_3}}}$$

$$CH_2{=}CH{-}\!\!\left\langle\!\!\bigcirc\!\!\right\rangle\!\!{-}\overset{O}{\overset{\|}{C}}H + n\ \underset{CH_3}{\overset{R}{\underset{|}{CH_2{=}CHO{-}Si{-}CH_3}}}$$

$$\xrightarrow{\text{Lewis Acid}} CH_2{=}CH{-}\!\!\left\langle\!\!\bigcirc\!\!\right\rangle\!\!{-}\underset{\underset{CH_3}{\overset{|}{R{-}Si{-}CH_3}}}{\overset{|}{\underset{|}{O}}}CH{-}(CH_2{-}\underset{\underset{CH_3}{\overset{|}{R{-}Si{-}CH_3}}}{\overset{|}{\underset{|}{O}}}CH)_{n\text{-}1}CH_2{-}\overset{O}{\overset{\|}{C}}H$$

Scheme 67

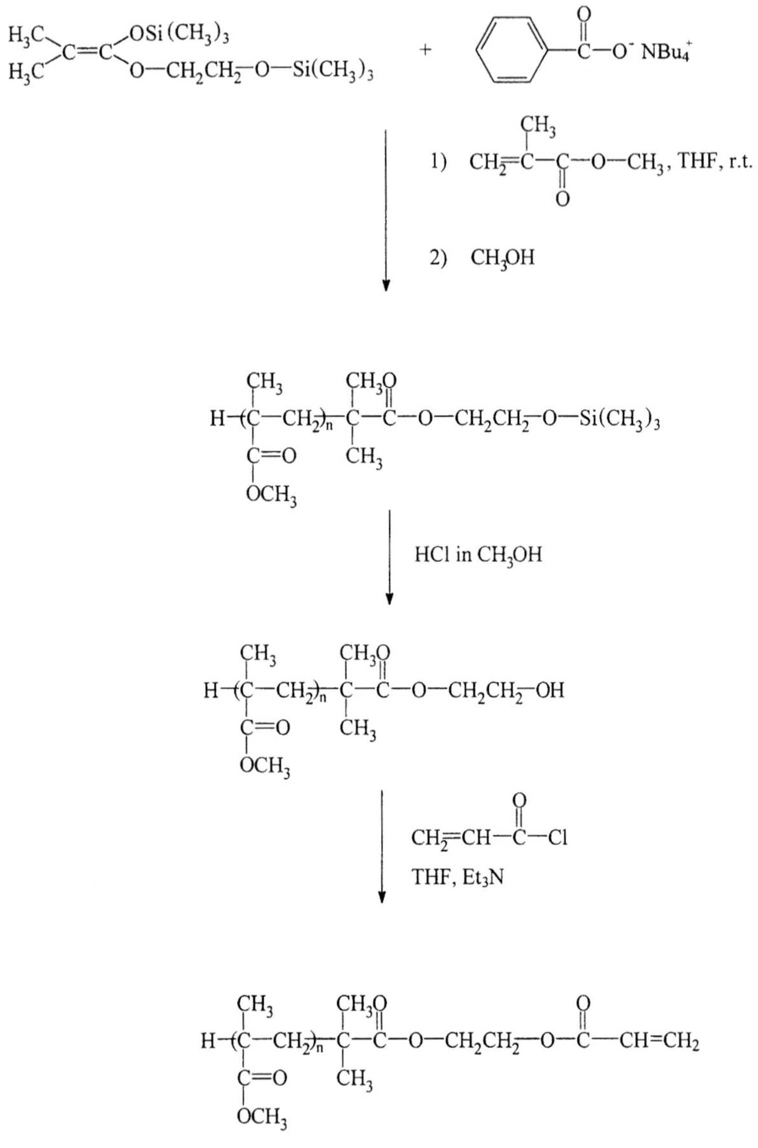

Scheme 68

The GTP of 2-phenyl-1,3,2-dioxaborole with *p*-formylstyrene as initiator in the presence of zinc halides as catalyst followed by hydrolysis gave styryl functionalized oligomeric monosaccharides [220]. The reaction sequence is the following (Scheme 69).

Polyacrylate macromonomers were prepared by GTP using zinc halide, triphenylphosphine, and trimethylchlorosilane. The resulting polymers, having end

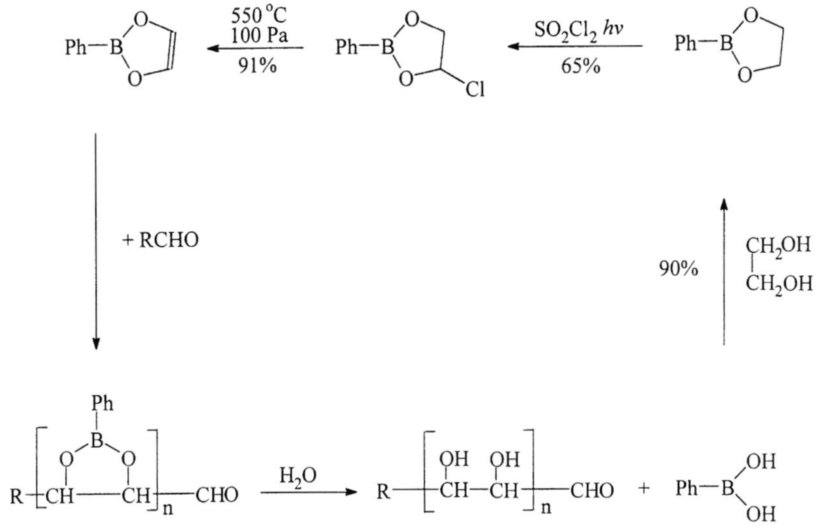

Scheme 69

trimethylphosphinium groups, were converted with sodium ethanolate to the corresponding macromolecular ylide, which undergoes the Wittig reaction to provide the macromonomer [221]. The process is outlined in the following scheme 70.

Table 6 summarizes the work reported in the literature on the synthesis of graft copolymers prepared from macromonomers synthesized by GTP.

Other methods have also been used for the synthesis of macromonomers. For example the synthesis of poly(D,L)-lactide macromonomers was reported [225]. It is known that aluminum alkoxides can be used as initiators for the polymerization of lactides via a coordination-insertion mechanism. A functional initiator was prepared by the reaction of triethylaluminum with 2-hydroxyethylmethacrylate (HEMA) as shown in the following Scheme 71.

The polylactides prepared by this initiator have a methacrylate end-group and possess a narrow molecular weight distribution (I ≤ 1.2). Subsequent radical copolymerization with HEMA provided poly(HEMA-g-(D,L)-lactide) graft copolymers according to the Scheme 72.

If the macromonomer produced from the functional initiator is reacted with methacryloyl chloride instead of being hydrolysed, then α,ω-dimethacryloyl polylactides can be prepared.

Polypropylene macromonomers having vinylidene end-groups were prepared by Ziegler-Natta polymerization using the Cp$_2$ZrCl$_2$/methylaluminoxane (MAO) and the silylene-bridged Brintzinger-type metallocene/MAO catalysts [226]. Atactic and isotactic polypropylenes, respectively, with narrow molecular weight distributions (1.5 ≤ I ≤ 2.5) were prepared. Addition of thioacetic acid to the vinylidene end-group introduced thiol functional groups, whereas hydroboration and subsequent reaction of the resulting hydroxyl group with methacrylic acid led to the formation of methacryloyl-terminated polypropylenes. Radical copolymerization with MMA provided PMMA-g-polypropylene graft copolymers.

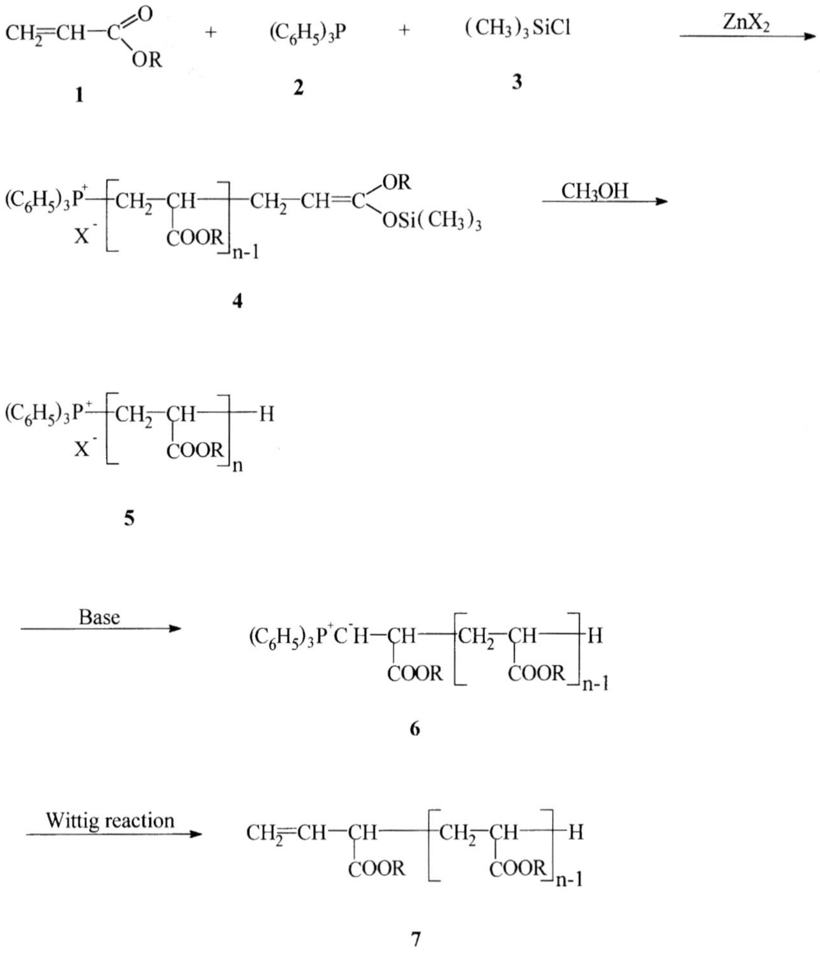

$CH_2=CH-C(=O)OR$ + $(C_6H_5)_3P$ + $(CH_3)_3SiCl$ $\xrightarrow{ZnX_2}$

1   **2**   **3**

$(C_6H_5)_3P^+ - [CH_2-CH(COOR)]_{n-1} - CH_2-CH=C(OR)(OSi(CH_3)_3)$ X^- $\xrightarrow{CH_3OH}$

4

$(C_6H_5)_3P^+ - [CH_2-CH(COOR)]_n - H$ X^-

5

$\xrightarrow{Base}$ $(C_6H_5)_3P^+\bar{C}H-CH(COOR) - [CH_2-CH(COOR)]_{n-1} - H$

6

$\xrightarrow{Wittig\ reaction}$ $CH_2=CH-CH(COOR) - [CH_2-CH(COOR)]_{n-1} - H$

7

1a, 4a, 5a: R=CH$_3$; 1b, 4b, 5b: R=C$_2$H$_5$; 1c, 4c, 5c: R=C$_4$H$_9$

Scheme 70

Table 6. Graft copolymers prepared by macromonomers synthesized by GTP

Macromonomer	End group	Comonomer	Copolymerization technique	Ref.
polyvinyl alcohol	styryl	p-oligodimethyl-siloxanylstyrene	radical	218
PMMA	acryloyl	ethyl hexyl acrylate	radical	219
polyvinyl alcohol	styryl	styrene	radical	222,223
poly(2-phenyl-1,3,2-dioxaborole)	styryl	styrene	radical	220
methyl, hexyl, butyl acrylate	vinyl	styrene	radical	221
PMMA	styryl	styrene	radical	224

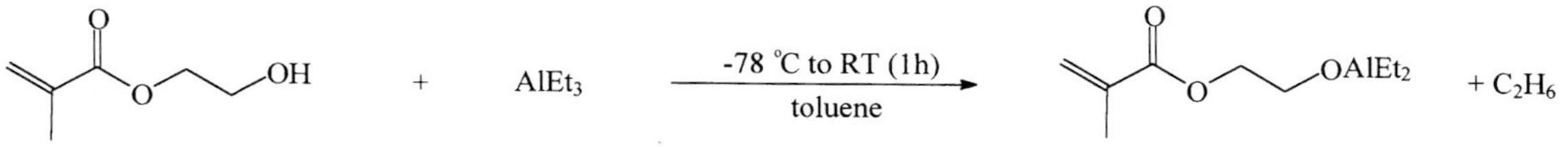

Scheme 71

Ring opening polymerization methods were developed for the synthesis of glycopeptide macromonomers [227]. *O*-(Tetra-*O*-acetyl-β-D-glucopyrasonyl)-L-serine *N*-carboxyanhydride (a) and *O*-(2-acetamido-3,4,6-tri-*O*-acetyl-2-deoxy-β-D-glucopyrasonyl)-L-serine *N*-carboxyanhydride (b) were polymerized using *p*-vinylbenzyl-amine as initiator. The macromonomers thus produced are characterized by narrow molecular weight distributions, and their molecular weights were controlled by the monomer/initiator ratio. Graft copolymers were prepared by radical copolymerization of the macromonomers with acrylamide using AIBN as initiator. The resulting products were subjected to deacetylation by treatment with hydrazine monohydrate in methanol. As an alternative method the

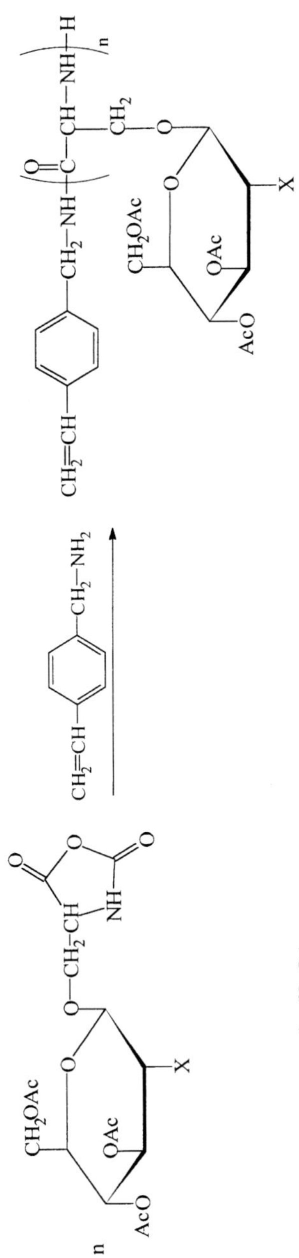

1a, X=OAc
1b, X=NHAc

2a, X=OAc
2b, X=NHAc

Scheme 73

macromonomers were first deacetylated and then copolymerized with acrylamide using 2,2′-azobis(2-amidinopropane) dihydrochloride (AAPD) as initiator. The synthetic procedure is given in the following Scheme 73.

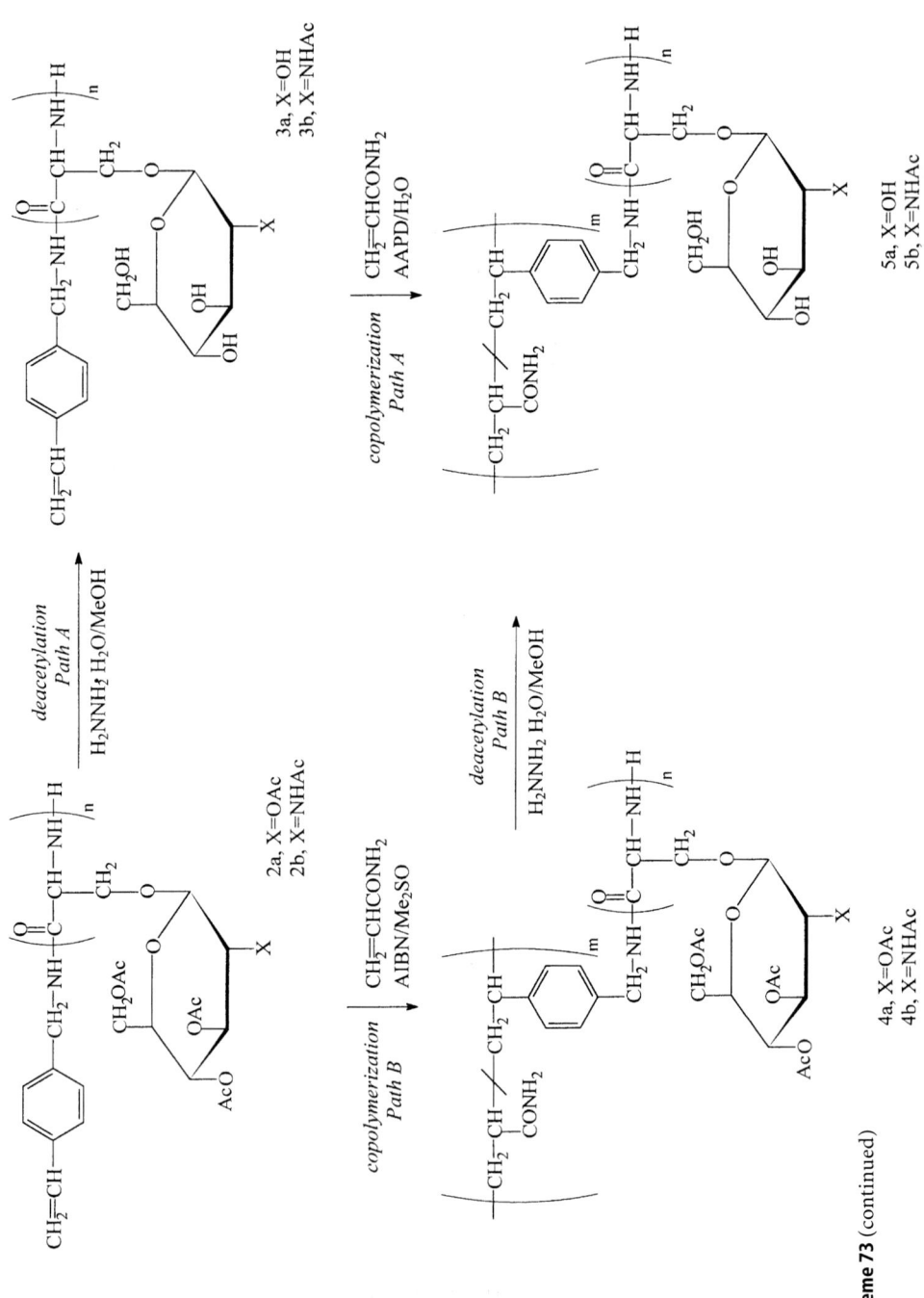

Scheme 73 (continued)

2.3
Miktoarm Star Polymers

Star polymers are the simplest branched species having only one branch point to which several linear chains are attached. Miktoarm stars (mikto from the greek word „μικτοσ" meaning mixed) are a special group of stars containing chemically different arms linked to the branch point. They differ from star-block copolymers, where all the arms are chemically identical and consist of block or triblock copolymers of the A-B or A-B-A type. Several methods have been utilized for the synthesis of miktoarm stars and most of them are based on anionic polymerization techniques. The living character of anionic polymerization in combination with several linking agents provide the means for the synthesis of several types of miktoarm stars. The most common examples found in the literature are stars of the type A_2B, A_3B, ABC, and A_2B_2 with other more extraordinary structures also available (e.g., ABCD, A_qB_q, q>2 etc.).

2.3.1
The Chlorosilane Method

In 1987, Xie and Xia reported the synthesis of A_2B_2 and A_2B miktoarm stars, where A is PS and B is PEO [228]. Their method involves the reaction of living polystyrene chains with $SiCl_4$ in a molar ratio 2:1. This reaction leads to the formation of a two arm star with the remaining Si-Cl bonds available for reaction with the living polyethylene oxide chains in order to produce the $(PS)_2(PEO)_2$ stars (Scheme 74).

This synthetic route takes advantage of the steric hindrance to linking of the living PS chains, which avoids the formation of three or four arm stars. Using CH_3SiCl_3 instead of $SiCl_4$, A_2B stars were prepared.

A near monodisperse 3-miktoarm star copolymer of the A_2B type with two PI arms and one PS arm was prepared by Mays [229]. The synthetic approach involves the reaction of living arm B with excess CH_3SiCl_3 followed, after the evaporation of the excess linking agent, by the addition of a slight excess of living arm A. The excess of A arm is removed by fractionation after the linking is complete (Scheme 75).

$$2 \text{ PS}^- \text{Li}^+ + \text{SiCl}_4 \xrightarrow{C_6H_6/THF} (PS)_2SiCl_2 \xrightarrow{\text{excess PEO}^- K^+} (PS)_2Si(PEO)_2$$

Scheme 74

$$PS^- Li^+ + \text{excess } CH_3SiCl_3 \longrightarrow PS\text{-}SiCH_3Cl_2 + LiCl + CH_3SiCl_3\uparrow$$

$$PS\text{-}SiCH_3Cl_2 + \text{excess } PI^- Li^+ \longrightarrow PS\text{-}SiCH_3(PI)_2$$

Scheme 75

Hadjichristidis and collaborators extended this work to the synthesis of various combinations of A_2B miktoarm stars containing PS, PI, or PB arms [230]. By using $SiCl_4$ as the linking agent instead of CH_3SiCl_3 and similar procedures, the synthesis of A_3B 4-miktoarm stars, where A is PI and B is PS, was reported [231]. SEC, used to monitor the reaction sequence, revealed the absence of any detectable amount of coupling after the reaction of living arm B with the excess silane, due to the special conditions employed for this reaction. The molecular characterization of the arms (A and B) and the stars and the compositional characterization by NMR and UV spectroscopy confirmed the molecular and compositional homogeneity of these miktoarm stars.

Using the chlorosilane approach, Iatrou and Hadjichristidis synthesized 3-miktoarm star terpolymers of the ABC type [232] and 4-miktoarm star copolymers and quaterpolymers of the A_2B_2 and ABCD type [233], respectively. The synthesis of the ABC star was based on the following reactions (Scheme 76).

The first step involved the reaction of living PI chains with excess CH_3SiCl_3, followed by the removal of the excess silane exactly as it was described in the case of the A_2B stars. For the incorporation of the second arm (PSLi) a slow stoichiometric addition (titration) was employed towards the formation of the two arm product [(PI)(PS)(CH$_3$)SiCl]. This procedure was monitored by SEC, taking samples during the addition. Finally, a slight excess of PBLi was added for the preparation of the ABC star. The success of this synthetic route is based on the steric hindrance of the PSLi, which prevents the complete reaction with the macromolecular difunctional linking agent.

Isoprene + sec BuLi $\longrightarrow$ PILi

PILi + excess CH_3SiCl_3 $\longrightarrow$ $PICH_3SiCl_2$ + LiCl + $CH_3SiCl_3 \uparrow$

Styrene + sec BuLi $\longrightarrow$ PSLi

$PICH_3SiCl_2$ + PSLi $\xrightarrow{\text{titration}}$ (PI)(PS)CH$_3$SiCl + LiCl

CH_2=CH—CH=CH_2 + sec BuLi $\longrightarrow$ PBLi

(PI)(PS)CH$_3$SiCl + excess PBLi $\longrightarrow$ (PS)(PI)(PB)SiCH$_3$ + LiCl

Scheme 76

Styrene + sec BuLi ⟶ PSLi

PSLi + excess SiCl$_4$ ⟶ PSSiCl$_3$ + LiCl + SiCl$_4$↑

PSSiCl$_3$ + PSLi $\xrightarrow{\text{titration}}$ (PS)$_2$SiCl$_2$ + LiCl

CH$_2$=CH—CH=CH$_2$ + sec BuLi ⟶ PBLi

(PS)$_2$SiCl$_2$ + excess 2 PBLi ⟶ (PS)$_2$(PB)$_2$Si + 2 LiCl

Scheme 77

The synthesis of the A$_2$B$_2$ 4-miktoarm star was accomplished in a similar way to that for the ABC star, but by using SiCl$_4$ as the linking agent (Scheme 77).

A similar procedure was employed for the synthesis of 4-miktoarm star copolymers of the type A$_2$B$_2$, where A=PS and B=PI, by Young et al. [234]. The living PS solution was reacted with SiCl$_4$ in a molar ratio 2:1 for the synthesis of the two arm star. Subsequent reaction with an excess of PI$^-$Li$^+$ solution led to the formation of the (PS)$_2$(PI)$_2$ stars. A disadvantage of this approach is that careful control has to be exercised over the stoichiometry of the reaction between the living PS chains and SiCl$_4$ in order to avoid the presence of macromolecular linking agents with different functionalities. Nevertheless, under the proper experimental conditions, well-defined miktoarm stars can be prepared. The difference between this procedure and the one reported by Hadjichristidis et al. is the method of addition of the first two arms. In the first case the living A chains were introduced in two steps (the second A arm by titration), whereas in the latter case a stoichiometric quantity was used. The first method provides the best control over the reaction, but on the other hand it requires the complete evaporation of the excess silane at the first step of the synthesis, which is a time consuming process.

Two methods have been reported for the synthesis of A$_2$B$_2$ stars, where A is PI and B is PB [235]. According to the first method, the polyisoprenyllithium chains were end-capped with 2-3 units of styrene in order to increase the steric hindrance of the active center, followed by titration with SiCl$_4$ and reaction with an excess of PBLi. The second method involved the reaction of living polyisoprenyllithium chains with SiCl$_4$ in a molar ratio 2:1, at −40 °C in order to lower the reactivity of the living chains and to avoid the presence of macromolecular linking agents with functionality other than two. Addition of an excess of PBLi produced the A$_2$B$_2$ stars.

The ABCD 4-miktoarm quaterpolymer was prepared according to the following Scheme 78.

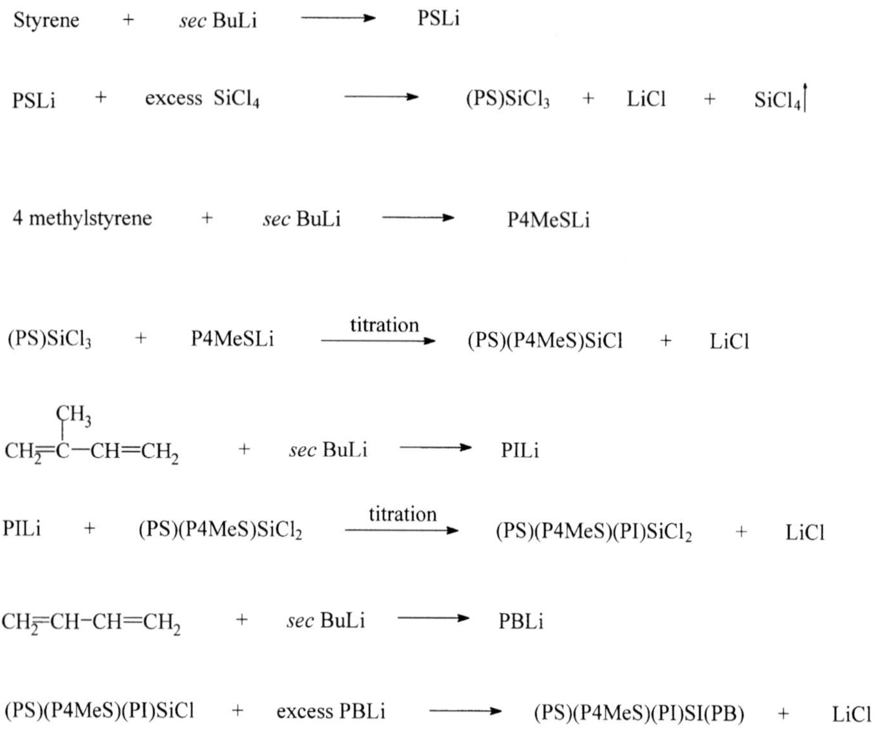

Styrene + *sec* BuLi ⟶ PSLi

PSLi + excess SiCl$_4$ ⟶ (PS)SiCl$_3$ + LiCl + SiCl$_4$↑

4 methylstyrene + *sec* BuLi ⟶ P4MeSLi

(PS)SiCl$_3$ + P4MeSLi $\xrightarrow{\text{titration}}$ (PS)(P4MeS)SiCl + LiCl

$\underset{\text{CH}_2=\text{C}-\text{CH}=\text{CH}_2}{\overset{\text{CH}_3}{|}}$ + *sec* BuLi ⟶ PILi

PILi + (PS)(P4MeS)SiCl$_2$ $\xrightarrow{\text{titration}}$ (PS)(P4MeS)(PI)SiCl$_2$ + LiCl

CH$_2$=CH−CH=CH$_2$ + *sec* BuLi ⟶ PBLi

(PS)(P4MeS)(PI)SiCl + excess PBLi ⟶ (PS)(P4MeS)(PI)SI(PB) + LiCl

Scheme 78

The characteristic of this procedure is that two of the arms are incorporated by titration. The difficulty for the introduction of the arms gradually increases and consequently the more sterically hindered polyanions have to react first. For this reason, P4MeSLi was chosen to react with the trifunctional macromolecular linking agent and the less sterically hindered PILi with the difunctional linking agent. It is obvious that the sequence of addition of the arms is crucial for this procedure.

The samples prepared by this chlorosilane route are characterized by high degrees of molecular and compositional homogeneities, as was confirmed by the exhaustive characterization data provided by SEC, membrane osmometry, low angle laser light scattering, differential refractometry, vapor pressure osmometry, UV-SEC, and NMR. The most important feature of the method is that every step of the reaction is monitored by taking small aliquots from the reaction mixture.

Polymers of the type (AB)B$_3$, A being PS and B being PB, were prepared by Tsiang [236]. The strategy followed was the same as the one adopted by Xie, Xia and Young et al. for the synthesis of A$_2$B$_2$ miktoarm stars. Living chains of arm B were reacted with SiCl$_4$ in a molar ratio 3:1, followed by addition of the living diblock AB (Scheme 79).

$$3 \text{ PB}^- \text{Li}^+ + \text{SiCl}_4 \longrightarrow (\text{PB})_3\text{SiCl} + 3 \text{ LiCl}$$

$$(\text{PB})_3\text{SiCl} + (\text{PS-b-PB})^- \text{Li}^+ \longrightarrow (\text{PB})_3\text{Si}(\text{PS-b-PB}) + \text{LiCl}$$

Scheme 79

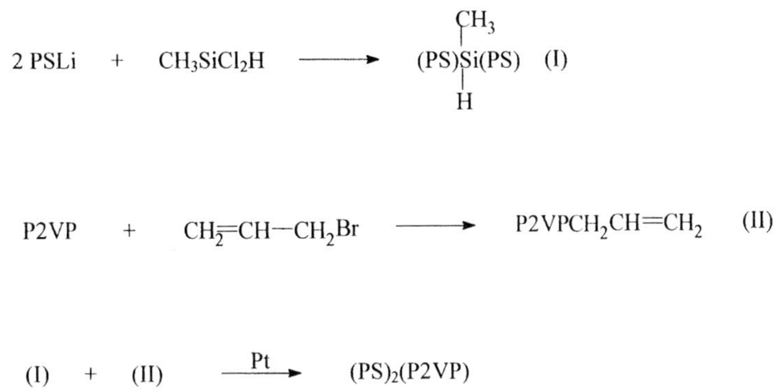

$$2 \text{ PSLi} + \text{CH}_3\text{SiCl}_2\text{H} \longrightarrow (\text{PS})\underset{\underset{H}{|}}{\overset{\overset{CH_3}{|}}{\text{Si}}}(\text{PS}) \quad (\text{I})$$

$$\text{P2VP} + \text{CH}_2{=}\text{CH}{-}\text{CH}_2\text{Br} \longrightarrow \text{P2VPCH}_2\text{CH}{=}\text{CH}_2 \quad (\text{II})$$

$$(\text{I}) + (\text{II}) \xrightarrow{\text{Pt}} (\text{PS})_2(\text{P2VP})$$

Scheme 80

The ability of the PBLi chains to react completely with the SiCl_4 may lead to several byproducts at the first step of the synthesis such as $(\text{PB})_2\text{SiCl}_2$, $(\text{PB})_4\text{Si}$, $(\text{PB})\text{SiCl}_3$. As a consequence, the purity of the products is questionable. Avgeropoulos and Hadjichristidis have prepared $(\text{AB})_2\text{A}$ and $(\text{AB})_3\text{A}$ where A is PS and B is PI by using the same techniques as in the case of A_2B and A_3B miktoarm stars [237].

A different approach for the synthesis of A_2B miktoarm stars, where A is PS and B is P2VP, was reported [238]. $\text{CH}_3\text{SiCl}_2\text{H}$ was used as the linking agent for the formation of the two arm star $(\text{A}_2\text{Si}(\text{CH}_3)\text{H})$. The living chains B were end-capped with a hydrocarbon moiety having a terminal double bond. Hydrosilylation reaction between the end-capped B chains and the Si-H bond of the two arm star leads to the formation of the A_2B stars (Scheme 80).

The final products were rather polydisperse (I=1.33-1.5) probably due to incomplete hydrosilylation, especially in the case of higher molecular weight arms.

Miktoarm star copolymers of the type $(\text{PS})_8(\text{PI})_8$ were also prepared using chlorosilane chemistry [239]. A silane with 16 chlorine atoms, $\text{Si}[\text{CH}_2\text{CH}_2\text{Si}(\text{CH}_3)(\text{CH}_2\text{CH}_2\text{Si}(\text{CH}_3)\text{Cl}_2)_2]_4$ was used as the linking agent. Living polystyrene chains were reacted with the linking agent in a stoichiometric ratio 8:1 (even a slight excess of living PS is used due to the steric hindrance) followed by the addition of excess PILi to produce the desired product.

Very recently, Hadjichristidis and co-workers [240] synthesized model 3-miktoarm star terpolymers of styrene, isoprene, and methyl methacrylate. This

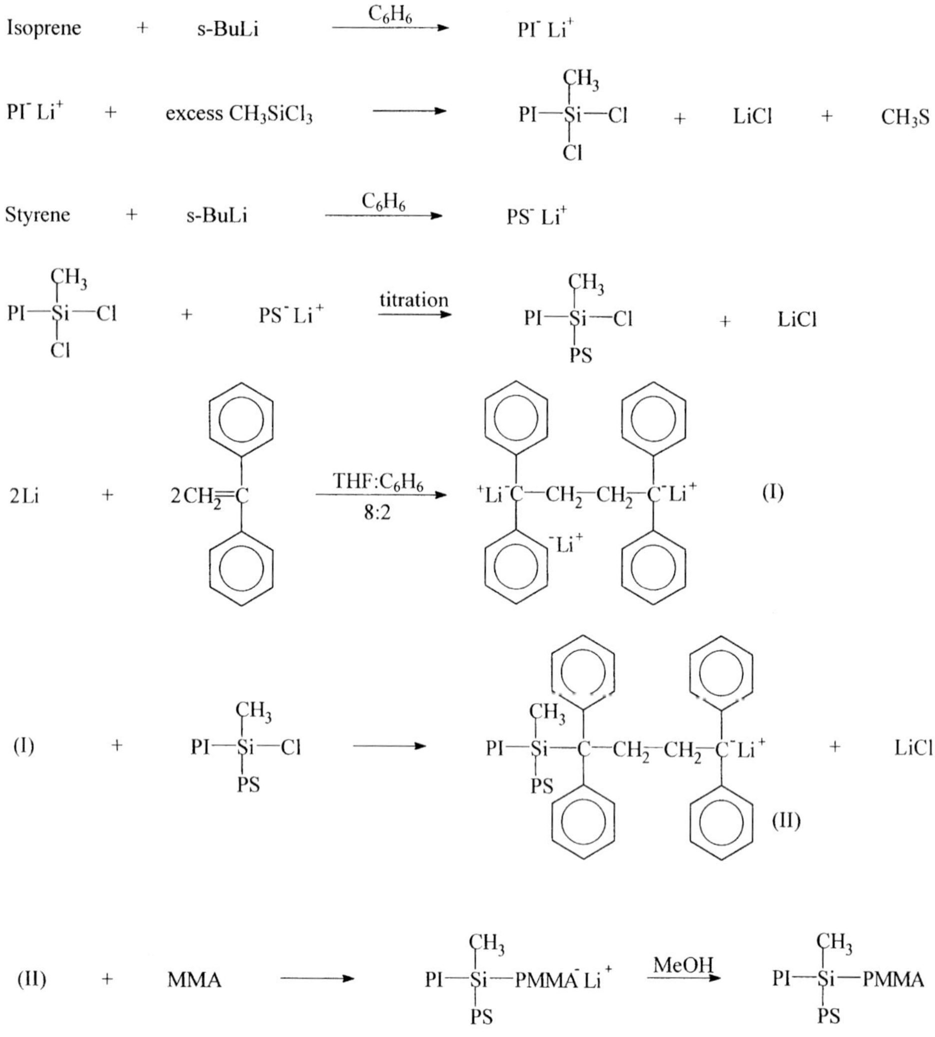

Scheme 81

involved the synthesis of (PS)(PI)Si(CH$_3$)Cl in hydrocarbon solvents (using methods described above), followed by reaction with the dianion formed by reaction of 1,1-diphenylethylene and Li, and polymerization of MMA in THF at -78 °C (Scheme 81).

The resulting polymers were rigorously characterized and found to have I values in the range of 1.06 to 1.12.

2.3.2
The Divinylbenzene Method

This method was first reported by Eschwey and Burchard [241] and developed by Rempp and coworkers [242-244]. Living polymer chains initiate the polymerization of divinylbenzene (DVB), leading to the formation of living star polymers composed of a polydivinylbenzene core from which emanate the arms, which have contributed to its formation. The core contains the same number of active sites as the arms of the star. These living anions can be used for the polymerization of a second monomer leading to the formation of miktoarm stars of the A_nB_n type. The polymers prepared by this approach have PS as A arms and as B arms PtBuMA, PBuMA, PEO or PtBuA [242-246] (Scheme 82).

The advantage of the DVB method is that it is easier to perform than the chlorosilane approach and can be used even on an industrial scale. Nevertheless this route suffers many disadvantages. Only stars of the A_nB_n type can be prepared by this method and even in this case there is no precise control over the reaction. Disregarding the case of accidental deactivation of the living chains, used to polymerize the DVB, other reasons like the higher molecular weight of these chains and/or the low molar ratio of DVB to living chains (due to stereochemical reasons) may impede the living polymers in reacting with the core. The remaining living arms can act as initiators after the addition of the second monomer. Another major disadvantage is that the number, n, of the branches incorporated into the core cannot be predicted and in fact there is a distribution of the number of branches. The average number of arms, n, is influenced mainly by the molecular weight of the precursor chains, by the overall concentration, and by the molar ratio of DVB to living chains. Other less important reasons influencing n are the rate of addition of DVB, whether styrene is diluted with solvent

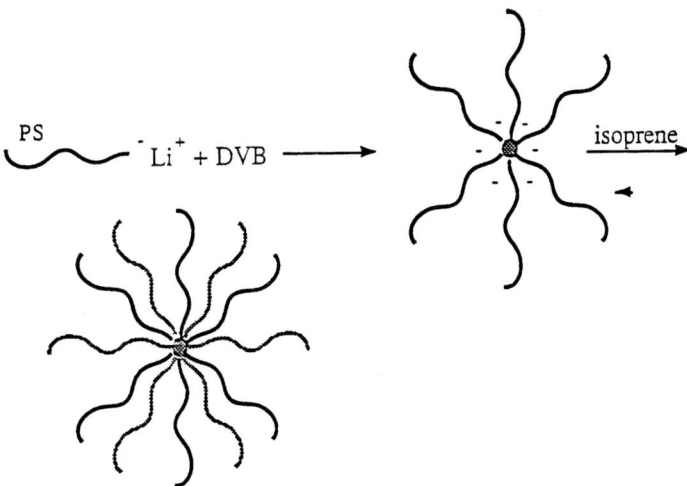

Scheme 82

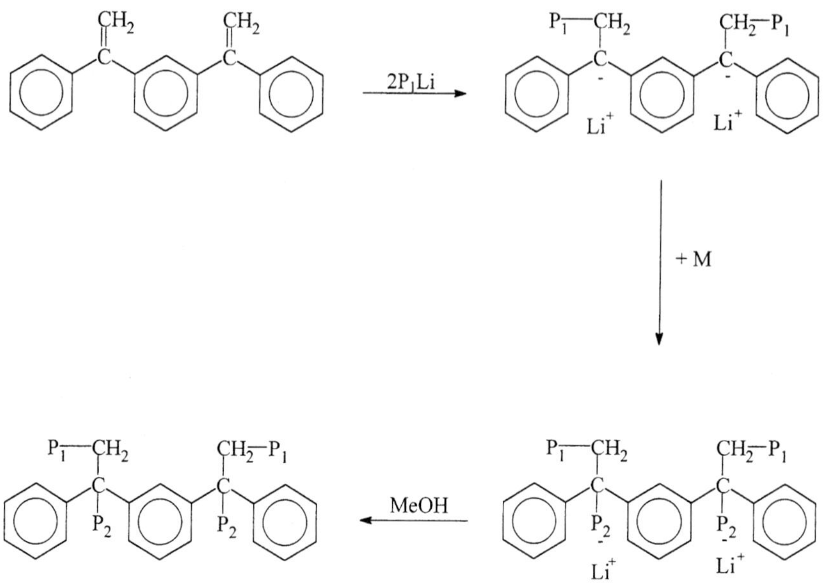

Scheme 83

or not, the stirring efficiency, and the amount of precursor which was not incorporated into the core. It is obvious that it is very difficult to predetermine the n value and that the samples prepared by this method are not characterized by high degrees of molecular and compositional homogeneity.

2.3.3
1,1-Diphenylethylene Derivative Method

This method uses a non-polymerizable divinyl compound to prepare miktoarm stars [246]. The reaction sequence is outlined in the following Scheme 83.

The first step involves the coupling reaction of living polymeric chains with 1,3-bis(1-phenylvinyl benzene) (DPE derivative). In the next step a second monomer is added and its polymerization starts from the living sites formed in the first step. As a result, an A_2B_2 miktoarm star is prepared. Using this approach, Quirk et al. synthesized A_2B_2 stars where A is PS and B is PI or PB or a PB-PS diblock [$A_2(BA)_2$] [248-250].

The method suffers from several disadvantages. Probably the most important is that the rate constants for the reaction of the first and second living polymeric chains with the DPE derivative are different. This leads to bimodal distributions, due to the formation of both the dianion and the monoanion during the first step. This problem can be avoided by the addition of polar compounds (e.g., THF), but these compounds influence dramatically the microstructure of the polydienes. However, it was found that the addition of *sec*-butoxide to the living coupled product, prior to the addition of the diene monomer, can produce monomodal miktoarm stars while maintaining high 1,4 polydiene microstructure.

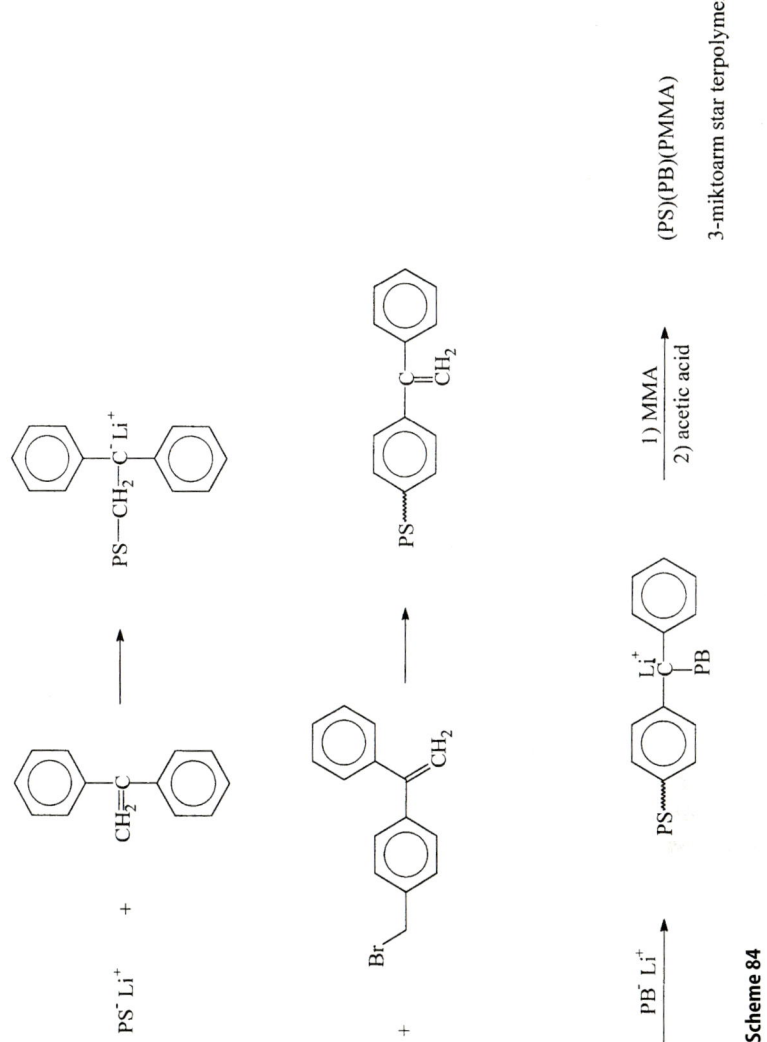

Scheme 84

Other limitations of the method are the following: 1) it is essential to control the stoichiometry of the reaction of the P_2 chains with the linking agent, otherwise a mixture of stars will be produced; 2) as in the case of the DVB method the P_2 arms cannot be isolated and therefore characterized independently. For this reason the method cannot provide unamibiguous proof for the formation of the claimed products.

Using the *meta* or *para* DPE derivatives under proper conditions it is possible to prepare the monoaddition product. Subsequent stoichiometric addition of the second living monomer leads to a living linking agent carrying two arms. The living center can be used for the polymerization of another monomer [251]. Potentially, A_2B_2 and ABC miktoarm stars can be prepared by this method. Extreme care has to be taken with the stoichiometry in all steps of the synthesis and with the conditions chosen for the reaction.

In fact, Stadler and co-workers [252] recently synthesized an ABC 3-miktoarm star terpolymer having arms of PS, PI, and PMMA using the approach shown in Scheme 84. After purification this polymer exhibited I=1.13 as measured by SEC. Unfortunately no absolute characterization was performed.

2.3.4
Other Methods Using Anionic Polymerization

Other approaches have also been developed for the preparation of miktoarm stars. Teyssie et al. used naphthalene chemistry for the preparation of AB_2 stars, A being PS, PI, poly(α-methylstyrene) or poly(*tert*-butylstyrene) and B being PEO [253] (Scheme 85).

The reaction of a living polymer chain with bromomethylated naphthalene leads to the formation of a fairly large amount of coupled product. The coupling is reduced to 5-10% by using the Grignard reagent, as shown in the reaction scheme. The crude product contains 90% of the desired AB_2 star, 10% of the starting homopolymer, and traces of PEO homopolymer. The miktoarm stars produced by this method have relatively high polydispersity indices (I=1.2-1.3).

Naka et al. used complexes of Ru(III) with bipyridyl terminated polymers to prepare A_2B miktoarm stars, where A is PEO and B is polyoxazoline (POX) [254], according to the following reaction (Scheme 86).

No characterization data were given in this communication.

Fujimoto et al. prepared the lithium salt of *p*-(dimethyl hydroxy) silyl-α-phenylstyrene and used it as initiator for the polymerization of hexamethyl cyclotrisilox-

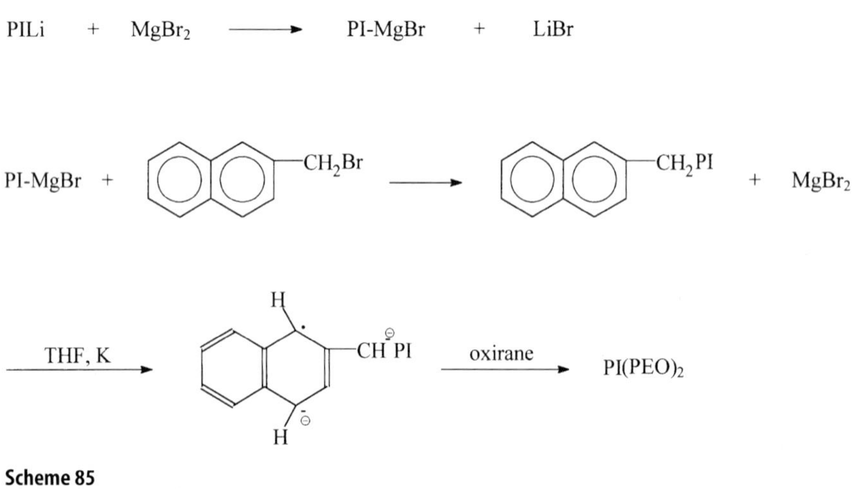

Scheme 85

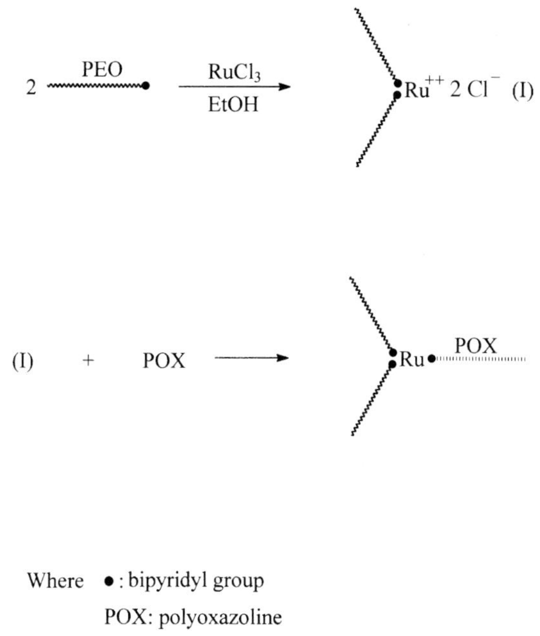

Scheme 86

2 PEO —• →(RuCl₃, EtOH)→)•Ru^{++} 2 Cl$^-$ (I)

(I) + POX →)•Ru—POX

Where •: bipyridyl group
POX: polyoxazoline

Scheme 86

ane [255]. The end-reactive poly(dimethyl siloxane) thus prepared was reacted with PSLi chains, followed by anionic propagation of t-butyl methacrylate, leading to the formation of an ABC miktoarm star copolymer (Scheme 87).

The polydimethylsiloxane prepared by this initiator has a polydispersity index as high as 1.4, and fractionation is therefore required before proceeding with the following steps of the synthesis. An additional problem is that the third arm cannot be isolated and characterized.

Takano et al. employed anionic block copolymerization of a monomer with a reactive functional group and an ordinary monomer [256]. The requirement for this procedure is that the functional group has to be unreacted. This is accomplished using THF as a solvent and short reaction times. Since the multifunctional linking agent has been prepared, subsequent reaction with living chains produces miktoarm stars of the type A_nB. In this study n=13, A is PS, and B is poly(vinylnaphthalene) (Scheme 88).

This approach does not provide the best control over the number of A arms. It was found that, due to steric reasons, only one of 3.6 silylvinyl groups reacted to produce the miktoarm star.

Ishizu and Kuwahara reported the synthesis of miktoarm star copolymers of styrene and isoprene [257] according to the following scheme (Scheme 89).

The copolymerization of the macromonomers produced comb-shaped copolymers but their solution and solid state properties were similar to those of miktoarm stars. The number of PS and PI arms incorporated in the final product

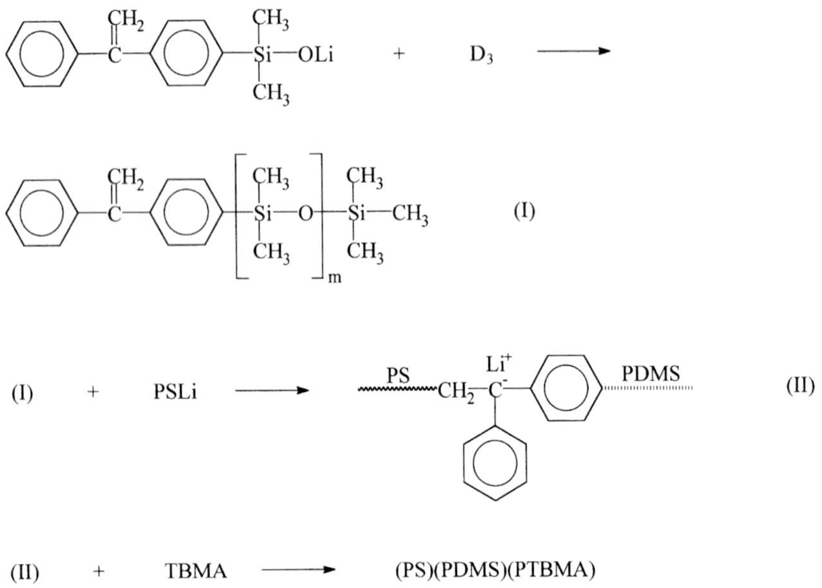

Scheme 87

depends on the reactivity ratios of the copolymerization system and thus it is not easily controlled.

This work was extended to the preparation of $(PS)_n(PtBuMA)_n$ miktoarm stars, using diblock macromonomers having central vinylbenzyl groups [258, 259]. The macromonomers were prepared by sequential polymerization of styrene 1,4-divinylbenzene (DVB), and t-BuMA. When living PS reacts with 1,4-DVB for short times (~5 min) only a few units of the difunctional monomer are incorporated at the end of the PS chain, avoiding the production of PS stars through the formation of polyDVB cores. The macromonomers were polymerized in solution using AIBN as initiator or in bulk using AIBN, tetramethylthiuram (photosensitizer), and ethyl glycol dimethacrylate (crosslinking agent). The SEC analysis showed that n=2.8. Similar work was presented using PS-b-P2VP macromonomers having central isoprene units [259]. Sulfur monochloride was used as crosslinking agent to form the final products.

Miktoarm stars with PS and Nylon 6 arms were prepared using phosphagene linking agents [260]. Hexachlorocyclotriphosphagene was reacted with 4-hydroxybenzoic acid ethyl ester to give the totally substituted cyclophosphagene. Subsequent hydrolysis with NaOH led to the corresponding acid, and reaction with $SOCl_2$ gave the acid chloride. This compound was used as linking agent to prepare PS stars. Two methods were employed for this reaction. Radical polymerization of styrene in the presence of aminoethenethiol produced PS with end-amine groups, capable of reacting with the linking agent to produce star polymers. The second method consisted of the reaction of living PS chains, prepared by anionic polymerization, with the linking agent. Hydrolysis of residual acid

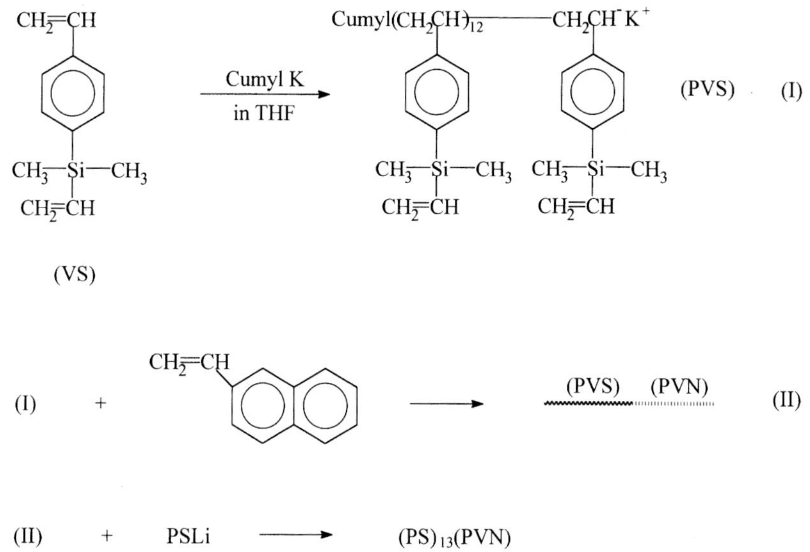

Scheme 88

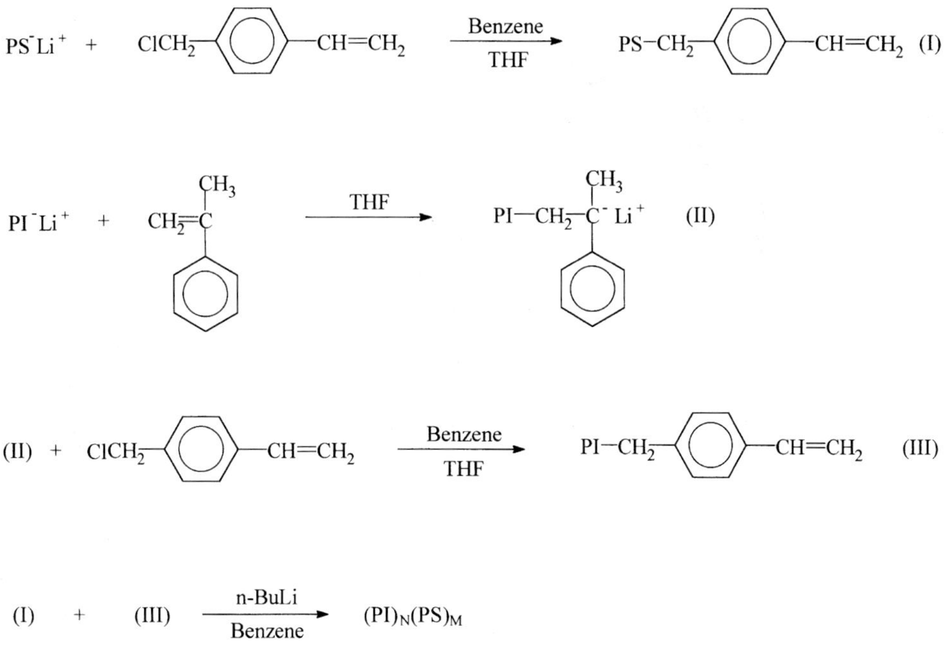

Scheme 89

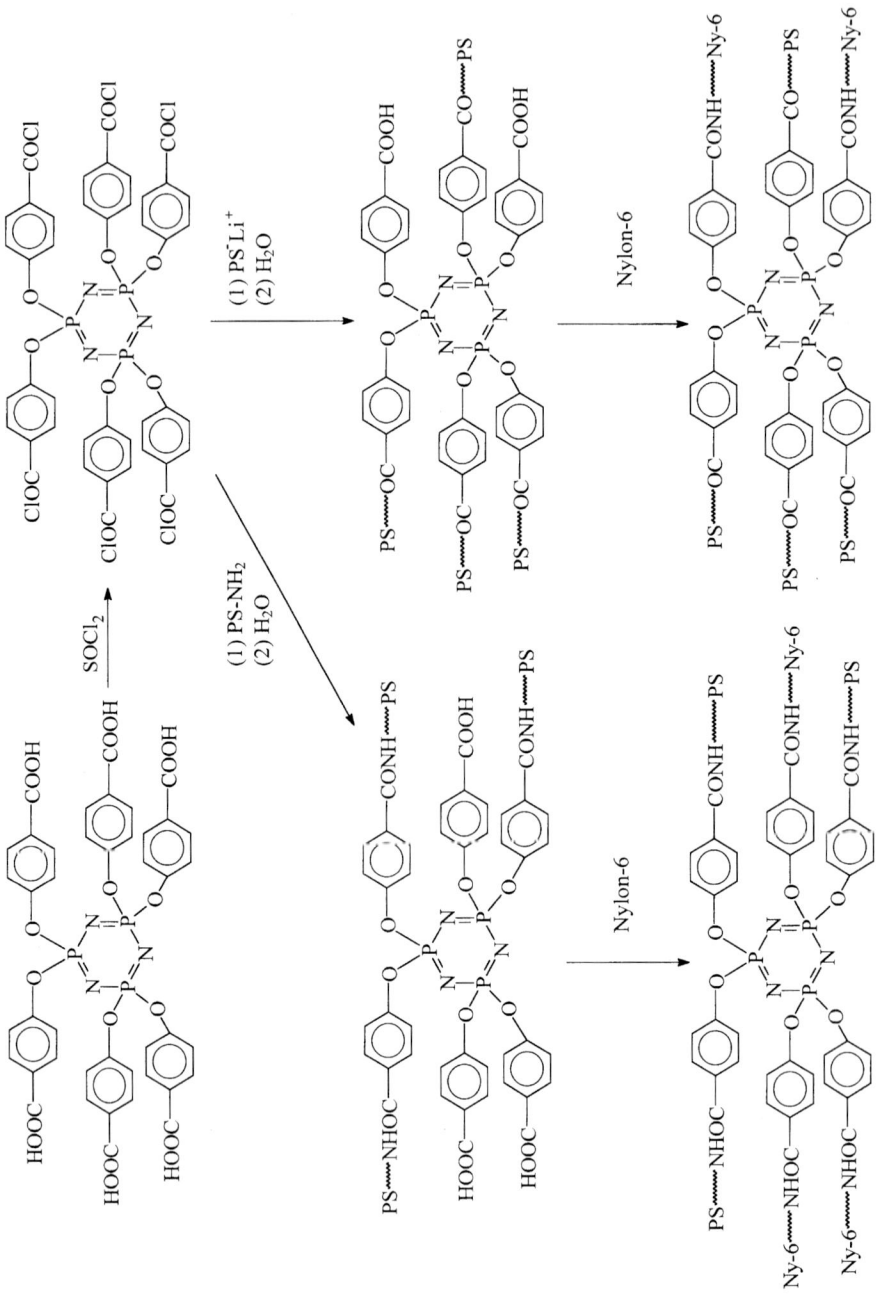

Scheme 90

chloride and titration of the resulting carboxyl groups showed that less than two groups remained unreacted. These carboxylic acids were used for the ring opening polymerization of ε-caprolactone to give the final miktoarm stars. The procedure is described in the following Scheme 90.

The main disadvantage with the method is that there is poor control over the number of PS arms linked to the phosphagene ring. It was observed that the number of arms decreased upon increasing molecular weight of the PS arms. This problem leads to the formation of stars with a broad distribution in the number of arms and thus to ill-defined structures.

2.3.5
Living Cationic Polymerization Method

Recent advances in living cationic polymerization have led to the preparation of miktoarm stars. The method involves the generation of living polymers and their reaction with a small amount of an appropriate divinyl compound leading to the formation of a star polymer with a central core, formed by polymerization of the divinyl compound, carrying active sites. These sites can be used for the polymerization of another monomer, thus producing miktoarm stars of the A_nB_n type, as follows (Scheme 91).

Vinyl ethers with isobutyl-, acetoxy ethyl-, and malonate ethyl- pendant groups have been used. Hydrolysis of the pendant groups in the last two cases led to the formation of amphiphilic polymers [50, 261–264]. It is obvious that this approach is analogous to the DVB method discussed above, and therefore it suffers the same disadvantages (poor control of the number of the arms, architectural limitations, etc.). Unfortunately detailed characterization data were not provided in these studies.

2.4
Other Architectures

The synthesis of complex architectures such as H-, super-H, umbrella type, dendrimers as well as cyclic and catenated types, will be discussed in this section.

H-shaped polystyrene homopolymers were prepared by Roovers and Toporowski using anionic polymerization techniques [265]. This work was extended by Mays and collaborators to the preparation of H-shaped copolymers, where the backbone is polyisoprene and the branches are polystyrenes [266]. A stoichiometric quantity of living polystyryllithium was reacted with CH_3SiCl_3 for the preparation of the two arm star $(PS)_2 SiCl(CH_3)$ (1). The steric hindrance of the active center prevents the formation of the three arm star. Using the difunctional initiator developed by Tung and Lo [267] and later Quirk and Ma [268], a difunctional living polyisoprenyllithium was prepared in benzene. Subsequent reaction with (1) produces the H-shaped copolymer. The advantage of using this difunctional initiator is that it can be used in benzene solutions, producing polydienes with high 1,4 contents. The disadvantage is that the polymers have somewhat broader distributions (I = 1.15-1.20) compared with the products of the classical difunc-

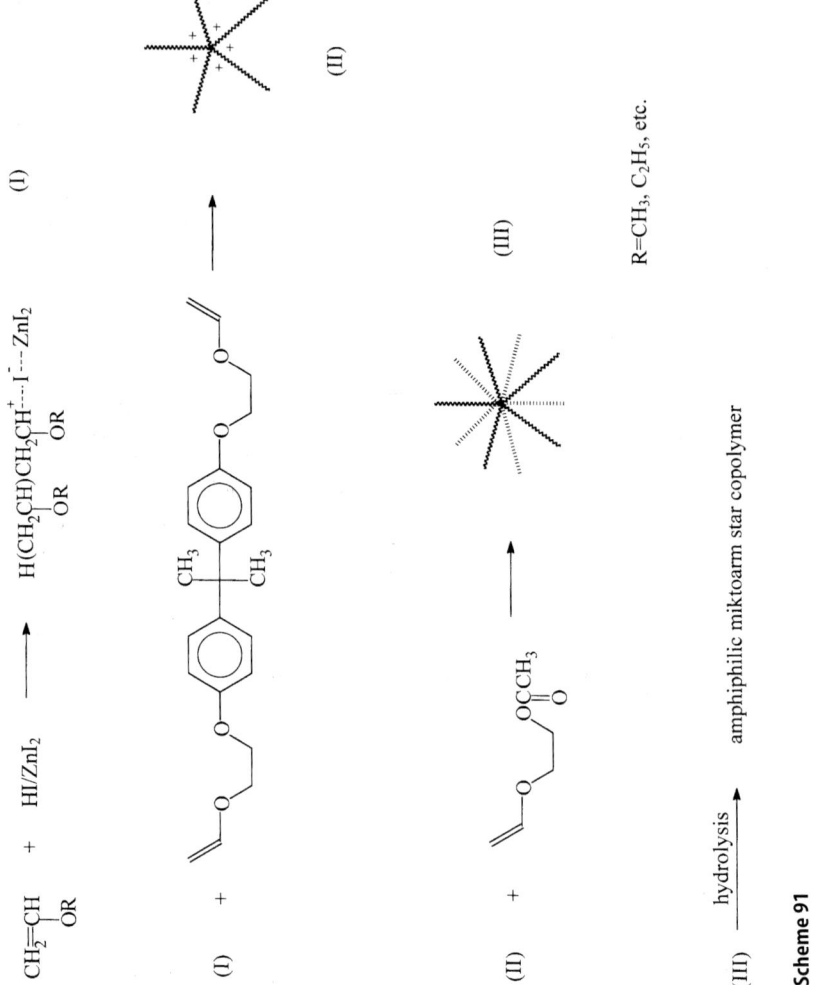

Scheme 91

tional initiator sodium/naphthalenide. Nevertheless, the final products after fractionation have narrow molecular weight distributions (I <1.1) (Scheme 92).

If $SiCl_4$ is used instead of CH_3SiCl_3 then super H-copolymers can be prepared. These are copolymers of the type B_3AB_3, where A is PI or PS. Super-H copolymers were first prepared by Hadjichristidis and collaborators using the chlorosilane method developed for the synthesis of miktoarm stars [269]. Styrene (or isoprene) was polymerized in THF at -78 °C using sodium/naphthalenide as initiator. The difunctional polymer was reacted with a very large excess of $SiCl_4$, leading to the formation of a polystyrene (or polyisoprene) chain end-capped at both ends with the -$SiCl_3$ trichlorosilyl group. Living polyisoprene (or polystyrene) chains

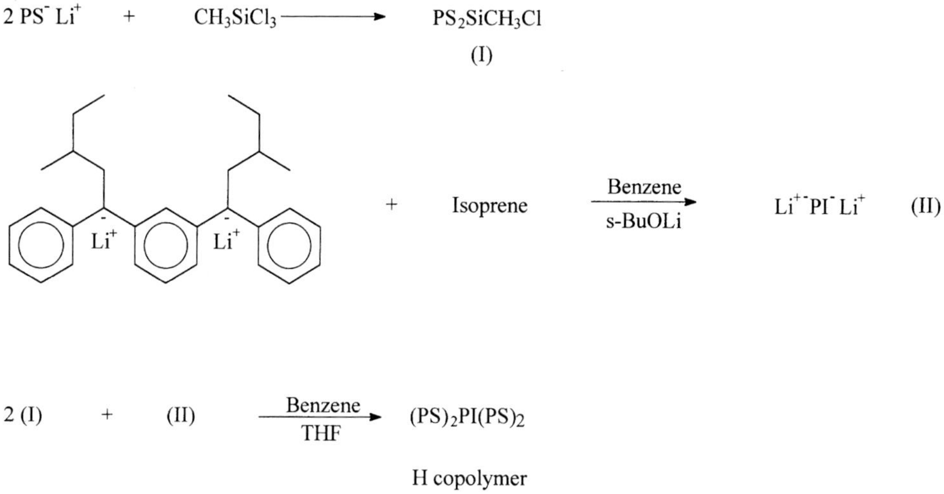

2 PS⁻ Li⁺ + CH₃SiCl₃ ⟶ PS₂SiCH₃Cl (I)

+ Isoprene $\xrightarrow[\text{s-BuOLi}]{\text{Benzene}}$ Li⁺⁻PI⁻Li⁺ (II)

2 (I) + (II) $\xrightarrow[\text{THF}]{\text{Benzene}}$ (PS)₂PI(PS)₂

H copolymer

Scheme 92

Naphthalene + Na $\xrightarrow{\text{THF}}$ Na Naphthalene (I)

St (or Is) + (I) ⟶ Na⁺⁻PS⁻Na⁺ (II)
(or Na⁺⁻PI⁻Na⁺)

(II) + excess SiCl₄ ⟶ (Cl₃SiPSSiCl₃) (III)
(or Cl₃SiPISiCl₃)

(III) + excess PI⁻ Li⁺ ⟶ (PI)₃PS(PI)₃
(or PS⁻ Li⁺) (or (PS)₃PI(PS)₃)

Scheme 93

were reacted with the remaining Si-Cl bonds for the preparation of the super-H copolymer (Scheme 93).

The use of a polar solvent (THF) for the preparation of the connector leads, in the case of polyisoprene, to high vinyl contents, but I values are very low (<1.1).

Mays and collaborators [266] have also used related chemistry, in benzene, to prepare a "π-copolymer", where two PS branches divide a PI backbone into three equal parts (Scheme 94).

The final product was rigorously characterized and found to have I = 1.09.

As an extension of the miktoarm polymers, Roovers et al. prepared the umbrella and the umbrella star polymers of the AB_x and $(AB_x)_y$ type respectively [270]. A is PS and is much larger than the B block which is PB or P2VP. In the presence of dipiperidinoethane (dipip) the polymerization of butadiene leads to 1,2 PBLi. Subsequent addition of styrene yields a diblock copolymer 1,2-PB-b-PS, for which the PB block is relatively short. Hydrosilylation of the remaining double bonds is performed for the introduction of chlorosilane groups $-Si(CH_3)Cl_2$ or $-Si(CH_3)_2Cl$. The coupling reaction of the hydrosilylated polymer with living PB or P2VP chains leads to the umbrella polymer, according to the following route (Scheme 95).

A similar method was used for the preparation of the umbrella star polymers. Prior to the hydrosilylation reaction the living diblock copolymers (1,2-PB-b-PS) was coupled to form a star polymer, using the appropriate chlorosilane linking agent. Subsequent hydrosilylation and coupling reaction with living PB or P2VP chains produces the following structure (Scheme 96).

This procedure suffers the disadvantage of the limited control by which the hydrosilylation reaction is characterized. The number of the Si-Cl bonds and consequently the number of arms of the umbrella polymers cannot be predicted accurately.

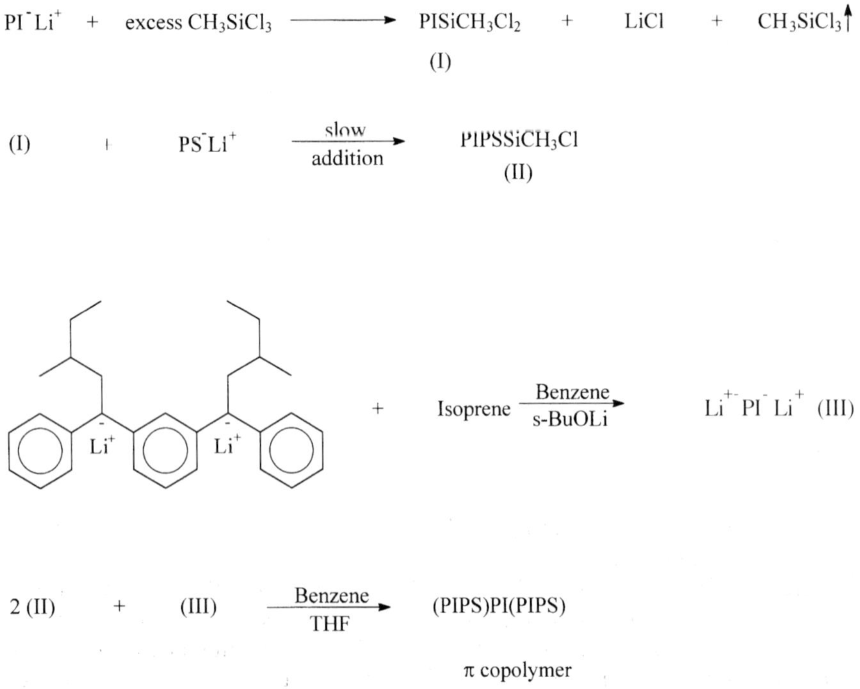

PI^-Li^+ + excess CH_3SiCl_3 ⟶ $PISiCH_3Cl_2$ (I) + LiCl + $CH_3SiCl_3\uparrow$

(I) + PS^-Li^+ —slow addition→ $PIPSSiCH_3Cl$ (II)

2 (II) + (III) —Benzene/THF→ (PIPS)PI(PIPS)

π copolymer

+ Isoprene —Benzene/s-BuOLi→ $Li^+\ PI^-\ Li^+$ (III)

Scheme 94

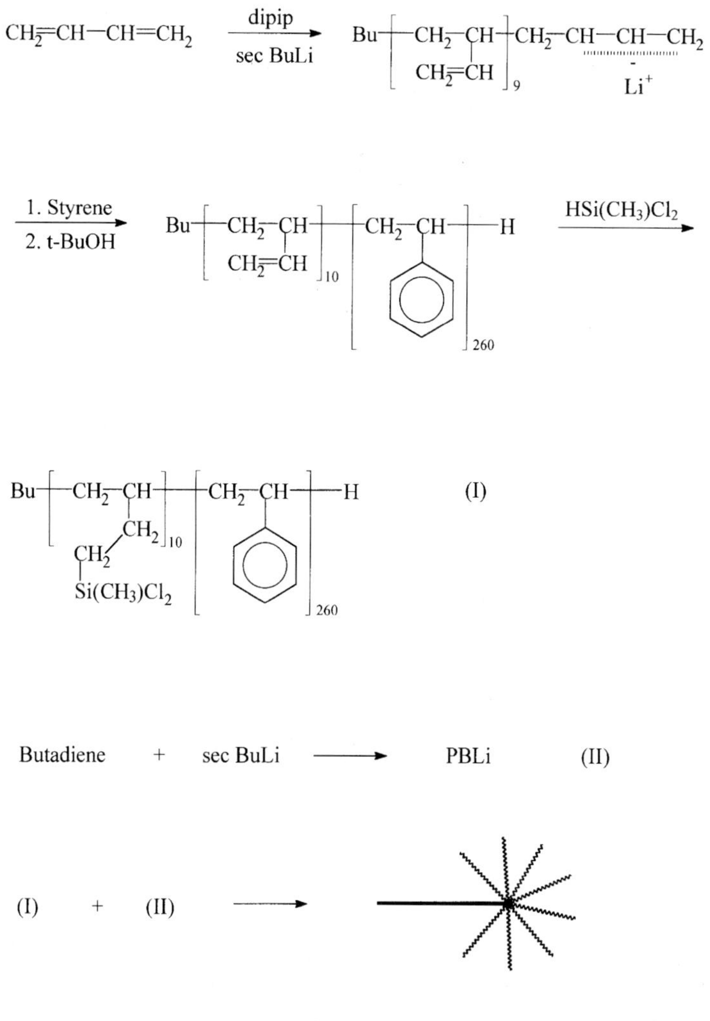

umbrella copolymer

Scheme 95

The synthesis of dumbbell polymers was reported in the literature [271]. These materials can be considered as double umbrella stars with functionality of umbrella polymers equal to two (Scheme 97).

The connector chain is PS, whereas the side branches are PEO. Using anionic polymerization techniques and potassium naphthalenide as initiator, a triblock copolymer, poly(B-*b*-S-*b*-PB) with short polybutadiene chains, was prepared. The PB blocks were subjected to hydroboration-oxidation reaction for the addition of H_2O to the pendant double-bonds of the 1,2-PB units, according to the reactions (Scheme 98).

Scheme 96

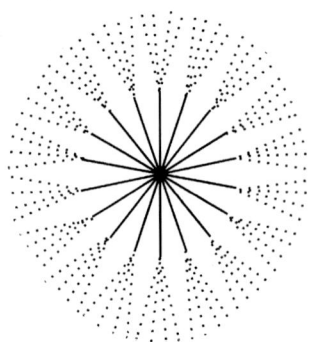

C: "dumbbell"

Scheme 97

The incorporated -OH groups were transformed to alkoxides using an organometallic base (cumylpotassium) and a suitable cryptand to avoid the precipitation of the polyfunctional initiator. Addition of ethylene oxide produces the dumbbell polymers, as the following reactions indicate (Scheme 99).

The disadvantage of this approach is that since the conversion of the hydroboration reaction is not quantitative it is very difficult to have precise control over the number of the PEO arms. Special care should be taken during the hydroboration to avoid side reactions and specifically the crosslinking of the hydroxylated polymer.

A structure resembling that of the dumbbell polymers was made by Frechet et al. In this case the connector is linked with polyether dendritic groups [272, 273]. The synthetic approach involved the preparation of a difunctional polystyrene chain in THF using potassium naphthalenide as initiator. The living polymer was end-capped with 1,1-diphenylethylene (DPE) to reduce its nucleophilicity and avoid side reactions with benzylic halomethyl groups. Addition of the fourth generation dendrimer [G-4]-Br led to the final product (Scheme 100).

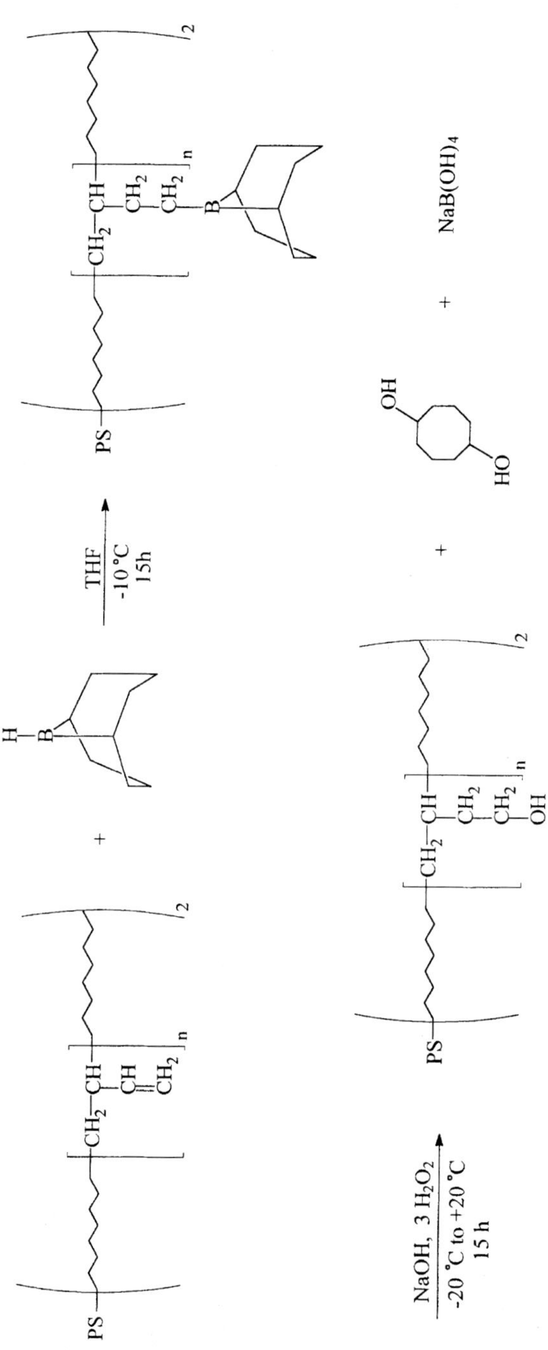
PS
CH_2–CH
CH
CH_2
n
2
+
H–B
THF
-10 °C
15h
PS
CH_2–CH
CH_2
CH_2
B
n
2
NaOH, 3 H_2O_2
-20 °C to +20 °C
15 h
PS
CH_2–CH
CH_2
CH_2
OH
n
2
+
OH
HO
+
$NaB(OH)_4$

Scheme 98

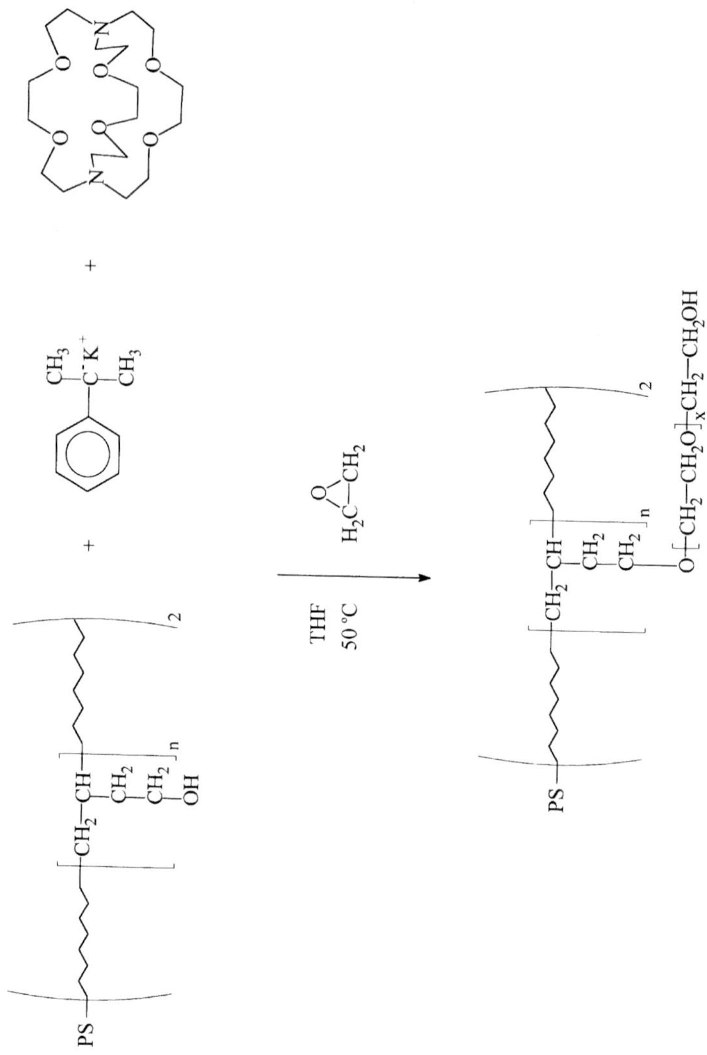

Scheme 99

Similar structures were prepared using poly(ethyleneglycol) as the connector. In this case the terminal -OH groups were transformed to alkoxides using NaH [274, 275] (Scheme 101).

PEO prepared by anionic polymerization using the monofunctional initiator ((diphenylmethyl)potassium) has one -OH end group leading to the formation of linear-dendritic diblock copolymers (Scheme 102).

Linear polymers with dendritic chain ends were also prepared by Chapman et al. [276]. A poly(ethylene oxide) chain served as the platform for the synthe-

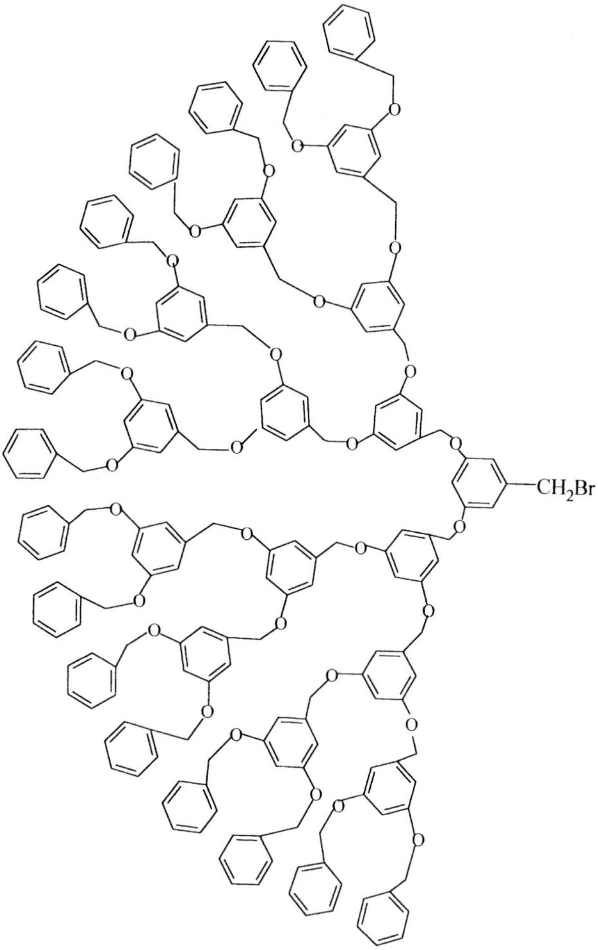

Scheme 100

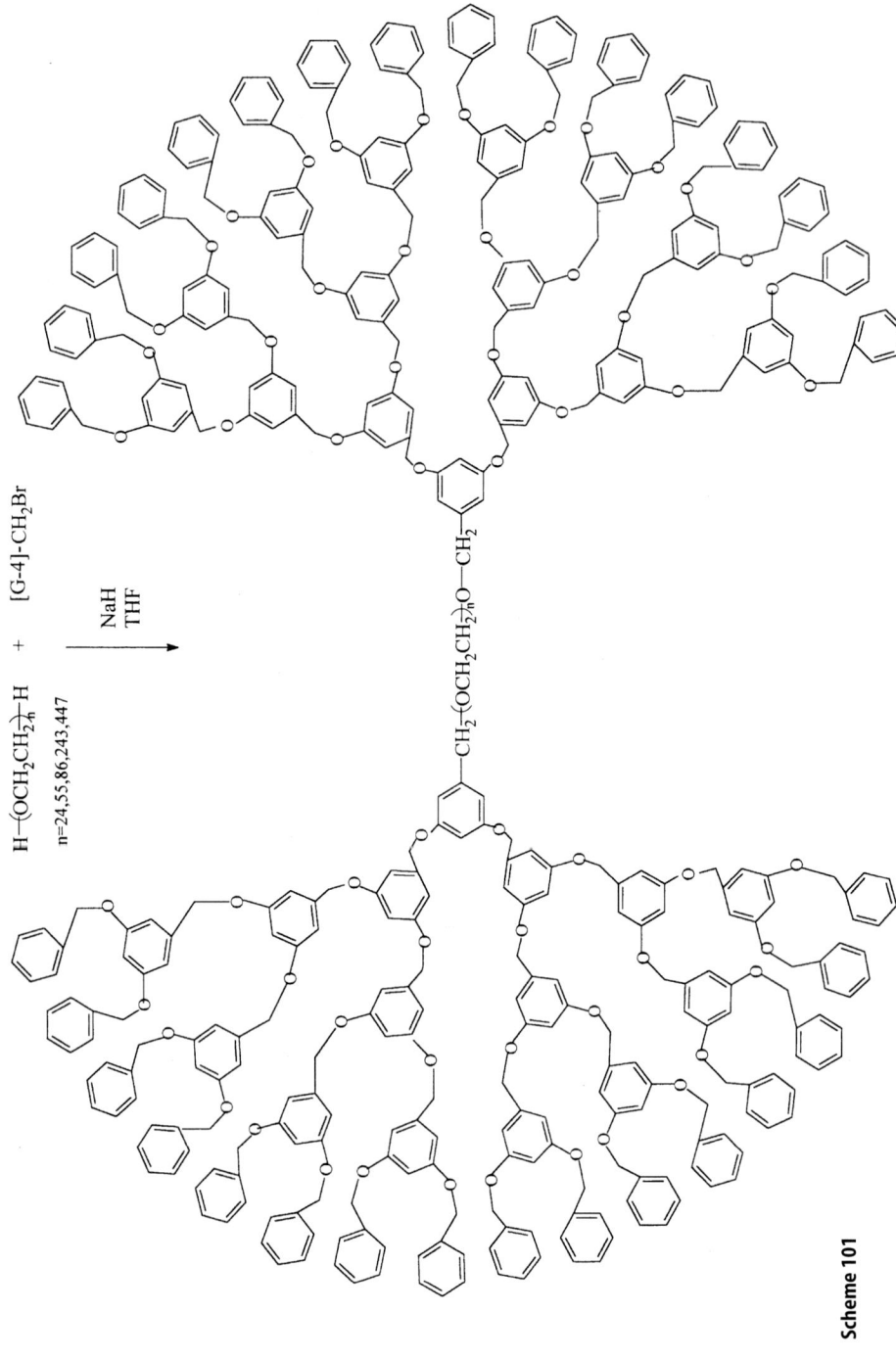

Scheme 101

sis of the dendrimer consisting of lysine groups. Methoxy-terminated PEO was esterified with Boc-glycine and then after the removal of the functional group pentafluorophenyl-N-α-N-ε-di-Boc-L-lysinate (l) was used for the introduction of the lysine group. Deprotection was performed with a 1:1 mixture of trifluoroacetic acid (TFA)-CH_2Cl_2. Repetition of this cycle generated a dendritic group at the end of the PEO chain (Scheme 103).

Comparing this synthetic procedure with the one developed by Frechet et al., in the former case the dendrimer is formed at the end of the polymer chain, whereas in the latter case the dendrimer is already prepared and is attached to the macromolecular chain end(s) using the proper reactions.

Molecular macrocylinders with functional groups were prepared in order to attach dendritic fragments to the polymer chain [277]. Random copolymers of [1.1.1] propellanes with different side groups were prepared by radical polymerization (Scheme 104).

Deprotection to the corresponding hydroxylic compounds with methanolic hydrochloric acid, deprotonation of the hydroxyl groups with sodium hydride, and the reaction of the resulted alkoxides with (G-1)-Br (the first generation dendrimer) produced polymers with incorporated dentritic fragments (Scheme 105).

Using Pd-catalyzed polycondensation reactions poly(p-phenylenes) with protected hydroxyl groups were prepared. These groups were used for the attachment of dendritic groups, in a similar way to the one referred to above. Dendritic monomers can also be used for the direct preparation of the desired structures (Scheme 106).

Due to their polymerization techniques (radical polymerization or polycondensation) the polymers are rather polydisperse and the control over the rod's molecular weight is limited, whereas the yield of the post-polymerization reactions is not quantitative.

Arborescent graft copolymers having a highly branched PS core and a PEO shell were prepared by a combination of anionic polymerization, grafting "onto" and "from" techniques [278]. The synthetic procedure is described in the following scheme (Scheme 107).

Styrene was polymerized with s-BuLi and the living polymer chains thus produced were end-capped with DPE, followed by a reaction with chloromethylated PS to provide a graft polymer. This sample was chloromethylated and was reacted with living PS chains to provide the second generation graft polymer. This procedure can be continued to give higher generation grafts. The PS used for the last grafting reaction was prepared using (6-lithiohexyl)acetaldehyde acetal, a bifunctional initiator containing a protected hydroxyl functionality. The graft polymers thus prepared have protected hydroxyl end-groups. Mild acidic hydrolysis deprotected the hydroxyl functionalities, followed by their titration with potassium naphthalide in order to generate the alcoholate anions. These groups serve as initiator for the polymerization of ethylene oxide leading to the formation of the final products.

In a series of papers, the synthesis of cyclic block copolymers was described. Using lithium naphthalenide as initiator and THF as solvent the difunctional living triblock copolymer $^+$Li$^-$P2VP-b-PS-b-P2VP$^-$Li$^+$ was prepared by sequential

Scheme 102

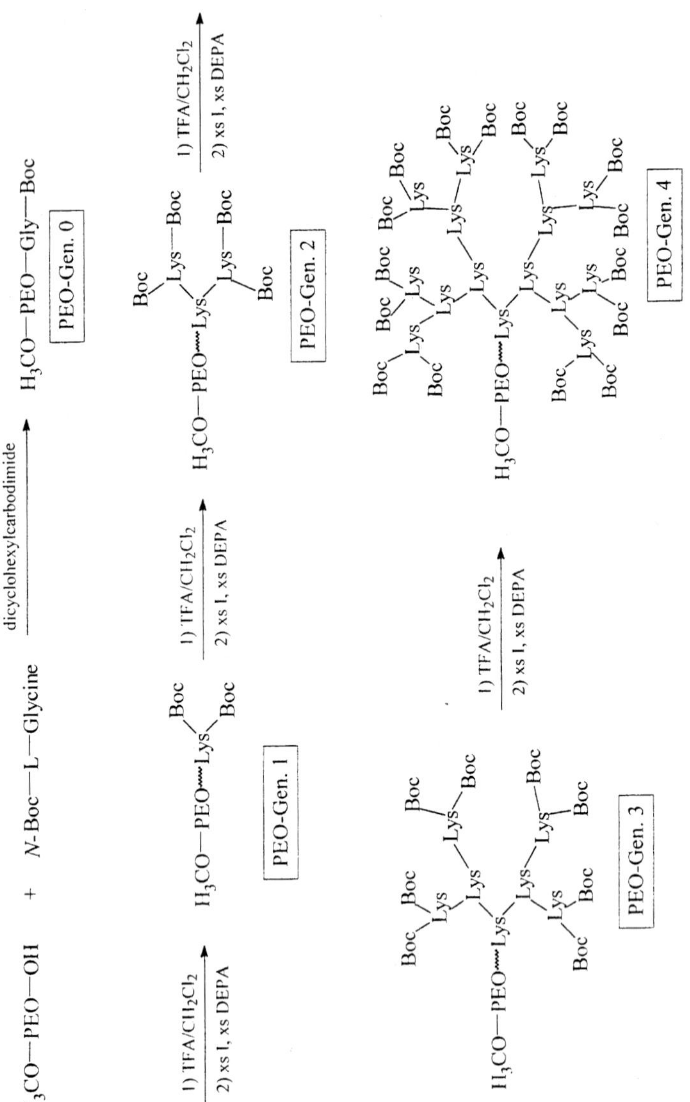

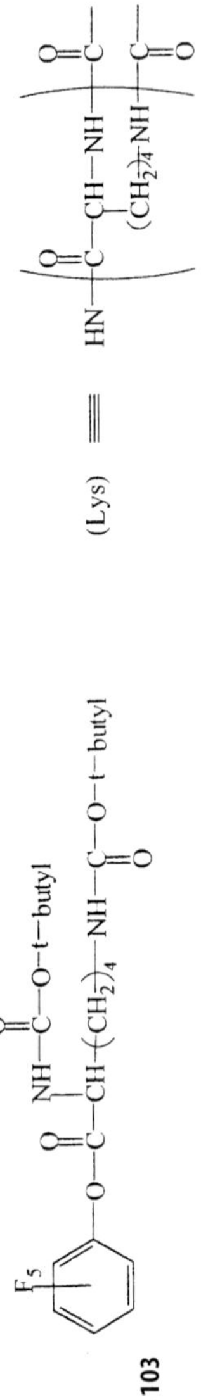

Scheme 103

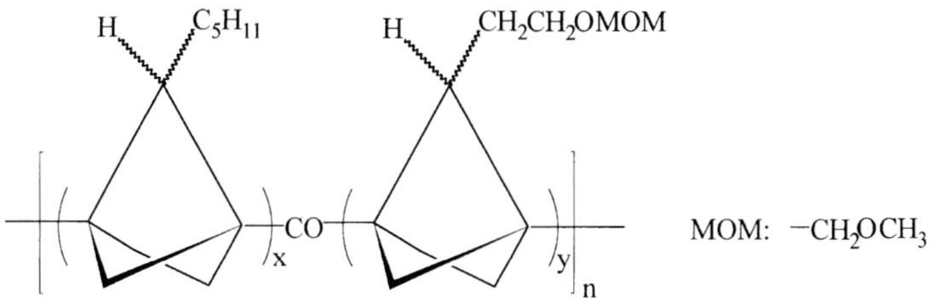

Scheme 104

MOM: $-CH_2OCH_3$

Scheme 105. (see next page) ──────────────▶

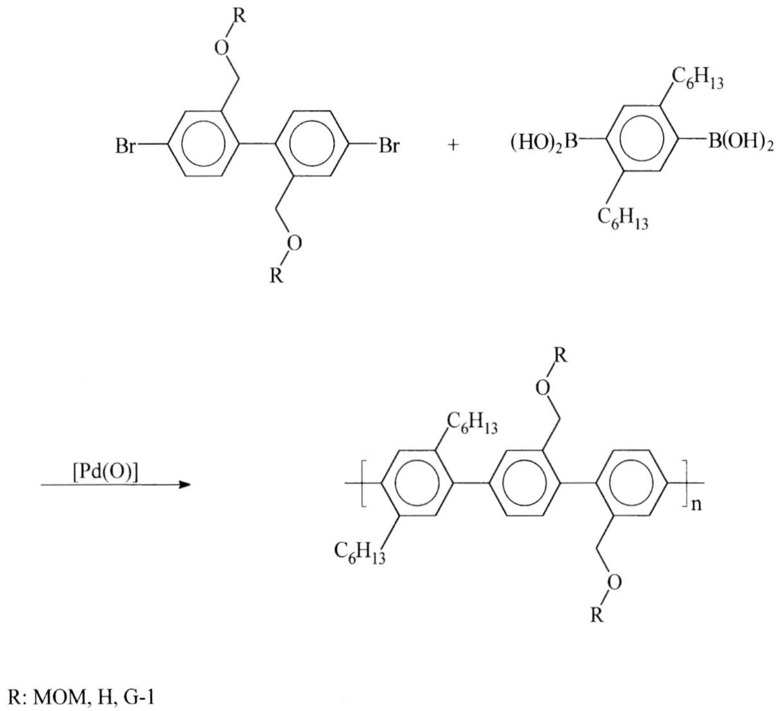

[Pd(O)]

R: MOM, H, G-1

Scheme 106

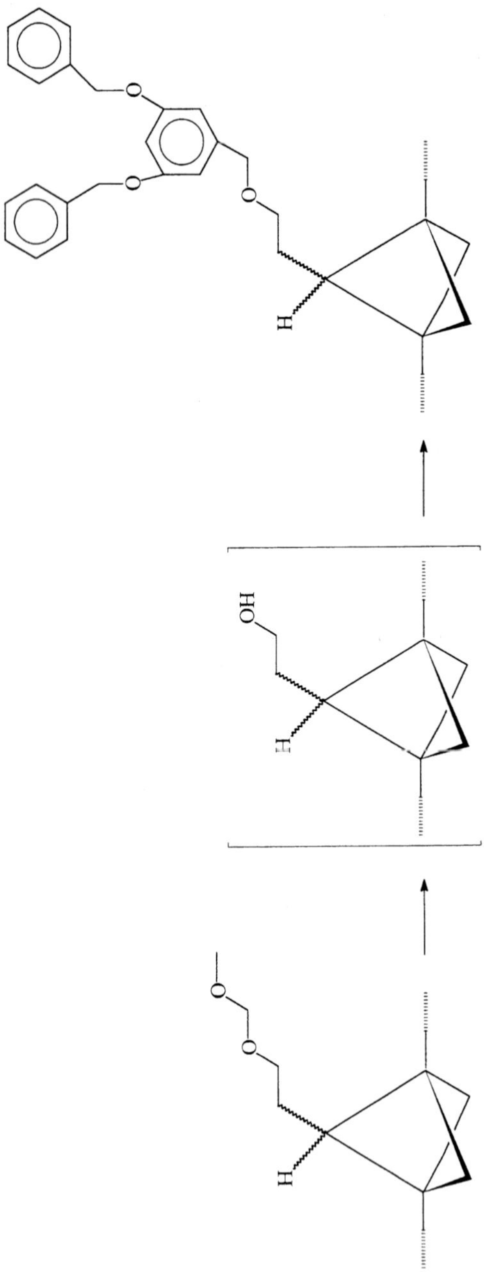

Scheme 105

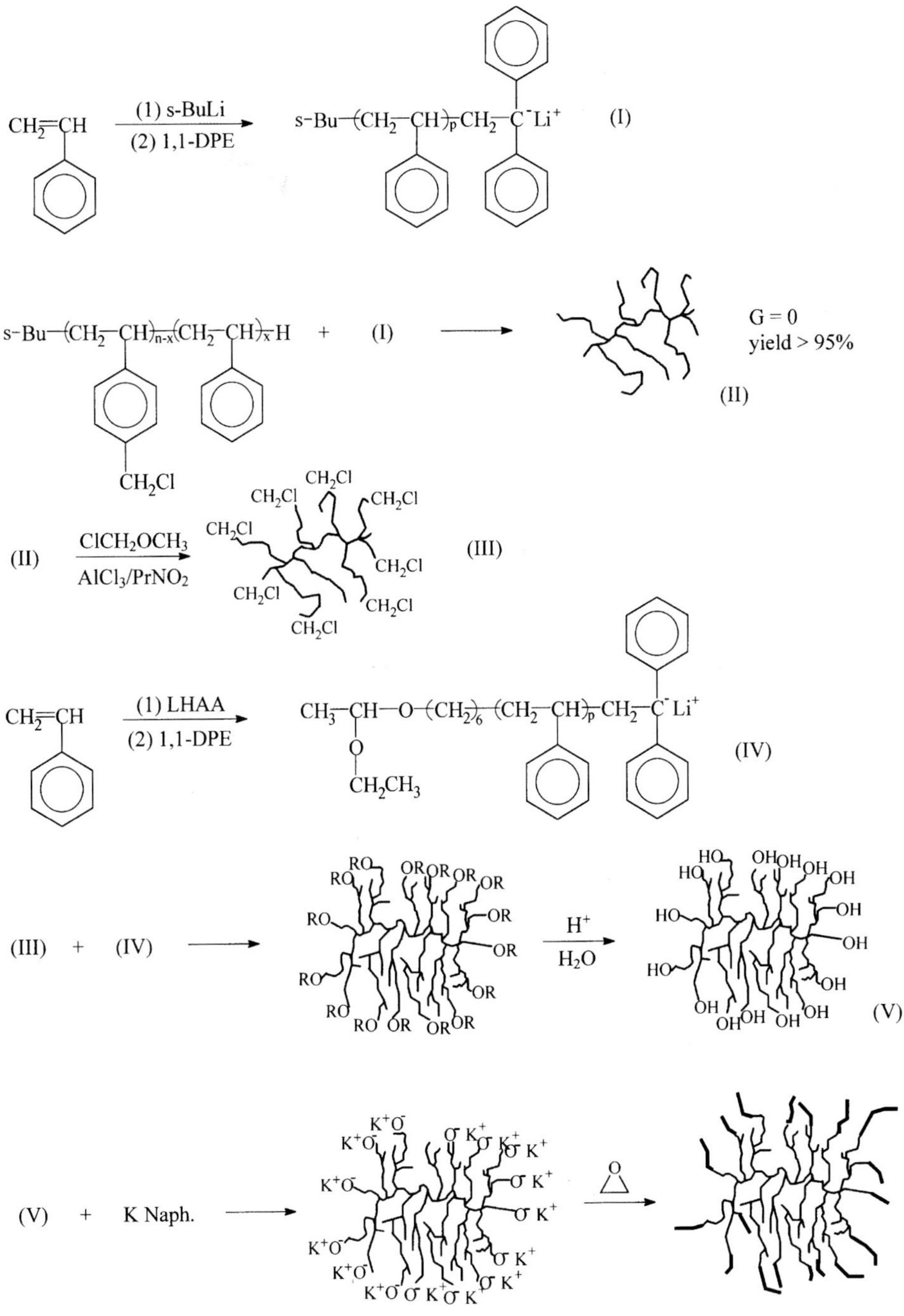

Nonlinear Block Copolymer Architectures

Scheme 107

addition of the two monomers. In very dilute solutions the coupling reaction was performed using 1,4- bis(bromomethyl)benzene [279]. Several techniques were employed to separate the cyclic polymers from the polycondensates and the linear block copolymer formed in the reaction mixture. Polymers with narrow molecular weight distributions were obtained after the fractionation (Scheme 108).

Using the same method, cyclic block copolymers of PS and PDMS were prepared [280, 281]. The cyclization reaction takes place in very dilute solution but using dichlorodimethylsilane as the coupling agent. Prior to the addition of D_3 the difunctional living PS chain was end-capped with 2,2,5,5-tetramethyl-1-oxa-2,5-disilacyclopentane. In this way, the initiation reaction of D_3 is much faster, leading to relatively narrow molecular weight distributions (Scheme 109).

Cyclic block copolymers of PS and PB were also prepared by a similar method [282]. sec-BuLi and 1,3-bis(1-phenylethenyl)benzene (DDPE) in a molar ratio 2:1 was the difunctional initiator used for the polymerization of B in the presence of sec-BuOLi. Subsequent addition of styrene forms the living triblock copolymer $^{(+)}Li^{(-)}$PS-b-PB-b-PS$^{(-)}Li^{(+)}$. Reaction with the linking agent (DDPE or $(CH_3)_2SiCl_2$) produced the cyclic block copolymers.

This work also described the preparation of cyclic polymers with two attached branches [282] using the following procedure (Scheme 110).

A four-arm star with two living chains is formed, and it reacts with the coupling agent to produce the desired structure. The critical step is the reaction of the living PS$^-$Li$^+$ chains with the DDPE in order to prepare the living intermediate from which the PB blocks emanate.

Catenated macromolecules were also prepared by similar techniques [283]. In the presence of PS macrocycles, 2-VP was polymerized anionically at –78 °C using Li dihydronaphthylide as initiator and THF as solvent. The cyclization reaction takes place in solution with high concentration of cyclic PS with 1,4-bis(bromomethyl)benzene as the coupling agent. Under these conditions, catenated PS-P2VP copolymers were prepared. Extraction techniques with cyclohexane and methanol were efficient enough, according to the authors, for the removal of the

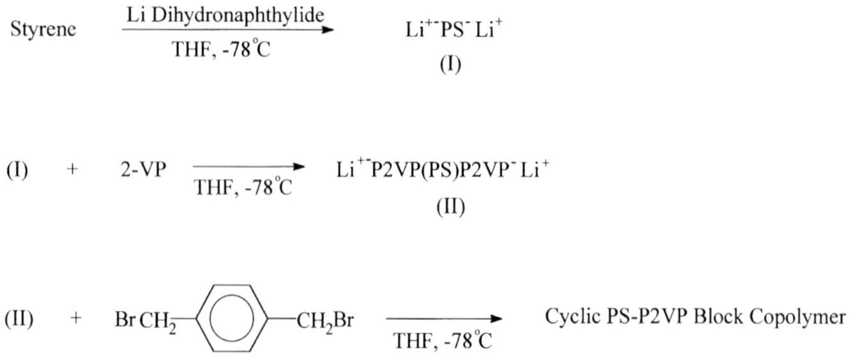

Styrene $\xrightarrow[\text{THF, -78°C}]{\text{Li Dihydronaphthylide}}$ Li$^+$$^-PS^-Li^+$ (I)

(I) + 2-VP $\xrightarrow[\text{THF, -78°C}]{}$ Li$^+$$^-$P2VP(PS)P2VP$^-Li^+$ (II)

(II) + BrCH$_2$–⌬–CH$_2$Br $\xrightarrow[\text{THF, -78°C}]{}$ Cyclic PS-P2VP Block Copolymer

Scheme 108

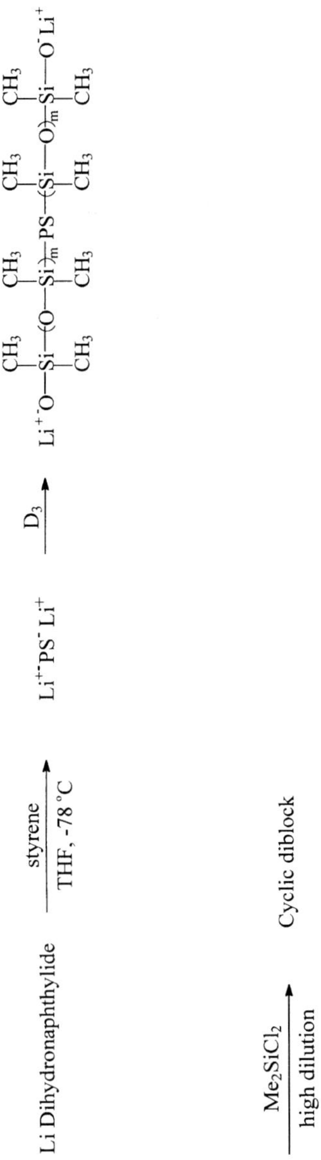

Scheme 109

cyclic PS, the cyclic P2VP, and the corresponding P2VP polycondensates. The remaining polymer was not soluble in either methanol or cyclohexane and its molecular weight equals the sum of the molecular weights of the PS and P2VP macrocyclic polymers. Nevertheless, the polydispersity index was rather high ($I = 1.30$).

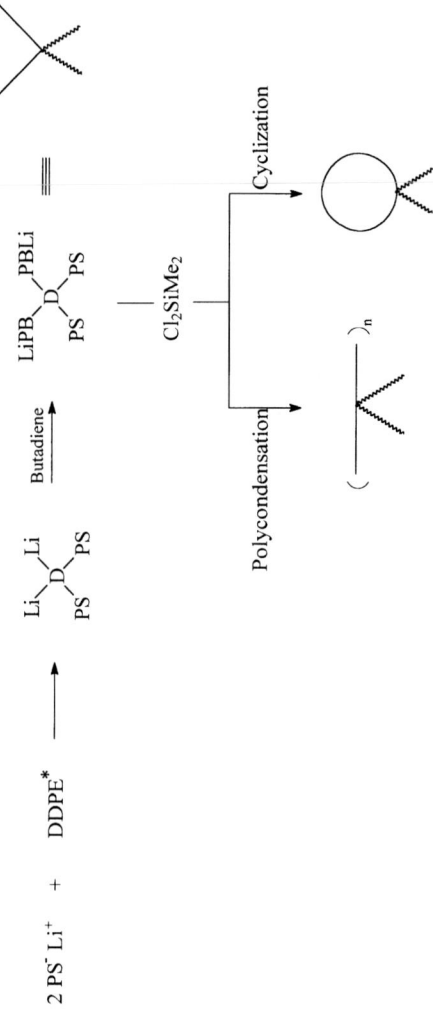

DDPE*: 1,3-bis(1-phenylethenyl)benzene

Scheme 110

3 Properties

3.1 In Solution

3.1.1 Theory

Relatively few theoretical studies have been devoted to the conformational characteristics of nonlinear block copolymers in different solvent environments. Burchard and coworkers [284] studied theoretically the behavior of the static and dynamic structure factors for regular star-block copolymers in dilute solutions. They considered different cases where the refractive index $(n)_s$ of the solvent takes certain values with respect to the corresponding refractive indices of the inner and outer blocks. A different dependence of the ratios

$$\frac{\langle S^2 \rangle app}{\langle S^2 \rangle true} \text{ and } \frac{\langle R_h^{-1} \rangle app}{\langle R_h^{-1} \rangle true}.$$

($<S^2>$: mean-square radius of gyration, R_h: hydrodynamic radius, app.: apparent) with the dn/dc (specific refractive index increment) of the copolymer was predicted for the case of high functionality star-block copolymers in comparison with linear diblock and triblock copolymers, the differences being more pronounced in the case of the ratio of the radii of gyration. The values of the radius of gyration, static structure factor, diffusion coefficient and the reduced and normalized first cumulant are close to the true ones when the refractive indices of both blocks differ considerably from the refractive index of the solvent (i.e., /n_i-n_o/<</n_i-n_s/), where i, o, s refer to inner and outer blocks and the solvent respectively. Figure 1 presents the predictions of Burchard and co-workers [284] for the dependence of the apparent radii of 18-arm star-blocks on refractive index. The angular dependence of the first cumulant of the dynamic structure factor, as well as the effects of heterogeneity, were also discussed.

Freire and coworkers [285, 286] studied the case of miktoarm star copolymers of the type A_xB_{f-x}, where f is the total functionality of the star copolymer. The conformational characteristics of these kinds of molecules were investigated as a function of molecular weight and number of the different branches, as well as the thermodynamic cross interactions between the arms and the solvent medium. Calculations based on the renormalization group and Monte Carlo methods allowed the estimation of the dimensions of each arm and of the whole molecule and the mean square distance between the two centers of mass of the different homopolymers. From these estimations different expansion factors relative to the homopolymer precursors could be calculated (Fig. 2). Different degrees of agreement were obtained by the two methods depending on the property under consideration.

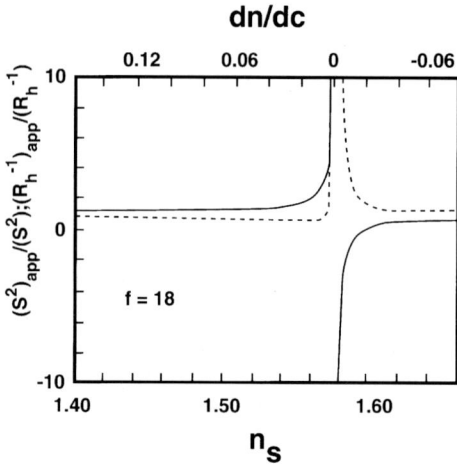

Fig. 1. Variation of the apparent mean-square radius of gyration $\langle S^2 \rangle_{app}$ (solid lines) and the apparent reciprocal hydrodynamic radius $\langle R_h^{-1} \rangle_{app}$ (dotted lines) as a function of the solvent refractive index n_s and refractive index increment dn/dc for a heterogeneous star molecule $(AB)_f$ with f=18: block B, polyisoprene, M=230 000; block A, polystyrene, M=150 000

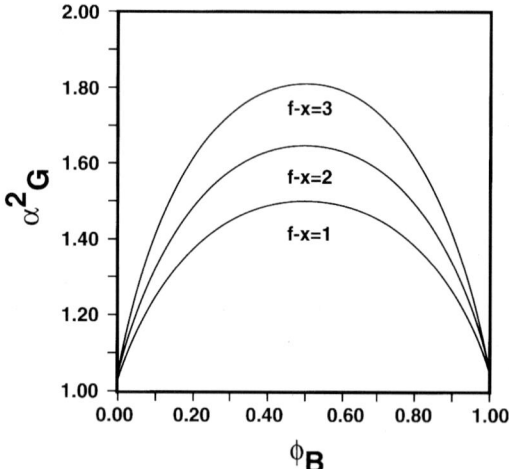

Fig. 2. Expansion factor α_G^2 of the mean square distance between the two centers of mass as a function of the length fraction of the B branch Φ_B for star copolymers of the A_xB_{f-x} type, where x=2 and f-x takes on various values (f is the total number of arms in the star)

The homopolymer arms in miktoarm stars are predicted, by the renormalization method, to expand monotonically, reaching a limit which is dictated by the increase in the size of the other kind of homopolymer arms. The whole molecule seems to increase in size monotonically with molecular weight and functionality. The same behavior is predicted for any part of the molecules consisting of two or more arms of different homopolymers. This work was expanded with the study of A_2B and A_3B miktoarm copolymers. The dimensionless ratio δ_G, which expresses quantitatively the effect of heterointeractions between unlike units on the conformational properties of copolymers, was calculated by an intrinsic viscosity analysis. It is defined as

$$\delta_G = \frac{\langle G^2_{AxBy} \rangle}{\langle G^2_{Ax} \rangle + \langle G^2_{By} \rangle}$$

where $\langle G^2_{AxBy} \rangle$ is the distance between the centers of mass of the two homopolymers A and B of the star and $\langle G^2_i \rangle$ is the distance between the star center and the center of mass of the i homopolymer branch in a symmetric star homopolymer with the same number of branches as the miktoarm star. Monte Carlo calculations were applied to calculate the Flory parameter ϕ for the asymmetric stars whereas, for the symmetric stars, results on 3 and 4 arm homopolymer stars were used. Higher values of the δ_G parameter were obtained in the case of A_3B stars, indicating the existence of stronger heterointeractions between unlike units. These theoretical predictions were also confirmed by experimental results [287].

The same group also investigated the case of ring diblock copolymers in a common θ solvent, a good solvent, and in selective solvents, using theoretical (renormalization group theory) and numerical simulation (Monte Carlo) methods [288]. In this way the average dimensions of each block and of the whole molecule were obtained and compared with results on linear diblock copolymers.

The basic difference between the ring copolymers and the linear ones is that the average dimensions of each block are affected by changing both the solvent conditions and the length of the other block. The effect of ring architecture vanishes as the molecular weight of both blocks increases. The cross interactions seem to have more effect on the average dimensions in the case of the ring copolymers than in the linear diblock chains. Calculations indicate that the greatest segregation between blocks appears in copolymers symmetric in composition. Simulation results agree relatively well with the theoretical predictions, especially in the asymmetric composition regimes. Comparatively speaking the chain architecture seems to affect mostly the block expansion while block segregation and global excluded volume effects contribute about the same in the average dimensions of ring and linear diblock chains.

In another study, Borsali and Benmouna discussed the static and dynamic scattering properties of ring diblock copolymers in solution [289]. They focused their attention on the semi-dilute region and the case of compositionally symmetric ring diblock copolymers. Differences in the scattering profiles were predicted, the main reason being the presence of connection between the two ends of the chain in the case of ring copolymers. Due to this connectivity effect, different conformations are possible for linear and cyclic chains. The elastic scattering intensity is reduced for cyclic diblocks and its maximum is shifted to higher q values. The dynamics of interdiffusion are predicted to speed up in ring diblocks and the minimum of the frequency, related to the interdiffusion mode, is predicted to shift to higher q in comparison to the linear material of the same molecular weight and composition. The kinetics of microphase separation in the Rouse regime were found to occur at $\chi N=18$, where χ is the Flory interaction parameter and N is the overall degree of polymerization, for the cyclic diblocks, indicating greater compatibility between the different blocks for this kind of macromolecular architecture.

At least three theoretical models for conformational characteristics of dendritic macromolecules have been proposed. De Gennes and Hervet predict a density minimum in the center [290], while Lescanec and Muthukumar argue for a density maximum in the center [291]. Properties such as radius and intrinsic viscosity differ substantially for the two models. Fundamental differences in these models include: 1) the de Gennes-Hervet analytical model does not account for backfolding of chains but is based on equilibrium structures; 2) the Lescanec-Muthukumar model (numerical approach) accounts for backfolding but is based on kinetically grown structures. Very recently, Boris and Rubinstein [292] developed a self-consistent mean field model for a starburst dendrimer. Their model conclusively shows that flexible dendrimers have dense, not hollow, cores. For all cases studied, they find that the density is greatest at the core and that the chain ends are distributed throughout the volume of the dendrimer.

No theoretical treatment exists so far, to our knowledge, which deals explicitly with the very interesting subject of the influence of chain architecture of nonlinear block copolymers on their micellization behavior in selective solvents.

3.1.2
Experiment

The behavior of nonlinear block copolymers with different macromolecular architectures has been studied extensively in different solvent environments. Most studies are concerned with the conformation of these complex molecules in good solvents and their micellization behavior in selective, for one part of the copolymer, solvents.

Fetters and coworkers [293] employed static light scattering in solvents of varying refractive index in order to investigate the conformation of star-block copolymers of styrene and isoprene having 18 diblock arms. The use of solvents isorefractive with either the outer or the inner part of the molecule allowed them to determine independently the radius of gyration of each part of the molecule and compare them with the size of the whole star. Their results support the idea that high functionality star-block copolymers adopt a more or less segregated, vesicle like, conformation in dilute solution. Their conclusions are in agreement with the theoretical treatment and experimental results of Burchard et al. [284].

Tuzar and coworkers [294] investigated the micellization behavior of styrene-butadiene star-block copolymers with four arms and polybutadiene inner blocks in the mixed solvent tetrahydrofuran/ethanol, selective for polystyrene blocks.

Spherical micelles with a polybutadiene core were formed for a certain range of compositions of the mixed solvent. The equilibrium between micelles and nonassociated macromolecules was found to be consistent with a closed association mechanism. In comparison with a linear triblock copolymer, the star-block sample showed a lower aggregation number.

Star-block copolymers with flexible inner blocks comprised of poly(dimethylsiloxane) and rigid-rod poly(p-benzamide) outer blocks were studied in solutions by wide and small angle X-ray scattering, polarized microscopy and rheology [295]. The ability of the polymers to form lyotropic liquid-crystalline phases was influenced by the length of the flexible block. For $M_{PDMS} > 1500$/arm, anisotropic solutions were formed whereas, for $M_{PDMS} < 1500$/arm, isotropic solutions were formed. Rheological measurements showed that both isotropic and mesophasic solutions were shear thinning, with no low-shear Newtonian plateau. Fibers spun from anisotropic solutions showed a greater degree of crystalline orientation, as determined by X-ray scattering.

The viscoelastic properties of concentrated solutions of styrene-butadiene star-block copolymers were studied by Masuda et al. [296] in good solvents for both blocks and in selective ones. A significant dependence of the loss and storage moduli on the strain amplitude was observed in the case of dibutylphthalate, a selectively good solvent for the PS blocks at temperatures below 60 °C, which indicates the presence of a microdomain structure due to self assembling of the insoluble blocks. At a certain value of the applied strain the microdomain structure in solution was disrupted.

Hadjichristidis and coworkers [230] studied the hydrodynamic behavior, in dilute solution, of miktoarm stars of the types A_2B and A_2B_2 where A, B = PS, PI, and PBD in solvents good for both segments or theta for one of the arms and good for the others. Analysis of the results suggests that the experimentally determined values of intrinsic viscosity, [η], viscometric radius, R_v, and R_h for the copolymers are higher than the ones calculated from homopolymer star data. The phenomenon was perceived as an indication of repulsive interactions between A and B chains, which tend to increase the sizes of the individual chains and of the star molecule as a whole [230]. A similar conclusion was reached from SEC experiments on polystyrene-poly-t-butylacrylate miktoarm stars with equal number of branches of the two components [243]. The phenomenon, in this case, was more pronounced as the molecular weight of the branches increased.

Teyssie and coworkers have studied the surface, interfacial, and emulsifying properties of AB_2 stars where A is a polydiene or polyvinyl block and B poly(ethylene oxide)(PEO) [297, 298]. The miktoarm stars were shown to be better emulsifying agents for water-organic solvent mixtures than linear block copolymers. Saturation of the interface is reached more quickly with the miktoarm stars. Their results are in agreement with the results of Xie and Xia on PS_2PEO_2 stars [228].

Higashimura et al. [299] used NMR to probe the interactions between amphiphilic star molecules (star-blocks and miktoarm stars) and small molecules and tried to evaluate the influence of macromolecular architecture on these interactions. No distinct differences were observed between star-blocks and miktoarms, both being efficient enough for accommodating hydrophilic molecules within their structure.

Tsitsilianis et al. recently published [245] preliminary results on the micellization behavior of anionically synthesized amphiphilic heteroarm star copolymers with polystyrene and poly(ethylene oxide) branches in THF and water. The former solvent is not very selective for one of the segments whereas the latter is strongly selective for PEO. The apparent molecular weights found for the micelles in THF were two orders of magnitude larger than the ones measured for the unimers. By increasing concentration an increase in the depolarization ratio was observed supporting the conclusion that multimolecular micelles are formed by this kind of miktoarm star copolymer.

The solution properties of graft copolymers have been studied for a relatively longer time in comparison with other copolymer structures. One of the earliest goals was to evaluate their conformational characteristics in good and selective solvents. Roovers and coworkers have studied the conformation of anionically prepared poly(styrene-g-isoprene) and poly(isoprene-g-styrene) in good solvents with varying refractive index [30, 39]. By utilizing static light scattering in isorefractive solvents for one of the components they have drawn the conclusion that in both cases the molecules acquire a more or less segregated structure with the core comprised of the backbone and the grafts being always on the outer part of the macromolecule.

Other groups focused their attention on the behavior of graft copolymers in solvents of varying selectivity towards each component [300-302]. Price and Woods studied the micellization behavior of poly(styrene-g-isoprene) anionically prepared copolymers in methylcyclohexane and n-decane (selective solvents for the polyisoprene grafts) [303, 304]. Compact intermolecular micelles were formed in decane, their aggregation number depending primarily on composition and temperature. Different trends were observed for the radius of gyration and the viscometric radius as a function of temperature and composition of the samples studied. In methylcyclohexane, apparent M_w values were almost constant upon varying temperature and equal to the molecular weight of the isolated molecule. The values for R_V, viscometric radius, were constant, whereas the radius of gyration increased slightly with increasing temperature. These results suggest the presence of intramolecular micelles of the graft copolymers having polystyrene cores in the latter solvent.

Tuzar and coworkers [305] studied the micellization of a poly(isoprene-g-styrene) copolymer in solvent mixtures selective either for the backbone or for the grafts. Their results from static and dynamic light scattering and sedimentation velocity experiments favor the closed association model for the description of the unimers-micelles equilibrium. The graft copolymer micelles were found to have lower aggregation numbers and to be less compact than the micelles formed by linear diblocks.

In another study, Selb and Gallot investigated the conformational properties of poly(styrene-g-4-vinyl-N-ethylpyridinium bromide) in water/methanol/LiBr mixtures [306]. The graft copolymers did not show intermolecular association in contrast to the linear block copolymers. Viscometric results showed that these graft copolymers also form compact, star-like monomolecular micelles with polystyrene cores and poly(4-vinyl-N-ethylpyridinium bromide) coronas, which resemble the polymolecular micelles of diblock materials.

The colloidal properties of anionically prepared poly(styrene-*g*-ethylene oxide) graft copolymers were studied by Candau et al. in different water/toluene/alcohol mixtures by light and neutron scattering, NMR, and viscometry [307-309]. Aggregation numbers depend on mixture compositions with the highest values attained for water-rich systems. The micelles formed seem to have a core and shell conformation, with PS cores, in all cases studied. Dialysis experiments showed that the enhanced water-oil solubility was due to preferential solvation of the two segregated components of the copolymers by the solvent mixture and not to one specific solvent entrapment as is the case of classical microemulsions.

In another study poly(acrylic acid-*g*-styrene) copolymers were also shown to have good emulsifying ability and high water absorbency [105]. Membranes produced from intermolecular complexes of the above materials with poly(ethylacrylate-*g*-ethylene oxide) copolymers behaved like chemical valves, whose permeativity could be controlled reversibly by changing the pH of the surrounding medium, since both graft copolymers behave as polyelectrolytes in aqueous solution.

Recently the study of the dilute solution behavior of polymacromonomer, a limiting case of graft copolymer where each repeat unit carries a grafted chain, has been initiated. The main interests are focused on the dependence of conformation and size of the whole molecule on factors such as nature of the backbone and side chain, molecular weight of backbone and grafts, solvent interactions with respect to both components, and chain stiffness induced on the backbone due to the high grafting density and its dependence on the aforementioned factors [310-313].

The availability of relatively more architecturally complicated macromolecules has also initiated investigation of their dilute solution properties. The behavior of dumbbell copolymers with a polystyrene connector and poly(ethylene oxide) end grafted arms has been studied by Bayer and Stadler [314]. In toluene the viscosity of these copolymers depends primarily on the degree of polymerization of the branches and not on the number of branches (a situation also observed for homostars and star-block copolymers in a good solvent). Aggregation is less pronounced than in the case of the linear precursors. Light scattering measurements in *N*,*N*-dimethylformamide gave classical Zimm plots and apparent molecular weights slightly higher than the unimer molecular weights.

Experiments on super-H shaped copolymers, [315] using small angle neutron scattering, low angle light scattering, and viscometry, with polystyrene connectors and polyisoprene branches in *n*-decane, also indicate decreased aggregation numbers for this kind of polymer in comparison with linear diblock copolymers. This is due to an increase in solubility induced by the macromolecular architecture, and it is related to the higher number of branches per molecule. It is worth mentioning that the samples with lower polystyrene content and molecular weight of the connector did not show intermolecular association. The experimental results from small-angle neutron scattering (SANS) agreed better with an intermediate scaling model, with characteristics between those of starlike and crew cut micelles, for micelles with polystyrene cores relatively free of solvent.

The dilute solution properties of cyclic diblock copolymers of styrene and dimethylsiloxane have been investigated by Amis and coworkers [316, 317]. The apparent temperature in cyclohexane of the cyclic diblocks was found to be about

10 °C lower than the one measured for the linear triblock analogs. As a consequence, at the temperature range investigated the cyclic copolymers do not show any evidence of aggregation, whereas micelles exist in the linear triblock solutions over the same range of temperatures. Comparison of the ratios of diffusion coefficients and second virial coefficients of the cyclic and linear samples (D_c/D_l and A_{2c}/A_{2l}) under theta and good solvent conditions, respectively, with theoretical values of these ratios for linear and cyclic homopolymers showed good agreement between theory and experiment.

Gitsov and Frechet [273] studied the solution properties of linear-dendritic hybrid triblock copolymers composed of a linear central polystyrene block and dendritic polyether end blocks. No dependence of the elution volume, in SEC experiments, on concentration was observed for these copolymers and the molecular weights determined by the universal calibration method were overestimated. For low molecular weights of the PS connector the intrinsic viscosity values, in THF, were larger than the PS precursor alone but for $M_{PS} > 40 000$ [η] values became increasingly lower than those of the PS block. However, the radii of gyration for the copolymers were always larger than the ones measured for the connector block. This was attributed to a shape transition as the molecular weight of the central block increases, and to the nonuniform density distribution of the dendritic hybrids.

Diblock copolymers comprised of a linear poly(ethylene oxide) block and a dendritic polyether block [318] have been found to form monomolecular micelles in THF (a good solvent for the dendritic block) and mono- and multimolecular micelles in methanol/water mixtures (good solvent for the linear blocks). The presence of mono- or multimolecular micelles in the latter system depends on the dendrimer generation and polymer concentration. Increasing concentration results in multimolecular micelles, where decreasing the dendritic generation favors the formation of monomolecular micelles. The amphiphilic character of linear PEO-dendritic polylysine diblocks has also been investigated through surface tension and solubility experiments [276].

In closing this section, we note that size exclusion chromatography with multiple detection has emerged as a powerful tool for the characterization and investigation of dilute solution properties of copolymers with complex architectures [319-321]. Aspects like molecular weight distribution, true molecular weight and size of macromolecules with branched architectures, and composition and polymerization mechanism determinations have attracted most of the interest of investigators. For example, Mourey et al. [321a] used an SEC system coupled with light scattering and viscometric detection to examine the solution behavior of polyether dendrimers. They observed low polydispersities ($I \approx 1.04$), which did not broaden appreciably with increasing dendrimer generation (G). Molecular weights were in agreement with values predicted based upon the synthetic strategy, and intrinsic viscosity passed through a maximum as a function of G, as predicted by Lescanec and Muthukumar [291] and confirmed by Boris and Rubinstein [292].

3.2
In Bulk

3.2.1
Theory

In contrast to the situation found for dilute solutions, the behavior of nonlinear block copolymers in the solid state seems to have attracted great attention. Many theoretical publications appeared in recent years, dealing mainly with the phase behavior of star-block, simple graft and comb copolymers. Issues like the nature of the phase diagram and the order-disorder transition have been studied in considerable detail. The compatibilizing effects of complex copolymers, in comparison to simple diblock copolymers, were also investigated.

Olvera de la Cruz and Sanchez [322] were the first to report theoretical calculations concerning the phase stability and the static structure factors of star-block copolymers, simple graft copolymers, and miktoarm $A_n B_n$ star copolymers with equal numbers of A and B chains. Using mean field theory and assuming the chains to be Gaussian, they predicted that a simple graft has no critical point for any composition. Irrespective of the position of the branch point, the minimum value at the spinodal appears at volume fraction $f=0.5$. For the special case of the symmetric graft (an A_2B type miktoarm star) $(\chi N)_s = 13.5$ (χ is the Flory-Huggins interaction parameter and N is overall degree of polymerization; see Fig. 3). This implies that it is more difficult for the chains to phase separate in this kind of architecture than it is for diblock copolymers. Star copolymers of the $A_n B_n$ type are predicted to have a critical point at $(\chi N)_s = 10.5$, the same value as for diblocks,

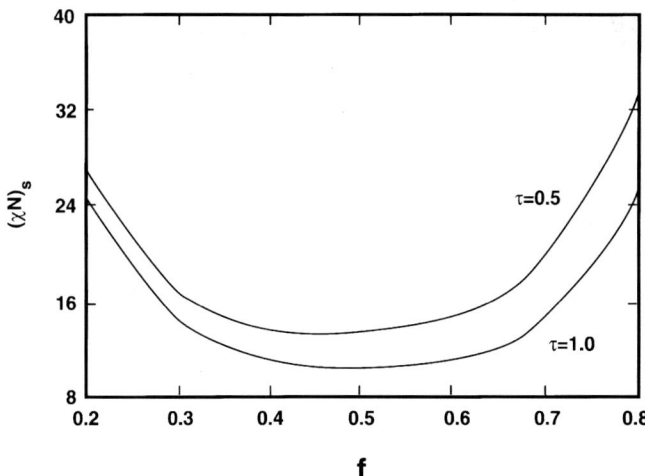

Fig. 3. Comparison of the variation of $(\chi N)_s$ with composition f for simple graft (or A_2B star; $\tau=0.5$) and diblock ($\tau=1.0$) copolymers. τ is the fractional position along the A chain backbone at which the B graft is chemically linked. The number of monomers (N) in each copolymer is the same

when $f=0.5$ (Fig. 4). When the branching point moves to the end of the diblock chains, resulting in the formation of star diblock copolymers, $(AB)_n$, again no critical point can be found and at $f=0.5$ the minimum value of $(\chi N)_s$ is decreasing by increasing the number of arms, in other words the tendency toward phase separation increases (Fig. 5). As far as the structure factor is concerned, theory predicts that the maximum in q (q is the scattering vector equal to $4\pi/\lambda \sin(\theta/2)$ where λ is the wavelength of the radiation and θ is the scattering angle) for graft copolymers (q^*_{graft}) should appear at larger q compared to the diblock case and q^* is not symmetric around $f=0.5$ due to the inherent asymmetry of the graft copolymer. For $(AB)_n$ star copolymers q^* passes through a minimum as n increases whereas for A_nB_n star $q^* (A_nB_n)$ $q^* (A_nB_n) \geq q^* (AB)$.

Benoît and Hadjiioannou [323] calculated the scattering functions in the homogeneous state for complex copolymer architectures like comb copolymers and star-blocks. They also considered changes in the scattering profiles as a function of the number N of blocks in the block copolymers as well as a function of composition, molecular weight, and polydispersity in molecular weight and composition. The scattering intensity at small angles was found to be independent of architecture. For larger angles the scattering intensity goes through a maximum, the position of which is independent of the interaction parameter χ but its magnitude depends strongly on χ. As the number N of repeating sequences increases both for star-blocks and comb copolymers, the maximum of intensity rapidly reaches an asymptotic value whereas the position of the maximum remains relatively constant. The maximum is more pronounced for compositions around $f=0.5$ and becomes less pronounced for asymmetric compositions. Polydispersity in molecular weight seems to change (increase) the intensity at low q without affecting the maximum but when it is increased substantially the q max can be shifted to lower values. The calculated phase diagram for regular comb copoly-

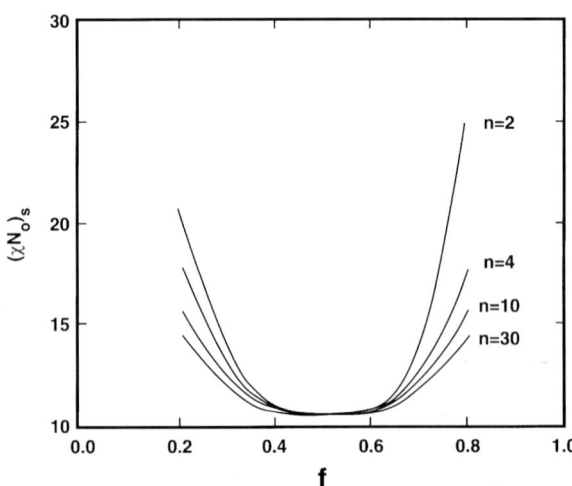

Fig. 4. Variation of $(\chi N_o)_s$ with composition and arm number for A_nB_n star copolymers. N_o is the number of monomers in the diblock (A_1B_1)

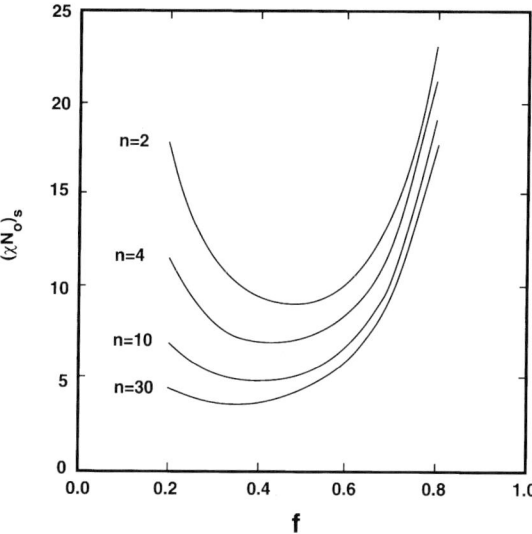

Fig. 5. Variation of $(\chi N_o)_s$ with composition and arm number for $(AB)_n$ star copolymers

mers is symmetric around f = 0.5, where the minimum appears. When $N \to \infty$ miscibility increases and the minimum shifts to higher values of f. The value for $(\chi N)_s$ tends to become higher. The phase diagram for star-block copolymers agrees with the one reported by Olvera de la Cruz and Sanchez.

Anderson and Thomas [324] studied the microphase separation of star-diblock copolymers in the strong segregation limit using mean field theory. They probed the thermodynamic stability of the bicontinuous morphology that was observed for star-block copolymers with a large number of arms (see next section). However their calculations for the free energy of the bicontinuous morphology, described as the ordered bicontinuous double diamond structure, failed to show that it becomes less than that of the cylindrical structure. They attributed this failure to the inadequacy of Gaussian statistics to describe the conformation of chains in the microphase separated state of high functionality stars. Fluctuation corrections to the mean field theory have been incorporated for star copolymers by Dobrynin and Erukhimovich [325].

A different approach was used by Milner [326] in order to predict the phase diagram for asymmetric copolymer architectures (for example A_2B, A_3B etc. types of miktoarm stars). The free energy of the system can be calculated by summing the free energies of the polymer "brushes" existing on the two sides of the interphase. Milner described the effects of both chain architecture (i.e., number of arms) and elastic (conformational) asymmetry of the dissimilar chains, in the strong segregation limit, by the parameter

$$\varepsilon = \frac{n_A}{n_B}\left[\frac{(\ell_A)}{(\ell_B)}\right]^{1/2}$$

where n_A, n_B is the number of arms of types A or B and l_A, l_B is given by the ratio V_i/R^2_{gi} where V_i is the volume displaced by arms A or B and R_{gi} the radius of gyration of the chains A or B in the "brush". In this way the phase diagram of ε vs volume fraction of B monomer was calculated, and the boundaries within which each morphology is likely to be observed were established (Fig. 6). There is a strong dependence of the phase boundaries on the number of each kind of arms.

More recently the microphase separation of regular and random comb copolymers was studied theoretically by Balazs and coworkers in the framework of the random phase approximation [327]. The regular combs were assumed to have symmetrically or asymmetrically placed teeth (grafts) on the backbone. For the regular case the structure factor rapidly approaches an asymptotic behavior as the number of teeth increases. The spinodal curves become more asymmetric as the number of teeth increases and the minimum value of χN shifts to lower volume fractions, f_A, of the backbone for the symmetric case, whereas for the asymmetric the opposite behavior is observed. However, both curves approach to the same limiting form for large numbers of teeth. The spinodal curve for the random combs is significantly different in the limit of large numbers of teeth. More detailed calculations for the case of random combs allowed for determination of the effects of teeth placement correlation and polydispersity in the number of teeth and the average number of teeth. The above factors determine whether the homogeneous phase is unstable and susceptible to microphase or macrophase separation.

The same group has also presented Monte Carlo calculations on the compatibilizing effect of comb copolymers [328]. Their interfacial behavior was compared with that of linear multiblock copolymers. By varying the numbers and lengths of the teeth it was demonstrated that combs with fewer and longer teeth localize themselves easier at the interface between two homopolymers, of the same nature as the backbone and teeth, than combs with a larger number of shorter teeth. In

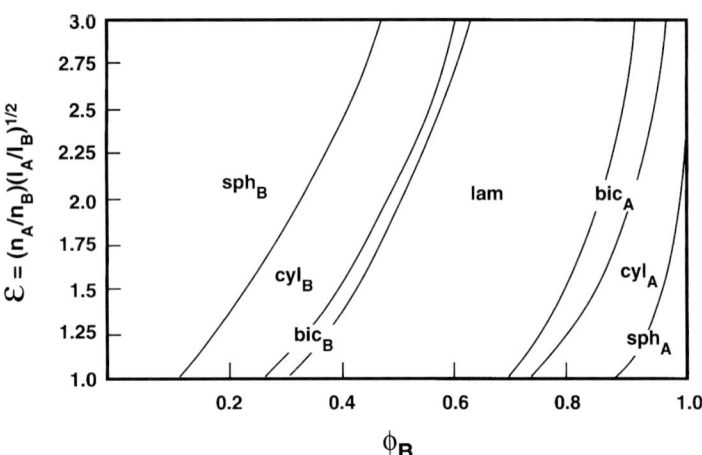

Fig. 6. Phase diagram in the strong segregation limit for star-block copolymers with n_A, A arms and n_B, B arms as a function of volume fraction of the B monomer

any case, comb copolymers are predicted to be better compatibilizers than multiblock copolymers.

In another case, a mixture of AC and BC combs, with backbones made up of A and B segments and teeth of C segments, and A, B homopolymers were studied [329]. All A, B, and C segments were taken to be incompatible with each other. It was found that AC and BC combs confine themselves at the interface for comb concentrations below the critical micelle concentration. The reduction in interfacial tension was calculated, taking into account the effect of micelle formation. The effect of number and length of C teeth was also studied. It was concluded that a critical tooth length is needed for the combs to absorb at the interface. On the other hand, combs with multiple short teeth are more efficient in lowering the interfacial tension between A and B homopolymers. When the interaction parameter between C and A, B was increased, a higher concentration of combs was necessary to lower the interfacial tension to the same extent.

The same group used analytical arguments and numerical self-consistent field calculations to compare the interfacial tensions for a system of A and B homopolymers, immiscible with each other, having AB diblocks and comb copolymers localized at the interface [330, 331]. In this case the overall molecular weight of the compatibilizer molecule was kept constant and the architecture was varied. At fixed N and χ, as the number of teeth increased the comb becomes less efficient at reducing the interfacial tension as compared to the diblock. At high values of N and χ, the difference between the architectures becomes more pronounced. The diblock produces the lowest interfacial tension at the lowest concentrations. On the other hand, long combs with multiple teeth were found to be more efficient compatibilizers than short diblocks.

3.2.2
Experiment

Much experimental work has been devoted so far to the study of the solid state and melt behavior of nonlinear block copolymers. Most of the studies are concerned with the microphase separation morphology of copolymers with various architectures as well as the phase transition temperature. Other solid state and melt properties have been examined in a few cases.

The most studied materials are the star-block copolymers since they were the first synthesized. The morphology of DVB linked poly(styrene-*b*-isoprene) star-blocks has been studied by Bi and Fetters [12]. Some more recent studies on these materials have been presented by other groups [332].

Thomas and coworkers made a very extensive investigation of the morphological characteristics for the better defined chlorosilane-linked star-blocks of styrene and isoprene with up to 18 arms. They studied the effect of arm number and arm molecular weight on the solid state morphology of the aforementioned star-blocks [16, 333, 334]. A transition from hexagonally-packed cylinders to the formation of a new ordered bicontinuous structure was observed as the number of arms was increased (above six), for constant arm molecular weight, and for compositions ~30 wt% of the outer segment, and also by increasing the arm molecular weight.

This type of morphology had not been observed previously for linear materials of similar composition. SAXS data supported the assignment from transmission electron microscopy (TEM) results of the new structure as the ordered bicontinuous double diamond (OBDD), which was shown afterwards by selective solvent casting to be an equilibrium morphology [334a]. The structure consists of two distinct, mutually interwoven but unconnected three-dimensional networks of polystyrene (minority component), each of which exhibits the symmetry of a diamond lattice, surrounded by a continuous polyisoprene phase (majority component). The cubic lattice of this particular arrangement exhibits the symmetry of the Pn3m space group. A detailed investigation of the dependence of morphology on the composition of 18-arm poly(styrene-b-isoprene) stars determined the window of stability of the OBDD structure (27-32vol.% of the outer segments). The new morphology was assumed to be a result of the molecular architecture of the star-block copolymers, where an increase in the arm number results in an increased crowding in the core region. Presumably OBDD is a more favorable phase-separated morphology that minimizes the interfacial area between the two phases. However, recent re-evaluation of the SAXS profiles from samples that were assigned the OBDD structure revealed that the cubic bicontinuous structure exhibits reflections that correspond more close to the Ia3d space group [335]. Bicontinuous structures have also been observed in linear diblock samples of certain composition [336].

The effect of surface constraints on the morphology of the star-block copolymers was studied by Thomas and coworkers [337]. Thin film droplets of samples with various functionalities were studied, and the ones that exhibited the OBDD structure in the bulk were found to be cylinders in this case. In an independent study, the lamellar domain spacings of 4-arm and 12-arm star-block copolymers of styrene and isoprene were found, by TEM and SAXS, to be the same as those of the arm material [338].

The order-disorder transition of anionically prepared, chlorosilane linked star-block copolymers of styrene and isoprene was examined by Hashimoto and coworkers, in respect to the effect of the arm number. In their first report [339], where a 6-arm star-block was compared with the precursor arm sample, they found that the position of maximum intensity, q_{max} in the SAXS profiles is independent of temperature at constant arm number ($f_A = 6$) but depends on the number of arms (f_A). The value of q_{max} for $f_A = 6$ is significantly smaller than that of $f_A = 1$ (the diblock arm). The χ parameter was higher for $f_A = 1$ than for $f_A = 6$ in the temperature range studied. In a later more detailed SAXS investigation [340] the same group reported that the order-disorder transition and the spinodal temperature decrease dramatically from $f_A = 1$ to $f_A = 2$ but remains almost constant for $2 \leq f_A \leq 18$. They reconfirmed the dependence of q_{max} on functionality (at constant arm molecular weight and composition) and its temperature independence at constant f_A. However, they observed a dependence of the effective interaction parameter χ_{eff} on the arm number (the greater the number of arms the smaller the value for χ_{eff}). Another study [341] on the microphase separation thermodynamics of star-block copolymers of constant functionality ($f_A = 18$) and arm molecular weight has revealed that the quantity $q_{max} R_G^2$, where q_{max} is the wave number in the dominant mode of the fluctuations and R_G

is the radius of gyration of the linear diblock arm, goes through a minimum as the composition of the copolymer changes.

SAXS and rheology were used to study the order-disorder transition and the ordering kinetics in four arm star-diblock copolymers consisting of PS-b-PI branches [342]. Both methods gave similar results regarding the T_{ODT}. When $T \gg T_{ODT}$ the SAXS profiles could be described by the mean-field structure factor. Due to the weak T dependence the phase diagram was explored in only a small region showing the existence of spherical PS microdomains in a bcc lattice. The ordering process was slower compared to linear diblocks due to the different architecture. This study was completed with computer simulations and theoretical predictions concerning the phase diagram of the star-diblock copolymers.

Thomas and coworkers [19] studied the morphological transition from lamellar to cubic microphase separated structure in compositionally symmetric inverse star-block copolymers. Samples having the general type $(PS\alpha_M - PI\alpha_M)_n - (PI\alpha_M - PS\alpha_M)_n$ where $M \sim 20000$, number of arms (n) = 1,2, and $\alpha = 1,2,4$ (α is the ratio of the outer block molecular weight to that of the inner block) were examined with TEM and SAXS. Lamellar structures were identified for $n = 1,2$ and $\alpha = 1,2$. For $\alpha = 4$ a transition from a lamellar structure (for $n = 1$) to a triconnected cubic structure (for $n = 2$) was observed. Comparison of TEM data with simulated images of three dimensional models of level surfaces confirmed that the cubic continuous structure is best described as an OBDD structure. This transformation was attributed to a preferred interfacial curvature induced by the change of the architecture in these asymmetric (in the arm diblock level) but symmetric (in overall composition) star copolymers.

Relatively few studies have appeared concerning the bulk properties of miktoarm star copolymers, mainly due to the fact that they have only recently been synthesized. Transmission electron microscopy and small angle X-ray scattering results for A_2B and A_3B miktoarm star copolymers, where A is PI or PS and PI, respectively, and B is PS, show that the observed morphologies are in qualitative agreement with the theory developed by Milner [231, 343]. Interestingly, an I_2S sample with 53 vol.% PS forms a tricontinuous cubic structure. On the other hand, the experimentally determined phase diagram for the I_3S miktoarm stars shows an absence of the spherical morphology in the high PS volume fraction region. The micrograph of a sample having 92 vol.% PS consists of PS cylinders in a PI matrix and order is apparent only on a local scale. Additionally, in the case of an I_3S copolymer with 39 vol.% polystyrene, PS cylinders were observed instead of PS spheres, and for an I_3S sample containing 86 vol.% PS, PI cylinders were present instead of lamellae. The theory can be brought "in line" with the experimental findings if the theoretical boundary curves could "bend back" towards the low volume fraction, ϕ_B, side of the phase diagram. In other words, multiple domain effects, i.e., effects of crowding of chains emanating from different but neighboring interfaces, should be taken into account by the theory, as the authors suggested. Similar effects are observed in a 3-miktoarm star terpolymer comprized of PS, PI, and PBd arms. With a 40 vol.% fraction of PS this sample would be expected to have a lamellar or a bicontinuous structure. However, micrographs show cylindrical morphology of the PS phase in a matrix of polydiene (PB and PI are miscible in this low molecular weight region).

In an independent study, Gido and coworkers [344] examined the morphology of A_2B miktoarms, where A is polyisoprene and B deuterated polystyrene, with PS volume fractions between 8 and 89% covering a wide range on the phase diagram. Their results show agreement with the theory of Milner, except in one case. A sample having 81 vol.% PS and expected to have the lamellar morphology, but situated close to the boundary between the lamellae and bicontinuous phases of the theoretical diagram, showed disordered PI spheres in a PS matrix. This peculiar microphase separated morphology of randomly oriented wormlike micelles, which may be a result of the frustration of the chains at the given volume fraction and molecular weights of the arms, was shown, by preferential solvent casting experiments, to be an equilibrium one [345].

The above are examples of the influence of architecture on the microphase morphology of miktoarm stars where, due to the crowding of the chains on one side of the interface, the interface becomes curved resulting in different morphological structures to those expected for linear diblocks of the same composition.

A thorough investigation of the static and kinetic aspects of the order-disorder transition in SI_2 and SIB miktoarm stars was presented by Floudas et al. [346], SAXS and rheology were employed for this purpose. The experimental results could be quantitatively described by mean field theory at temperatures higher than the ODT. However, near the ODT, SAXS profiles showed the existence of fluctuations. The q_{max} discontinuities in the SAXS peak intensity and the storage modulus near the ODT are more pronounced for the miktoarm stars than in the case of diblocks. The χN values of sample I_2S near the ODT were higher than a diblock of the same composition and almost independent of temperature. After quenching the samples to final temperatures near the ODT the rheological results indicate that ordering proceeds via heterogeneous nucleation. The width of the kinetically accessible metastable region was larger for the star copolymers.

Dielectric spectroscopy was also employed in order to study the local and global dynamics of the PI arm in these miktoarm samples [347]. Measurements were made in the ordered state and the dynamics of PI chains tethered on PS cylinders were observed in different environments (one of pure PI in the $(PI)_2PS$ case and one of a mixture of PI and PB for the (PS)(PI)(PB) case). At low temperatures the PI segmental dynamics are different due to the introduction of the faster PB chains in the terpolymer with the effect diminishing by increasing temperature. The global PI dynamics resemble the dynamics of PI stars with additional entanglement effects when PB is present in the system.

Gallot and coworkers studied the effect of macromolecular architecture on the lamellar structure of the poly(ethylene oxide) crystallizable block in linear poly-(4-*tert*-butylstyrene-*b*-ethylene oxide) diblocks and PtBS $(PEO)_2$ miktoarm stars by comparing results from SAXS and DSC [348]. At the same total molecular weight and composition, the melting temperature, the degree of crystallinity, and the number of folds of PEO chains are lower for the branched samples.

The microphase separation of (styrene-*g*-isoprene) graft copolymers with a large number of grafted chains was investigated by Price and coworkers [349]. Those films were cast from benzene, a nonselective solvent, and a relatively well-defined microphase separated morphology was observed, but the structure was less regular compared with dispersions in hexane on carbon films [350]. In the latter case, the morphology was described as polystyrene spheres in a polyisoprene matrix (at 26 vol% PS), with diffuse domain boundaries, probably due to the packing difficulties involved in the case of graft copolymers.

Fine microphase separation was also observed for model anionically prepared block-graft copolymers having polystyrene backbones with polyisoprene or poly-(ethylene oxide) grafts [351]. In the first case, for samples having 47 and 17 wt% PI, cylinders of PS in a PI matrix and lamellar structures were observed, respectively. In the second case, for ~30 wt% PEO, a lamellar microphase separated morphology was observed. The results were interpreted by assuming that the apparent volume fractions of the grafts are higher than the real ones. The cause of this behavior can be attributed to the more extended conformation of the grafted chains due to crowding around the backbone.

In another study, Kennedy and Delvaux reported that graft copolymers of polystyrene (grafts) and polybutadiene (backbone) did not show a well-defined microphase separated morphology [352]. The phenomenon was attributed to the frustration of the side chains, due to the increased grafting density on the backbone. It seems that molecular parameters like side chain and backbone length and incompatibility, as well as grafting density, are the major factors affecting the morphology of graft copolymers.

Some studies dealing with the solid state properties of graft copolymers with liquid-crystal side chains have appeared in the literature [353-355]. The main attention is focused on the influence of molecular parameters (nature of the backbone and the side chains, molecular weight, etc.) on the mesomorphic properties of the liquid-crystal side chain. Salt complexes of PEO grafted onto various backbones were also considered [356] and the influence of the nature of the backbone and salts on the conductivity of the resulting materials was studied.

Recently the morphology of microphase separated cyclic diblock copolymers of PS-P2VP and PS-PDMS was investigated by Thomas and coworkers [357] and comparisons with their linear analogs were made. All the samples studied showed the lamellar morphology. A decrease in the domain spacing was observed for the cyclic copolymers due to the fact that only looped conformations are allowed for these materials. It was pointed out, by scaling arguments, that χN values are identical for cyclic and linear chains and thus each pair is at the same degree of segregation. Therefore the dependence of the ratio of domain spacing D_{cyclic}/D_{linear} on χN was also studied.

The rheological properties of 3 and 4 arm star-block copolymers of styrene and butadiene, having either monomer as the outer segment, were studied by Krause et al. [358]. Steady flow and dynamic viscosities were greater for polymers with styrene outer blocks at constant molecular weight and hard segment composition. No dependence on functionality was observed. Compared at constant total molecular weight, the viscosity decreased with increasing functionality irrespective of the nature of the outer blocks. It was concluded, from comparisons at equal block lengths, that it is the length of the outer blocks and not the total molecular weight that dominates the viscoelastic behavior of these materials.

Bi and Fetters studied the rheological and mechanical properties of DVB linked star-block copolymers of styrene and dienes [12]. The dynamic viscosities were found to be independent of the star functionality but to depend on the molecular weight of the arm (a situation similar to that of the intrinsic viscosity values in dilute solution in good solvents). The absolute values were roughly equal to the ones measured for linear triblock copolymers. The tensile strengths for the star-block copolymers were greater than the ones observed for linear materials, for a wide range of temperatures, but their permanent set after break was lower.

Recently, Ma and coworkers (at 3 M Co.) studied the use of asymmetric star-block copolymers as pressure sensitive adhesives [358a]. Through a DVB-linking approach, they prepared stars having a PI core, with both "long" and "short" PS outer segments. The presence of both high and low molecular weight endblocks on the same molecule gave materials with removable adhesion and good resistance to lifting in the absence or presence of crosslinking.

McLeish and coworkers have published results on the rheological behavior of S_2I_2 miktoarm star copolymers [359]. For the temperature range between 100 and 150 °C it was evident that the rheology of a polymer with 20 wt%. PS was independent of temperature, implying a particular molecular mechanism for stress relaxation for this architecture. For the sample having 35 wt% PS, a failure of the superposition principle was observed, a fact that was attributed to the temperature sensitive effective modulus of the polymer.

A study of the phase behavior of (poly-t-butylmethacrylate)$_n$ – (polystyrene)$_n$ and (poly-t-butylacrylate)$_n$ – (polystyrene)$_n$ miktoarm stars by differential scanning calorimetry has shown a remarkable decrease in the Tg of the polystyrene phase, in comparison with linear precursors, and an increase in the width of the glass transition [360]. An intermediate Tg was observed in some cases for microphase separated samples. The above phenomenon was attributed to an extended interphase region in these materials, due to partial mixing of the different chains around the cores of the stars.

Much interest has been devoted towards the investigation of the behavior of blends comprised of homopolymers and copolymers of various architectures, the main goal being to determine the influence of macromolecular architecture on miscibility and compatibilizing effectiveness. A number of papers by Feng and coworkers have dealt with the miscibility of blends containing polystyrene and 4-arm poly(styrene-b-butadiene) star-block copolymers [361-363]. They concluded that miscibility depends on the molecular weight of the homopolymers and not on the architecture of the copolymer. The interdomain distance was found to increase with PS molecular weight for miscible blends whereas the interfacial thickness was independent of PS molecular weight. The molecular mobility of the polystyrene and polybutadiene blocks of the star depends strongly on the molecular weight of homopolystyrene added due to differences of solubility of the PS into the PS and PB block microdomains. On the other hand, the molecular diffusivity of the homopolymer increases as its molecular weight decreases.

For the case of homopolyisoprene/styrene-isoprene copolymer mixtures [364], it was shown that the miscibility increases in the order four-arm star-block < triblock < diblock. Increased incompatibility was observed in the pair poly(isoprene-g-styrene)/polyisoprene [365] even when the molecular weight of the homopolymer was much lower than the PI segment length between junction points of the graft copolymers.

In view of the available theoretical calculations, the ability of graft copolymers to act as compatibilizers for homopolymer blends was experimentally investigated by Balazs and coworkers [366] and good agreement with the theoretical predictions was observed.

Several studies have been published where different aspects of the solid state behavior of nonlinear block copolymers (like the adhesive properties, viscoelastic properties after aging and solvent treatment, thermal properties, etc.) vary with varying architectures [367-373]. Much work has still to be done in order to understand the structure-property relationship of these complex macromolecules.

4
Concluding Remarks

Advancements in synthetic polymer chemistry have allowed a remarkable range of new nonlinear block copolymer architectures to be synthesized. The result is a wide variety of new materials with the capacity to form self-assembled phases in bulk and in solution. At present our synthetic capabilities exceed our understanding, both theoretical and experimental, of the properties of such macromolecular systems. We anticipate that a better understanding of structure-property relationships for these materials will lead to impressive new polymers with applications such as structural plastics, elastomers, membranes, controlled release agents, compatibilizers, and surface active agents. From the synthetic standpoint it seems likely that recent advances in living free radical polymerization will make the syntheses of many non-linear block copolymers more commercially appealing.

Acknowledgements. *MP*, SP, and JWM are grateful to the U.S. Army Research Office for supporting our research on nonlinear block copolymers (Grant # DAAH04-94-G-0245) and making possible the preparation of this review. NH, MP, and SP also wish to thank the Greek General Secretariat of Research and Technology for financial support. We thank Lujia (Luke) Bu for the preparation of reaction schemes.

5
References

1. Allport DC, Janes WH (1973) Block copolymers. Wiley, New York
2. Noshay A, McGrath JE (1977) Block copolymers: overview and critical survey. Academic, New York
3. Riess G, Hurtrez C, Badahur P (1985) Block copolymers. In: Kroschwitz JI (ed) Encyclopedia of polymer science and engineering, 2nd edn, vol 2. Wiley-Interscience, New York, p 324
4. Cowie JMG (1989) Block and graft copolymers. In: Allen G, Bevington JC (eds) Comprehensive polymer science, vol 3. Pergamon, Oxford, p 33
5. Quirk RP, Kinning DJ, Fetters LJ (1989) Block copolymers. In: Allen G, Bevington JC (eds) Comprehensive polymer science, vol 7. Pergamon, Oxford, p 1.
6. Kennedy JP, Ivan B (1992) Designed polymers by carbocationic macromolecular engineering: theory and practice. Hanser, Munich

7. Hsieh H, Quirk RP (1996) Anionic polymerization: principles and practical applications. Marcel Dekker, New York
8. Dreyfus P, Quirk RP (1987) Graft copolymers. In: Kroshwitz JI (ed) Encyclopedia of polymer science and engineering, 2nd edn. Wiley-Interscience, New York, 7: 551
9. (a) Clark RJ, Henkee CS (1995) Macromol Symp 91: 27; (b) Keszler B, Kennedy JP (1984) J Macromol Sci Chem A21(3): 327
10. Bi LK, Fetters LJ, Morton M (1974) Polym Prepr (Am Chem Soc Div Polym Chem) 15 (2): 157
11. Bi LK, Fetters LJ (1975) Macromolecules 8: 90
12. Bi LK, Fetters LJ (1976) Macromolecules 9: 732
13. McGrath JE, Wang I, Martin MK, Crane KS (1979) Polym Prepr (Am Chem Soc Polym Div) 20(2): 114, 524
14. Lutz P, Rempp P (1988) Makromol Chem 189: 1051
15. Rein D, Lamps JP, Rempp P, Lutz P, Papanagopoulos D, Tsitsilianis C (1993) Acta Polymerica 44: 225
16. Alward DB, Kinning DJ, Thomas EL, Fetters LJ (1986) Macromolecules 19: 215
17. (a) Storey RF, George SE (1988) Polym Mater Sci Eng 58: 985; (b) Storey RF, George SE, Nelson ME (1991) Macromolecules 24: 2920
18. Dickstein WH (1987) Rigid rod star-block copolymers: synthesis and characterization of star-block liquid crystalline copolymers. Dissertation, Univ of Massachussets
19. Tselikas Y, Hadjichristidis N, Lescanec RL, Honeker CC, Wohlgemuth M, Thomas EL, (1996) Macromolecules 29: 3390
20. Kanaoka S, Sawamoto M, Higashimura T (1991) Macromolecules 24: 5741
21. Feger C, Cantow H-J (1980) Polym Bull 3: 407
22. Storey RF, Chisholm BJ, Lee Y (1993) Polymer 34: 4330
23. Shohl H, Sawamoto M, Higashimura T (1991) Polym Bull 25: 529
24. Fukui H, Sawamoto M, Higashimura T (1995) Macromolecules 28: 3756
25. Saunders RS, Cohen RE, Wong SJ, Schrock RR (1992) Macromolecules 25: 2055
26. Ishizu K (1993) Polym-Plast Technol Eng 32(6): 511
27. Ishizu K, Shimomura K, Saito R, Fukutomi T (1991) J Polym Sci: Part A: Polym Chem 29: 607
28. Ishizu K, Uchida S (1994) Polymer 35: 4712
29. Roovers J, Toporowski P, Martin J (1989) Macromolecules 22: 1897
30. Cameron GC, Quresh MY (1981) Makromol Chem Rapid Commun 2: 287
31. Cooper W, Vaughan G, Miller S, Fielden M (1959) J Polym Sci 34: 651
32. Oster G, Shibata O (1957) J Polym Sci 26: 233
33. Maadhah G, Amin M, Usmani A (1985) Polym Bull 14: 433
34. Brockway C, Moser K (1963) J Polym Sci A-1: 1025
35. Franta E, Reibel L, Rempp P (1979) Polym Prepr 20(2): 102
36. Falk J, Schlott R, Hoeg D (1973) J Macromol Sci Chem 7(8): 1647
37. Falk J, Hoeg D, Schlott R, Pendelton J (1973) J Macromol Sci Chem A 7(8): 1669
38. Falk JC, Schlott R, Hoeg DF, Pendelton JF (1973) Rubber Chem Technol 46: 1044
39. Hadjichristidis N, Roovers J (1978) J Polym Sci Polym Phys Ed. 16: 851
40. Gupta S, Kennedy J (1979) Polym Bull 1: 253
41. Nguyen H, Kennedy J (1983) Polym Bull 10: 74
42. Rempp P, Franta E (1984) Adv Polym Sci 58: 1
43. Corner T (1984) Adv Polym Sci 62: 95
44. Saegusa T (1982) Topics Curr Chem 100: 75
45. Kawakami Y (1987) In: Kroschwitz JI (ed) Encyclopedia of polymer science and engineering, vol 9. Wiley Interscience, New York, p 195
46. Rempp P, Franta E (1987) In: Hogen-Esch TE, Smid J (eds) Recent advances in anionic polymerization. Elsevier, New York, p 353
47. Schulz GO, Milkovich R (1982) J Appl Polym Sci 27: 4773
48. Sierra-Vargas J, Zilliox JG, Rempp P, Franta E (1980) Polym Bull 3: 83

49. Tsuruta T (1985) Makromol Chem Suppl. 13: 33
50. Kawakami Y (1994) Prog Polym Sci 19: 203
51. Sawamoto M (1991) Prog Polym Sci 16: 111
52. Quirk RP (1984) Rubber Chem Technol 57: 557
53. Altares T Jr, Wyman DP, Allen VR, Meyersen KJ (1965) J Polym Sci A 3: 4131
54. Candau F, Rempp P (1969) Makromol Chem 15: 122
55. Pepper KW, Paisley HM (1953) J Chem Soc 4097
56. Itsuno S, Uchikoshi K, Ito K (1990) J Am Chem Soc 112: 8187
57. Candau F, Afchar-Taromi F, Rempp P (1977) Polymer 18: 1253
58. Selb J, Gallot Y (1979) Polymer 20: 1259
59. Selb J, Gallot Y (1979) Polymer 20: 1273
60. George MH, Majid MA, Barrie JA, Rezaian I (1987) Polymer 28: 1217
61. Ishikawa S (1995) Macromol Chem Phys 196: 485
62. Narayan R, Shay M (1986) Polym Prepr 27(1): 204
63. Narayan R, Shay M (1987) In: Hogen-Esch TE, Smid J (eds) Recent advances in anionic polymerization. Elsevier, New York, p 441
64. Watanabe H, Ameniya T, Shimura T, Kotaka T (1994) Macromolecules 27: 2336
65. Derand H, Wesslen B (1995) J Polym Sci Polym Chem Ed 33: 571
66. Se K, Watanabe O, Shibamoto T, Fujimoto T (1988) Polym Prepr 29(2): 110
67. Se K, Watanabe O, Isono Y, Fujimoto T (1988) ACS Polymer Preprints 29(2): 110
68. Wesslen B, Wesslen KB (1989) J Polym Sci Polym Chem Ed 27: 3915
69. Ishizu K, Fukutomi T (1987) J Polym Sci Polym Chem Ed 25: 23
70. Eckert AR, Webber SE (1996) Macromolecules 29: 560
71. Moraes MAR, Moreira ACF, Barbosa RV, Soares BG (1996) Macromolecules 29: 416
72. Sinai-Zingde G, Verma A, Liu Q, Brink A, Bronk J, Allison D, Goforth A, Patel N, Marand H, McGrath JE, Riffle JS (1990) Polym Prepr 31(1): 63
73. Schulz RC, Dworak A (1994) Macromol Symp 85: 203
74. Jiang Y, Frechet JMJ (1989) Polym Prepr 30(1): 127
75. Pary B, Tardi M, Polton A, Sigwalt P (1985) Eur. Polym J 21: 393
76. Hrkach J, Ruehl K, Matyjaszewski K (1988) Polym Prep. 29(2): 112
77. Crivello JV, Fan M (1994) Macromol Symp 77: 413
78. Al-Jarrah MMF, Al-Kafaji JKH, Apikian RL (1986) Brit Polym J 18: 256
79. Jannasch P, Wesslen B (1995) J Polym Sci Polym Chem Ed 33: 1465
80. Jannasch P, Wesslen B (1993) J Polym Sci Polym Chem Ed 31: 1519
81. Inoki M, Akutsu F, Yamaguchi H, Naruchi K, Miura M (1994) Macromol Chem Phys 195: 2799
82. Roha M, Wang H-T, Harwood HJ, Sebenik A (1995) Macromol Symp 91: 81
83. Dong X, Geuskens G, Wilkie CA (1995) Eur Polym J 31: 1165
84. Gleria M, Bolognesi A, Porzio W, Catellani M, Destri S, Audisio G (1987) Macromolecules 20: 469
85. Meister JJ, Chen M-J (1991) Macromolecules 24: 6843
86. Hrkach JS, Ou J, Lotan N, Langer R (1995) Macromolecules 28: 4736
87. Norton RL, McCarthy TJ (1987) Polym Prepr 28(1): 174
88. Feast WJ, Gibson VC, Johnson AF, Khosravi E, Moshin MA (1994) Polymer 35: 3542
89. Tanaka S, Uno M, Teramachi S, Tsukahara Y (1995) Polymer 36: 2219
90. Tsukahara Y, Tsai H-C, Yamashita Y, Muroga Y (1987) Polymer J 19: 1033
91. Takano A, Furutani T, Isono Y (1994) Macromolecules 27: 7914
92. Teramachi S, Hasegawa A, Matsumoto T, Kitahara K, Tsukahara Y, Yamashita Y (1992) Macromolecules 25: 4025
93. Feinberg SC (1985) Polym Prepr 26(2): 296
94. Yamashita Y, Tsai HC, Tsukahara Y (1985) Makromol Chem Suppl 12: 51
95. Breuning S, Heroquez V, Gnanou Y, Fontanille M (1995) Macromol Symp 95: 151
96. Heitz T, Rohrbach P, Höcker H (1989) Makromol Chem 190: 3295
97. Xie H-Q, Wu X-D, Guo J-S (1994) J Appl Polym Sci 54: 1079
98. Xie H-Q, Wu X-D, Guo J-S (1993) Polym Prepr 34(2): 598

99. Arnold M, Frank W, Reinhold G (1990) Polym Bull 24: 1
100. Sato M, Mangyo T, Nakadera K, Mukaida K-I (1994) Macromol Rapid Commun 15: 243
101. Peiffer DG, Rabeony M (1994) J Appl Polym Sci 51: 1283
102. Xie H, Fong Y (1988) Polym Bull 19: 179
103. Sato M, Kobayashi T, Komatsu F Takeno N, (1991) Macromol Rapid Commun 12: 269
104. Yamashita Y (1986) Polym Prepr 27(2): 27
105. Xie H-Q, Liu Y (1991) Eur. Polym J 27: 1339
106. Xie H, Feng Y (1988) Polymer 29: 1216
107. Xie H-Q, Liu Y (1993) Polymer 34: 182
108. Laivins G, Worsfold S (1990) J Polym Sci Polym Chem Ed 28: 1413
109. DeSimone JM, Hellstern AM, Siochi EJ, Smith SD, Ward TC, Gallagher PM, Krukonis VJ, McGrath JE (1990) Makromol Chem Macromol Symp 32: 21
110. De Simone JM, York GA, Wilson GR, Smith SD, Marand H, Gozdz AS, Bowden MJ, McGrath JE (1989) Polym Prepr 30(2): 134
111. Itoh M, Mita I (1994) J Polym Sci Polym Chem Ed 32: 1581
112. Li W, Huang B (1989) Makromol Chem 190: 2373
113. Kitayama T, Nakagawa O, Kishiro S, Nishiura T, Hatada K (1993) Polym J 25: 707
114. Hatada K, Kitayama T, Ute K, Masuda E, Shinozaki T, Yamamoto M (1988) Polym Prepr 29(2): 54
115. Hatada K, Nakanishi H, Ute K, Kitayama T (1986) Polymer J.18: 581
116. Hellstern AM, Smith SD, McGrath JE (1987) Polym Prepr 28(2): 328
117. DeSimone JM, York GA, McGrath JE, Gozdz AS, Bowden MJ (1991) Macromolecules 24: 5330
118. Serre B, Worsfold DJ (1987) Polymer 28: 881
119. Hashimoto K, Shinoda H, Okada M, Sumitomo H (1990) Polymer J 22: 312
120. Hatada K, Shinozaki T, Ute K, Kitayama T (1988) Polym Bull 19: 231
121. DeSimone JM, Hellstern AM, Ward TC, McGrath JE, Smith SD, Gallagher PM, Krukonis VJ, Stejskal J, Strakova D, Kratochvil P (1989) Contemp Top Polym Sci 6: 227
122. DeSimone JM, Hellstern AM, Ward TC, McGrath JE, Smith SD, Gallagher PM, Krukonis VJ, Stejskal J, Strakova D, Kratochvil P (1988) Polym Prepr 29(2): 116
123. DeSimone JM, Smith SD, Hellstern AM, Ward TC, McGrath JE, Gallagher PM, Krukonis VJ (1988) Polym Prepr 29(1): 361
124. Smith SD, McGrath JE (1986) Polym Prepr 27(2): 31
125. Stejskal J, Strakova D, Kratochvil P, Smith SD, McGrath JE (1989) Macromolecules 22: 861
126. Hashimoto K, Sumitomo H, Kawasumi M (1985) Polymer J 17: 1045
127. Hashimoto K, Sumitomo H (1985) Makromol Chem Suppl 12: 39
128. DeSimone JM, Maury EE, Combes JR, Menceloglu YZ (1993) Polym Mater Sci Eng 68: 41
129. Schulz RC, Muhlbach K, Perner T, Ziegler P (1986) Polym Prepr 27(2): 25
130. Hatada K, Kitayama T, Ute K, Masuda E, Shinozaki T, Yamamoto M (1989) Polym Bull 21: 165
131. Masuda E, Kishiro S, Kitayama T, Hatada K (1991) Polymer J 23: 847
132. Radke W, Stein HM, Muller AHE (1993) Polym Prepr 34(2): 652
133. Hatada K, Shinozaki T, Ute K, Kitayama T (1988) Polym Bull 19: 231
134. Omeis J, Pennewib H (1994) Polym Prepr 35(1): 714
135. Hefft M, Springer J (1992) Makromol Chem 193: 329
136. Smith SD, DeSimone JM, Huang H, York G, Dwight DW, Wilkes GL, McGrath JE (1992) Macromolecules 25: 2575
137. Schunk TC, Long TE (1995) J Chromat A 692: 221
138. Ragunath R.P., Lutz P, Lamps JP, Masson P, Rempp P (1986) Polym Bull 15: 69
139. Ishizu K, Minematsu S, Fukotomi T (1991) J Appl Polym Sci 43: 2107
140. Schulz GO, Milkovich R (1986) Ind Eng Chem Prod Res Dev 25: 148
141. Nagase Y, Mori S, Egawa M, Matsui K (1990) Makromol Chem Rap Commun 11: 185
142. Xie H-Q, Liu Z, Li H (1990) J Macromol Sci Chem A27(6): 725
143. Cameron GG, Chisholm MS (1986) Polymer 27: 1420

144. Pratt LM, Waugaman M, Khan IM (1995) Polym Prepr 36(2): 263
145. Gordon GG, Chisholm MS (1985) Polymer 26: 437
146. Suzuki T, Okawa T (1988) Polymer Commun 29: 225
147. Akashi M, Wada M, Yanase S, Miyauchi N (1989) J Polym Sci Polym Lett 27: 377
148. Tezuka Y, Nobe S, Shiomi T (1995) Macromolecules 28: 8251
149. Akashi M, Chao D, Yashima E, Miyauchi N (1990) J Appl Polym Sci 39: 2027
150. Cui M-H, Guo J-S, Xie H-Q (1995) JMS.-Pure Appl Chem A32(7): 1293
151. Ishizu K, Mitsutani K (1988) Makromol Chem 189: 2875
152. Tezuka Y, Fukushima A, Imai K (1985) Makromol Chem 186: 685
153. Quinot P, Bryant L, Chow TY, Saegusa T (1996) Macromol Chem Phys 197: 1
154. Heroquez V, Breunig S, Gnanou Y, Fontanille Y (1996) Macromolecules 29: 4459
155. Chujo Y, Kobayashi H, Yamashita Y (1989) J Polym Sci Polym Chem Ed 27: 2007
156. Yamada B, Kato E, Kobatake S, Otsu T (1991) Polym Bull 25: 423
157. Rimmer S, George MH (1993) Eur. Polym J 29: 205
158. Ishizu K, Ono T, Fukutomi T (1987) J. Polym Sci Polym Lett 25: 131
159. Fukutomi T, Ishizu K, Shiraki K (1987) J Polym Sci Polym Lett 25: 175
160. Xie H-Q, Xu G-C, Guo J-S (1992) J Macromol Sci-Pure Appl Chem A29(3): 263
161. Cacioli P, Hawthorne DG, Laslett RL, Rizzardo E, Solomon DH (1986) JMS Chem A23(7): 839
162. Chujo Y, Kobayashi H, Yamashita Y (1988) Polymer J 20: 407
163. Ito K, Sabao K, Kawaguchi S (1995) Macromol Symp 91: 65
164. Xie H-Q, Zhou S-B (1991) J Appl Polym Sci 42: 199
165. Yoshinaga K, Yokoyama T, Kito T (1993) Polymers for Advanced Technologies 4: 38
166. Chujo Y, Hiraiwa A, Kobayashi H, Yamashita Y (1988) J Polym Sci Polym Chem Ed. 26: 2991
167. Chujo Y, Shishino T, Tsukahara Y, Yamashita Y (1985) Polymer J 17: 133
168. Bonardi C, Boutevin B, Pietrasanta Y, Taha M (1985) Makromol Chem 186: 261
169. Abdel-Razik EA (1993) J Photochem Photobiol A Chem 73: 53
170. Xie H-Q, Zhou S-B (1990) J Macromol Sci-Chem A27(4): 491
171. Akashi M, Yanagi T, Yashima E, Miyauchi N (1989) J Polym Sci Polym Chem Ed 27: 3521
172. Akashi M, Kirikihira I, Miyauchi N (1985) Die Angew Makromol Chem 132: 81
173. Tsukahara Y, Ito K, Tsai H-C, Yamashita Y (1989) J Polym Sci Polym Chem Ed 27: 1099
174. Ishihara K, Tsuji T, Sakai Y, Nakabayashi N (1994) J Polym Sci Polym Chem Ed 32: 859
175. Sugiyama K, Shiraishi K (1989) Makromol Chem 190: 2381
176. Nicholas PP (1993) ACS Polym Prepr 34(2): 650
177. Niwa M, Hayashi T, Matsumoto T (1986) J Macromol Sci-Chem A23(4): 433
178. Sato M, Yanahara M, Kobayashi T, Mukaida K (1992) Makromol Chem 193: 991
179. Mylonakis SG, Mrozack S (1989) US Pat 4866126
180. Tezuka Y, Okabayashi A, Imai K (1989) Makromol Chem 190: 753
181. Kobayashi S, Uyama H, Shirasaka H (1990) Makromol Chem Rapid Commun 11: 11
182. Kobayashi S, Masuda E, Shoda S, Shimano Y (1989) Macromolecules 22: 2878
183. Biedron T, Brzezinska K, Kubisa P, Penczek S (1995) Polym International 36: 73
184. Schulz RC, Schwarzenbach E (1988) Makromol Chem Macromol Symp 13/14: 495
185. Takacs A, Faust R (1996) JM.-Pure Appl Chem A33(2): 117
186. Kennedy JP, Carter JD (1986) Polym Prepr 27(2): 29
187. Kennedy JP, Carter JD (1991) US Pat 5075389
188. Kunisada H, Yuki Y, Kondo S, Nishimori Y, Masuyama A (1991) Polymer J 23: 1455
189. Niebner N, Heitz W (1990) Makromol Chem 191: 1463
190. Ishiyama N, Chung D, Maeda M, Inoue S (1989) Makromol Chem 190: 2417
191. Muruyama A, Senda E, Tsuruta T (1986) Makromol Chem 187: 1895
192. Zang Q, Li Z (1994) Macromolecules 27: 526
193. Wilson D, George MH (1992) Polymer 33: 3723
194. Wilson D, George MH (1990) Polymer Commun 31: 90
195. Geetha B, Baran Mandal A, Ramasami T (1993) Macromolecules 26: 4083
196. Kawaguchi S, Winnik MA, Ito K (1995) Macromolecules 28: 1159
197. Wicker M, Heitz W (1991) Makromol Chem 192: 1371

198. Akiyama E, Takamura Y, Nagase Y (1992) Makromol Chem 193: 1509
199. O'Shea MS, George GA (1994) Polymer 35: 4190
200. Eisenbach CD, Heinemann T (1995) Macromolecules 28: 2133
201. Berlinova IV, Panayotov IM (1989) Makromol Chem 190: 1515
202. Percec V, Wang JH (1991) Polym Bull 25: 41
203. Grutke S, Hurley JH, Risse W (1994) Macromol Chem Phys 195: 2875
204. Qiu Y, Yu X, Feng L, Yang S (1992) Makromol Chem 193: 1377
205. Levesque G, Moitie V, Bacle B, Depraetere P (1988) Polymer 29: 2271
206. Yan F, Dejardin P, Galin JC (1990) Polymer 31: 736
207. de Vos SC, Moller M (1993) Makromol Chem Macromol Symp 75: 223
208. Sanches NB, Oliveira CMF (1994) Polym Testing 13: 289
209. Gramain P, Frere Y (1986) Polymer Commun 27: 16
210. Bo G, Wesslen B, Wesslen KB (1992) J Polym Sci Polym Chem Ed 30: 1799
211. Kitayama T, Kishiro S, Hatada K (1991) Polym Bull 25: 161
212. Cunha HT, Oliveira CMF (1994) Eur Polym J 30: 1489
213. Oliveira CMF, Andrade CT, Delpech MC (1991) Polym Bull 26: 657
214. Wilson D, George MH (1990) Polymer Commun 31: 355
215. Ohta T, Yamaoka H, Gotoh J, Sano S (1994) US Pat 5319045
216. Ryntz RA, Kurple KR (1987) US Pat 673718
217. Sogah DY, Webster OW (1986) Macromolecules 19: 1775
218. Kawakamie Y, Aoki T, Yamashita Y (1987) Polym Bull 18: 473
219. Sheridan MS, Verma A, McGrath JE (1992) Polym Prepr 33(1): 904
220. Wulff G, Birnbrich P (1993) Makromol Chem 194: 1569
221. Shen W-P, Jin H-M (1992) Makromol Chem 193: 743
222. Charleux B, Pichot C, Llauro MF (1993) Polymer 34: 4352
223. Charleux B, Pichot C (1993) Polymer 34: 195
224. Asami R, Kondo Y, Takaki M (1987) In: Hogen-Esch TE, Smid J (eds) Recent advances in anionic polymerization. Elsevier, New York, p 381
225. Barakat I, Dubois P, Jerome R, Teyssie P, Goethals E (1994) J Polym Sci Polym Chem Ed 32: 2099
226. Duschec T, Mulhaupt R (1992) Polym Prepr 33(1), 170
227. Aoi K, Tsutsumiuchi K, Aoki E, Okada M (1996) Macromolecules 29: 4456
228. Xie H, Xia J (1987) Makromol Chem 188: 2543
229. Mays JW (1990) Polym Bull 23: 247
230. Iatrou H, Siakali-Kioulafa E, Hadjichristidis N, Roovers J, Mays JW (1995) J Polym Sci Polym Phys Ed 33: 1925
231. Tselikas Y, Hadjichristidis N, Iatrou H, Liang KS, Lohse DJ (1996) J Chem Phys 105: 2456
232. Iatrou H, Hadjichristidis N (1992) Macromolecules 25: 4649
233. Iatrou H, Hadjichristidis N (1993) Macromolecules 26: 2479
234. Wright SJ, Young RN, Croucher TG (1994) Polym International 33: 123
235. Algaier J, Young RN, Efstratiadis V, Hadjichristidis N (1996) Macromolecules 29: 1794
236. Tsiang RCC (1994) Macromolecules 27: 4399
237. Avgeropoulos A, Hadjichristidis N (1997) J Polym Sci Poly Chem Ed 35: 813
238. Khan IM, Gao Z, Khougaz K, Eisenberg A (1992) Macromolecules 25: 3002
239. Avgeropoulos A, Poulos Y, Hadjichristidis N, Roovers J (1996) Macromolecules 29: 6076
240. Sioula S, Tselikas Y, Hadjichristidis N (1997) Macromolecules 30: 1518
241. Eschwey H, Burchard W (1975) Polymer 16: 180
242. Tsitsilianis C, Chaumont P, Rempp P (1990) Makromol Chem 191: 2319
243. Tsitsilianis C, Graff S, Rempp P (1991) Eur Polym J 27: 243
244. Tsitsilianis C, Lutz P, Graff S, Lamps J-P, Rempp P (1991) Macromolecules 24: 5897
245. Tsitsilianis C, Papanagopoulos D, Lutz P (1995) Polymer 36: 3745
246. Tsitsilianis C, Boulgaris D (1995) Macromol Reports A32 (Suppls 5 and 6): 569
247. Yamagishi A, Szwarc M, Tung L, Lo GY-S (1978) Macromolecules 11: 607
248. Quirk RP, Hoover FI (1987) In: Hogen-Esch TE, Smid J (eds) Recent advances in anionic polymerization. Elsevier, New York, p 393

249. Quirk RP, Lee B, Schock LE (1992) Makromol Chem Macromol Symp 53: 201
250. Quirk RP, Yoo T, Lee B (1994) JM.-Pure Appl Chem A31: 911
251. Quirk RP, Yoo T (1993) Polym Bull 31: 29
252. Huckstadt H, Abetz V, Stadler R (1996) Macromol Rapid Commun 17: 599
253. Ba-Gia H, Jerome R, Teyssie P (1980) J Polym Sci Polym Chem Ed 18: 3483
254. Naka A, Sada K, Chujo Y, Saegusa T (1991) Polym Prepr Jp. 40(2): E116
255. Fujimoto T, Zhang H, Kazama T, Isono Y, Hasegawa H, Hashimoto T (1992) Polymer 33: 2208
256. Takano A, Okada M, Nose T, Fujimoto T (1992) Macromolecules 25: 3596
257. Ishizu K, Kuwahara K (1994) Polymer 35: 4907
258. Ishizu K, Ykimasa S, Saito R (1991) Polym Commun 32: 386
259. Ishizu K, Ykimasa S, Saito R (1992) Polymer 33: 1982
260. Miyata K, Watanabe Y, Itaya T, Tanigaki T, Inoue K (1996) Macromolecules 29: 3694
261. Kanaoka S, Sawamoto M, Higashimura T (1991) Macromolecules 24: 2309
262. Kanaoka S, Sawamoto M, Higashimura T (1991) Macromolecules 24: 5741
263. Kanaoka S, Omura T, Sawamoto M, Higashimura T (1992) Macromolecules 25: 6407
264. Kanaoka S, Sawamoto M, Higashimura T (1993) Macromolecules 26: 254
265. Roovers J, Toporowski PM (1981) Macromolecules 14: 1174
266. Gido SP, Lee C, Pochan DJ, Pispas S, Mays JW, Hadjichristidis N (1996) Macromolecules 29: 7022
267. Tung LH, Lo GY-S (1994) Macromolecules 27: 2219
268. Quirk RP, Ma J-J (1991) Polym International 24: 197
269. Iatrou H, Avgeropoulos A, Hadjichristidis N (1994) Macromolecules 27: 6232
270. (a) Wang F, Roovers J, Toporowski PM (1995) Macromol.Symp 95: 205; (b) Wang F, Roovers J, Toporowski PM (1995) Macromolecular Reports A32 (Suppls 5 and 6): 951
271. Bayer U, Stadler R (1994) Macromol Chem Phys 195: 2709
272. Gitsov I, Wooley KL, Hawker CJ, Frechet JMJ (1991) Polym Prepr 32(3): 631
273. Gitsov I, Frechet JMJ (1994) Macromolecules 27: 7309
274. Gitsov I, Wooley KL, Frechet JMJ (1992) Angew Chem Int Ed Engl 31: 1200
275. Gitsov I, Wooley KL, Hawker CJ, Ivanova PT, Frechet JMJ (1993) Macromolecules 26: 5621
276. Chapman TM, Hilyer GL, Mahan EJ, Shaffer KA (1994) J Am Chem Soc 116: 11195
277. Freudenberger R, Claussen W, Schluter A-D, Wallmeier H (1994) Polymer 35: 4496
278. Gauthier M, Tichagwa L, Downey JS, Gao S (1996) Macromolecules 29: 519
279. Gan Y-D, Zoller J, Yin R, Hogen-Esch TE (1994) Macromol Symp 77: 93
280. Yin R, Hogen-Esch TE (1993) Macromolecules 26: 6952
281. Yin R, Amis EJ, Hogen-Esch TE (1994) Macromol Symp 85: 217
282. Ma J (1995) Macromol Chem Phys Macromol Symp 91: 41
283. Gan Y, Dong D, Hogen-Esch TE (1995) Polym Prepr 36(1): 408
284. Burchard W, Kajiwara K, Nerger D, Stockmayer WH (1984) Macromolecules 17: 222
285. Vlahos CH, Horta A, Freire JJ (1992) Macromolecules 25: 5974
286. Vlahos CH, Horta A, Hadjichristidis N, Freire JJ (1995) Macromolecules 28: 1500
287. Vlahos C, Tselikas Y, Hadjichristidis N, Roovers J, Rey A, Freire J J (1996) Macromolecules 29: 5599
288. Vlahos CH, Hadjichristidis N, Kosmas MK, Rubio AM, Freire JJ (1995) Macromolecules 28: 6854
289. Borsali R, Benmouna M (1994) Macromol Symp 79: 153
290. de Gennes PG, Hervet H (1983) J Phys Lett 44: L-351
291. Lescanec RL, Muthukumar M (1990) Macromolecules 23: 2280
292. Boris D, Rubinstein M (1996) Macromolecules 29: 7251
293. Nguyen AB, Hadjichristidis N, Fetters LJ (1986) Macromolecules 19: 768
294. Prochazka K, Glockner G, Hoff M, Tuzar Z (1984) Makromol Chem 185: 1187
295. Ohta Y, Kojima T, Takigawa T, Masuda T (1987) J Rheology 31: 711
296. Cited in Campbell MM (1989) Rheological properties of liquid-crystalline star-block copolymers. Dissertation, Univ of Massachusetts

297. Gia H-B, Jerome R, Teyssie Ph (1981) J Appl Polym Sci 26: 343
298. Gia H-B, Jerome R, Teyssie Ph (1980) J Polym Sci B: Polym Phys 18: 2391
299. Kanaoka S, Sawamoto M, Higashimura T (1992) Macromolecules 25: 6414
300. Dondos A, Rempp P, Benoit H (1966) J Polym Sci D: Polym Letters 4: 293
301. Gosnell AB, Woods DK, Gervasi JA, Williams JL, Stannet V (1968) Polymer 9: 561
302. Evans DC, George MH, Barrie JA (1975) Polymer 16: 690
303. Price C, Woods D (1973) Polymer 14: 82
304. Price C, Woods D (1974) Polymer 15: 389
305. Tuzar Z, Kratochvil P, Prochazka K, Contractor K, Hadjichristidis N (1989) Makromol Chem 190: 2967
306. Selb J, Gallot Y (1981) Makromol Chem 182: 1775
307. Candau F, Boutillier J, Tripier F, Wittmann JC (1979) Polymer 20: 1221
308. Candau F, Guenet JM, Boutillier J, Picot C (1979) Polymer 20: 1227
309. (a) Candau S, Boutillier J, Candau F (1979) Polymer 20: 1237; (b) Ballet F, Candau F (1983) J Polym Sci Polym Chem Ed 21: 155
310. Ito K, Tomi Y, Kawaguchi S (1992) Macromolecules 25: 1534
311. Tsukahara Y, Kohjiya S, Tsutsumi K, Okamoto Y (1994) Macromolecules 27: 1662
312. Wintermantel M, Schmidt M, Tsukahara Y, Kajiwara K, Kohjiya S (1994) Macromol Rapid Commun 15: 279
313. Nemoto N, Nagai M, Koike A, Okada S (1995) Macromolecules 28: 3854
314. Bayer U, Stadler R (1994) Macromol Chem Phys 195: 2709
315. Iatrou H, Willner L, Hadjichristidis N, Richter D, Halperin A (1996) Macromolecules 29: 581
316. Wu W, Amis EJ (1995) Polym Prepr (Am Chem Soc Polym Div) 36 (1): 369
317. Amis EJ, Hodgson DF, Wenjun W (1993) J Polym Sci B: Polym Phys 31: 2049
318. Gitsov I, Frechet JMJ (1993) Macromolecules 26: 6536
319. Ambler MR (1981) Chromatographic Science 19: 29
320. (a) Jordan RC, Silver SF, Schon RD, Rivard RJ (1984) Size exclusion chromatography. ACS Symposium Series 245, Washington DC, p. 295-320; (b) Jordan RC, Silver SF, Schon RD, Rivard RJ (1983) Org Coat Appl Polym Sci Proc 48: 755
321. (a) Gores F, Kilz P (1993) In: Provder T (ed) Chromatography of polymers. ACS Symposium Series 521, Washington DC; (b) Mourey TH, Turner SR, Rubinstein M, Frechet JMJ, Hawker CJ, Wooley KL (1992) Macromolecules 25: 2401
322. Olvera de la Cruz M, Sanchez IS (1986) Macromolecules 19: 2501
323. Benoit H, Hadjiioannou G (1988) Macromolecules 21: 1449
324. Anderson DM, Thomas EL (1988) Macromolecules 21: 3221
325. Dobrynin AV, Erukhimovich IY (1991) J Phys II France 1: 1387
326. Milner ST (1994) Macromolecules 27: 2333
327. Shinozaki A, Jasnow D, Balazs AC (1994) Macromolecules 27: 2496
328. Shinozaki A, Jasnow D, Balazs AC (1995) Macromolecules 28: 3450
329. Gersappe D, Harm P, Irvine D, Balazs AC (1994) Macromolecules 27: 720
330. Israels R, Foster DP, Balazs AC (1995) Macromolecules 28: 218
331. Lyatskaya Y, Gersappe D, Balazs AC (1995) Macromolecules 28: 6278
332. Ishizu K, Uchida S (1994) Polymer 35: 4712
333. Herman DS, Kinning DJ, Thomas EL, Fetters LJ (1987) Macromolecules 20: 2940
334. (a) Kinning DJ, Thomas EL, Alward DB, Fetters LJ, Handlin DL (1986) Macromolecules 19: 1288; (b) Thomas EL, Alward DB, Kinning DJ, Martin DC, Handlin DL, Fetters LJ (1986) Macromolecules 19: 2197
335. Hajduk DA, Harper PE, Gruner SM, Honeker CC, Thomas EL, Fetters LJ (1995) Macromolecules 28: 2570
336. Hasegawa H, Tanaka H, Yamasaki K, Hashimoto T (1987) Macromolecules 20: 1651
337. Henkee CS, Thomas EL, Martin DC, Fetters LJ (1988) Polym Prepr (Am Chem Soc Polym Div) 29(1): 462
338. Matsushita Y, Takasu T, Yagi K, Tomioka K, Noda I (1994) Polymer 35: 2862

339. Hashimoto T, Ijichi Y, Fetters LJ (1988) J Chem Phys 89(4): 2463
340. Ijichi Y, Hashimoto T, Fetters LJ (1989) Macromolecules 22: 2817
341. Fetters LJ, Richards RW, Thomas EL (1987) Polymer 28: 2252
342. Floudas G, Pispas S, Hadjichristidis N, Pakula T, Erukhimovich I (1996) Macromolecules 29: 4142
343. Hadjichristidis N, Iatrou H, Behal SK, Chludzinski JJ, Disko MM, Garner RT, Liang KS, Lohse DJ, Milner ST (1993) Macromolecules 26: 5812
344. Pochan DJ, Gido SP, Pispas S, Mays JW, Ryan AJ, Fairclough P, Terrill N, Hamley IW (1996) Macromolecules 29: 5091
345. Pochan DJ, Gido SP, Pispas S, Mays JW (1996) Macromolecules 29: 5099
346. Floudas G, Hadjichristidis N, Iatrou H, Pakula T, Fischer EW (1994) Macromolecules 27: 7735
347. Floudas G, Hadjichristidis N, Iatrou H, Pakula T (1996) Macromolecules 29: 3139
348. Gervais M, Gallot B, Jerome R, Teyssie Ph (1986) Makromol Chem 187: 2685
349. Price C, Lally TP, Watson AG, Woods D (1972) Br Polym J 4: 413
350. Price C, Singleton R, Woods D (1974) Polymer 15: 117
351. Se K, Watanabe O (1989) Makromol Chem Macromol Symp 25: 249
352. Kennedy JP, Delvaux JM (1981) Adv Polym Sci 38: 141
353. Sato M, Kobayashi T, Komatsu F, Takeno N (1991) Makromol Chem Rapid Commun 12: 269
354. Williams G, Nazemi A, Karasz FE (1994) Macromolecular Reports A31 (Suppls 6 and 7): 911
355. Eisenbach CD, Heinemann T (1995) Macromolecules 28: 2133
356. Xie H-Q, Xie D, Liu J (1989) Polym-Plast Technol Eng 28(4): 355
357. Lescanec RL, Hajduk DA, Kim GY, Gan Y, Yin R, Gruner SM, Hogen-Esch TE, Thomas EL (1995) Macromolecules 28: 3485
358. (a)Krans G, Naylor FE, Rollmann KW (1971) J Polym Sci: Part A-2 9: 1839; (b) Ma J-J, Nestegard MK, Majumdar BD, Sheridan MM (1996) Polymer Preprints 37(2): 716
359. Johnson J, Young RN, Wright SJ, McLeish T (1994) Polym Prepr (Am Chem Soc Polym Chem Div) 35(2): 470
360. Tsitsilianis C (1993) Macromolecules 26: 2977
361. Feng H, Feng Z, Yuan H, Shen L (1994) Macromolecules 27: 7830
362. Feng H, Feng Z, Shen L (1994) Macromolecules 27: 7835
363. Feng H, Feng Z, Shen L (1994) Macromolecules 27: 7840
364. Jiang M, Cao X, Yu T (1986) Polymer 27: 1923
365. Jiang M, Cao X, Yu T (1986) Polymer 27: 1917
366. Gersappe D, Irvine B, Balazs AC, Liu Y, Sokolov J, Rafailovich M, Schwarz S, Peiffer DG (1994) Science 265: 1072
367. Meyer GC, Widmaier JM (1977) Polym Eng Sci 17(11): 803
368. LeBlanc JL (1977) J Appl Polym Sci 21: 2419
369. Pedemonte E, Dondero G, De Candia F, Romano G (1976) Polymer 17: 72
370. Takigawa T, Ohta Y, Masuda T (1990) Polymer J. 22: 447
371. George MH, Majidt MA, Barrie JA, Rezaian I (1987) Polymer 28: 1217
372. Lau FP, Silver SF(1987) Nippon Setchaku Kyokaishi 23(10): 382
373. Granger AT, Wang B, Krause S, Fetters LJ (1986) Adv Chem Ser 211, p.127-138

Polymer Solid Electrolytes: Synthesis and Structure

V. Chandrasekhar

Department of Chemistry, Indian Institute of Technology Kanpur – 208016, India

Solid state materials that exhibit high ion transport properties are of interest from both academic as well as applied points of view. Polymer solid electrolytes are materials of high technological promise in several electrochemical applications such as high energy density batteries, gas sensors, electrochemical devices etc. These polymeric materials have attracted much attention and hold great promise in this area.
   This review deals with several types of polymer hosts that have been investigated. These include polyethylene oxide and its several modified forms, comb like polymers such as polyacrylates and inorganic polymers such as polyphosphazenes and polysiloxanes. Various instrumental techniques have been employed in the structural characterization of polymer electrolytes. The structural information obtained from methods such as Extended X-ray Absorption Fine Structure (EXAFS), X-ray diffraction methods, vibrational spectroscopy and nuclear magnetic resonance (NMR) have also been discussed.

List of Abbreviations 140

1   Introduction 141

2   General Features 143

2.1   Requirements for a Polymer Electrolyte 143
2.2   Measurement of Ionic Conductivity in Polymer Electrolytes 145
2.3   Treatment of Conductivity Data 146

3   Poly(ethylene oxide) and Related Systems 147

3.1   Poly (ethylene oxide) (PEO) 147
3.2   Modification of PEO 151
3.2.1 Addition of Plasticizers 151
3.2.2 Plasticizing Salts 155
3.2.3 Other Modifications of PEO 158

4   Other PEO Related Polymers 160

4.1   Some Linear Polymers 160
4.2   Polymer 'Gel' Electrolytes 162
4.3   'Polymer-in-salt' Electrolytes 163

5	**Pendant Polymer Systems**	163
5.1	Poly Acrylates and Itaconates	163
5.2	Poly Crown-ether	168
5.3	Poly Phosphazenes	168
5.3.1	Structural Features of Polyphosphazenes	171
5.3.2	Oligo Etheroxy Substituted Polyphosphazenes	172
5.3.2.1	Methoxy Ethoxy Ethoxy Polyphosphazene (MEEP)	172
5.3.2.2	Modified MEEP Systems	174
5.3.2.3	Surfactant Substituted Polyphosphazenes	176
5.3.2.4	Other Etheroxy Side Chain Containing Polyphosphazenes	179
5.3.2.5	Mixed Substituent Polyphosphazenes	179
5.3.3	Polymers containing Pendant Oligo(oxyethylene) Cyclotriphosphazenes	181
5.3.4	Ionic Polyphosphazenes	183
5.4	Poly Siloxanes	183
6	**Structure of Polymer Electrolytes**	185
6.1	EXAFS of Polymer Electrolytes	186
6.2	X-ray Structures of Polymer Electrolytes	188
6.3	Infrared and Raman Spectroscopic Studies	193
6.4	NMR Studies on Polymer Electrolytes	196
7	**Summary**	199
	References	199

List of Abbreviations

AC	Anion conductor
AN	Acceptor number
CC	Cation conductor
DEP	Di ethyl phthalate
DOS	Di octyl sebacate
DN	Electron pair donicity
DSC	Differential scanning calorimetry
DTA	Differential thermal analysis
EC	Ethylene cabonate
EXAFS	Extended x-ray absorption fine structure
Li TFSI	Lithium bis(trifluoromethane sulfonyl) imide
Li TriTFSM	Lithium tris(trifluoromethyl sulfonyl) methanide
MC-3	Modified propylene carbonate
MEEP	Methoxy ethoxy ethoxy polyphosphazene
MEEMA	Methoxy ethoxy ethyl methacrylate
MPEG	Mono methyl poly ethylene glycol

NBR	Acrylonitrile-butadiene rubber
NMR	Nuclear magnetic resonance
PAAM	Poly acryl amide
PAN	Poly acrylo nitrile
PC	Propylene carbonate
PEG	Poly ethylene glycol
PEGDA	Poly ethylene glycol diacrylate
PEO	Poly ethylene oxide
PP	Surfactant substituted poly phosphazene
PPO	Poly propylene oxide
PVC	Poly vinyl chloride
PVP	Poly vinyl pyrrolidine
PVS	Poly vinyl sulfone
SBR	Styrene butadiene rubber
T_g	Glass transition temperature
T_m	Melting temperature
TEG	Tetra ethylene glycol
TEGDME	Tetra ethylene glycol dimethyl ether
THF	Tetrahydrofuran
VTF	Vogel-Tamann-Fulcher
VTXRD	Variable temperature powder x-ray diffraction

1
Introduction

In recent years there has been vigorous research activity in industrial and academic laboratories all over the world on solid materials which possess high ion-transport properties [1-5]. Generally ionic conduction is associated with liquids, either solvents with high dielectric constants or molten salts. However, solids that can function as electrolytes also known as solid ionic conductors, fast ion conductors or solid electrolytes (typical conductivity $10^{-6} \leq \sigma \leq 10^{-1} \text{S cm}^{-1}$) are exciting because of their wide ranging applications such as gas sensors [6-7], electrochemical display devices [8-9] high temperature heating elements [10], intercalation electrodes [11], power sources [12], fuel cells [13], solid state high energy density batteries [6, 14] and so on. Several solid electrolytes are known such as silver iodide, AgI [1-5] which transports Ag^+, β-alumina $((Na_2O)_x \cdot 11\, Al_2O_3)$ [15-16] and NASICON [17-18] $Na_{1+x}Zr_2Si_xP_{3-x}O_{12}$ which transport Na^+, modified zirconia (Ca^{2+} or Y^{3+} doped ZrO_2) which is an oxide ion conductor [1], or lithium ion conductors such as single crystal Li_3N [19-20] and glasses based on Li_2S [21-23]. Practical devices based on β-alumina have been realised such as a sodium-sulfur battery [24-25]. However, the operation of this device requires an impractically high temperature. In this context, all solid state rechargeable lithium batteries operating at room temperature are highly desirable because of several advantages such as high energy density (150 Wh/g), high voltage 4.0 V/cell) and longer charge retention characteristics [1, 26].

These advantageous features result in part from a high standard potential and a low electrochemical equivalent weight of Li.

In general, desirable battery properties are: energy content per unit volume and weight, discharge and charge characteristics at different rates and temperature, internal resistance, Ah and Wh efficiency, charge retention, life and mechanical stability. If not all most of these properties depend on the electrolytes that a battery is made up of. The choice of electrolyte for rechargeable batteries is governed by the following characteristics: (1) the electrolyte has to have negligent electronic conductivity (to prevent short circuiting) and favorable ionic conductivity, (2) the electrolyte should have a uni ion conduction, otherwise a concentration polarization in the cell may result, (3) the electrolyte must be electrochemically stable at least in the working potential range of the battery, (4) the electrolyte apart from being thermally stable should be compatible with other cell components. A recent review summarizes the progress in ceramic solid electrolytes in general and Li^+ conducting solid electrolytes in particular [1].

In spite of the attractive features of conventional solid electrolytes in various applications, one of the main difficulties in their use in all solid state batteries is the loss of contact between electrodes and electrolyte during the charge-discharge cycles of the battery. This is primarily as a result of dimensional changes occurring at the electrodes during the charging or discharging mode. With conventional liquid electrolytes such dimensional changes in the electrodes do not pose a problem, but with solid electrolytes, this leads to a loss of interfacial contact between the electrode and the electrolyte. In order to overcome this difficulty, batteries have to be operated at high temperatures so that the electrodes are molten (e.g. sodium-sulfur batteries using β-alumina as the electrolyte) [24-25]. Alternatively the solid electrolyte should be a material that is flexible and therefore can deform with the electrodes to suit the dimensional changes that occur so that interfacial contact is maintained throughout the operation of the battery. It is in this context that high molecular weight polymers with specially designed architectures are being investigated as solid electrolytes [27-38].

Polymers that function as solid electrolytes are a subclass by themselves and are known as polymer electrolytes [27, 29]. Besides the advantage of flexibility, polymers can also be cast into thin films and since thin films while minimizing the resistance of the electrolyte also reduces the volume and the weight, use of polymer electrolytes can increase the energy stored per unit weight and volume. In view of these attractive features, there has been considerable focus in recent years on the development of both inorganic and organic polymers as electrolytes for ion transport. This article deals with the recent developments in this area with emphasis on the new types of polymeric systems that have been used as polymer electrolytes.

In 1973, Peter Wright and coworkers first reported [39-41] the ionic conductivity of poly(ethylene oxide), $[CH_2CH_2O]_n$, (PEO), with alkali metal salts. This was followed by the visionary suggestion of M. Armand for the use of PEO as a solid electrolyte system for the transport of ions [42-43]. Since then, the area of polymer electrolytes has attracted considerable interest. In the following account, first a discussion is presented on the general features applicable to polymer electrolytes. This is followed by an account on individual polymer electrolytes, particularly on modifications of PEO and related systems, and etheroxy side chain containing organic and inorganic polymers. Some developments on the structural aspects of polymer electrolytes are also reviewed. This review does not deal with the applications of polymer electrolytes. These have been dealt with elsewhere [29, 31, 32].

2
General Features

2.1
Requirements for a Polymer Electrolyte

Since the polymer and the metal salt involved are both solid materials, the preparation of a polymer salt complex is achieved by the dissolution of the two materials in a common solvent such as acetonitrile, methanol or THF followed by a slow removal of the solvent in vacuum. This results in either the bulk polymer-metal salt complex or a thin film depending upon the method of preparation. It is essential to ensure that no traces of moisture are present and hence the operations are carried out by using Schlenk techniques or glove box methods. The essential reaction that occurs in the formation of a polymer-metal salt complex can be written as

$$[A - B]_y + MX \rightarrow [A - B]_y \cdot MX \tag{1}$$

where $[A - B]_y$ is a polymer chain and MX is an alkali metal salt or a transition metal halide. Divalent metal salts have also been used [27].

Just as the dissolution of ionic salts in a solvent system requires that the solvation energy of the ions in solution overcome the lattice energy of the ionic salt, similarly, polymer-metal salt complex formation proceeds, provided the polymer matrix effectively solvates the ions and overcomes the lattice energy of the ionic salt. Three essential criteria for this process have been identified [37]:

(a) Electron pair donicity (DN)
(b) Acceptor number (AN) and
(c) an Entropy term.

The DN term measures the effectiveness of the solvent to function as a Lewis base in its ability to solvate the cation, a Lewis acid. Thus, the polymer which should function as a host in the polymer electrolyte should posess donor sites such as oxygen, sulfur or nitrogen either in the backbone or in a group attached in the form of a side chain to the polymer. Similarly, the AN term describes the solvation of the anion, the Lewis base. PEO, a polyether, can be considered similar to 1,2-dimethoxy ethane (DN, 22; AN, 10.2) or even THF (DN, 20; AN, 8). Thus PEO can effectively solvate cations possessing counter anions that are bulky delocalized anions such as I^-, ClO_4^-, BF_4^- or $CF_3SO_3^-$ which require little or no solvation. The third term (entropy) has been related to the spatial disposition of the solvating unit and it has been shown that ethylene oxy (CH_2CH_2O) containing polymers such as PEO have the most favorable spatial orientation of the solvating units.

While small ions such as Li^+ which can be strongly solvated, lead to formation of polymer salt complexes even up to LiCl (lattice energy 853 KJ mol^{-1}) other larger cations such as Na^+, K^+ etc., require bulky counter anions such as I^-, SCN^-, or $CF_3SO_3^-$ in order to be solvated by PEO [38].

In addition to the above factors, it has also been recognized that the polymer should possess a low cohesive energy and a high flexibility in order to effectively solvate the ions. The former is characterized by lack of intermolecular interactions such as hydrogen bonding while the latter feature is indicated by a low glass transition temperature (T_g). Thus although polymers like polyamides contain oxygen and nitrogen atoms as donor sites in their backbone, these polymers are quite unsuitable as polymer hosts in polymer electrolytes because of the presence of extensive intermolecular hydrogen bonding. Metal complexation with these polymers would lead to the disruption of this energetically favorable situation. The second factor, viz. the high torsional flexibility of the polymer, is indicated by a low T_g and is crucial for ion transport. Thus large segmental motions of the polymer (either the backbone or the side chain) which is possible above its T_g can lead to fast ion movement.

In view of the above requirements, the polymers that have been studied as polymer electrolytes are either oxygen-, nitrogen-, or sulfur-containing materials. The heteroatoms are either part of the backbone of the polymer or are present in the side chain attachments. Some important polymers include

(a) poly (ethylene oxide)
(b) poly (ethylene glycol)
(c) poly (propylene oxide)
(d) poly (siloxanes)
(e) poly(phosphazenes)
(f) poly (vinyl pyrrolidine)
(g) poly (acrylates)
(h) poly (ethylene succinate)
(i) poly (vinyl alcohol)
(j) poly (ethylene imine)
(k) poly (alkylene sulphides).

While oxygen-containing polymers have received more attention other heteroatom-containing polymers have also been studied. In addition to homopolymers, copolymers containing more than one monomer has also received attention. Further, modifications of homopolymers by plasticizers, or crosslinking, or grafting to improve the properties of the polymers towards polymer-salt complex formation or increasing the dimensional stability of the materials has also been a focus of research.

In addition to the "salt in polymer" approach as described above, Angell and coworkers have described preparation of "polymer-in-salt" materials [44] (vide infra). Lithium salts are mixed with small amounts of poly propylene oxide and poly ethylene oxide to afford rubbery materials with low glass transition temperatures. This new class of polymer electrolytes showed good lithium ion conductivities and a high electrochemical stability.

2.2
Measurement of Ionic Conductivity in Polymer Electrolytes

In polymer electrolytes the electronic conductivity is minimal and the conductivity observed is due to migration of ions. Measurement of this ionic conductivity, however, is not straightforward. This aspect has been dealt with elsewhere in detail [45-47]. However, a brief description of the method of measurement of ionic conductivity is given below.

Because of the resistance to ion flow at the electrode-electrolyte interface, 'normal' measurement of total ionic conductivity is not possible in polymer electrolytes. In order to overcome this problem the conductivity measurements are carried out by the ac impedance spectroscopy method, which minimizes the effects of cell polarization. The measurements are often made with the electrolyte sandwiched between a pair of electrochemically inert electrodes made of platinum or stainless steel. The detailed methodology of impedance spectroscopy is reviewed thoroughly elsewhere [45-47].

Briefly, impedance spectroscopy is a powerful method of characterizing many of the electrical properties of materials. The dynamics of bound or mobile charge carriers in the bulk or interfacial regions of any kind of solid or liquid materials such as ionic, semiconducting, mixed electronic-ionic and even insulators can be derived from impedance spectroscopy. In this technique the impedance is directly measured in the frequency domain by applying a single frequency voltage to the interface and measuring the phase shift and amplitude or real and imaginary parts of current at that frequency. When a sinusoidal potential is applied, the magnitude I_m and the phase shift (θ) of the current (i) which are measured with time (t) are given by

$$i(t) = I_m \sin(\omega t + \theta) \tag{2}$$

where ω is the frequency and θ is the phase difference between the voltage and current. These measurements are repeated from very low (10^{-4} Hz) to very high frequencies. From an analysis of this data it is possible to arrive at the ac current vector (i_t^*) which is expressed in terms of real (i'_t) and imaginary parts (i''_t)

$$i_t^* = i'_t + j\, i''_t \;;\; j = \sqrt{-1} \tag{3}$$

Similarly ac potential is given by

$$v_t^* = v'_t + v''_t \tag{4}$$

and ac impedance is expressed as

$$Z^* = Z' + j Z'' \;;\; Z^* = v_t^* / i_t^* \tag{5}$$

In the impedance spectrum, also known as the cole-cole plot the real part of impedance is plotted against the imaginary part for the data collected at various

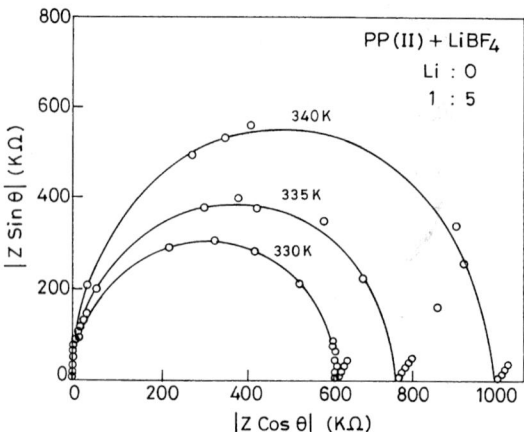

Fig. 1. Representative impedance spectra of [(PP-II)-LiBF$_4$](O:Li, 5:1) complex at three different temperatures (taken from Ref. 211)

frequencies (Fig. 1). From this plot the bulk resistance of the electrolyte (R_b) is obtained. The conductivity, σ is then obtained from

$$\sigma = g/R_b \tag{6}$$

$$R_b = Z^* \cos\theta \tag{7}$$

where g is the geometric factor of the electrolyte sample (thickness + area) in cm^{-1}, R_b is the bulk resistance, value at 100% resistance in ohms, Z^* is the impedance of the cell in ohms, and θ is the phase shift in degrees.

2.3
Treatment of Conductivity Data

The conductivity data can be treated by the use of Arrhenius equation

$$\sigma = A \exp(-E_A/kT) \tag{8}$$

or the VTF (Vogel-Tamann-Fulcher) equation

$$\sigma = A \exp[-B/K(T-T_o)] \tag{9}$$

The Arrhenius plot in which log σT is plotted against T^{-1} show straight lines. Generally it has been observed that this plot is followed by conventional solid electrolytes as well as crystalline polymer electrolytes.

The VTF equation [48-49] where T_o is a parameter to be determined (in many cases, however, it is found that T_o is very close to T_g, the glass transition temperature), B is a constant called the pseudo activation energy and is different from the activation energy E_A that appears in the Arrhenius equation, is generally obeyed by amorphous polymer solid electrolytes.

After the above discussion on the general features of the polymer electrolytes, an account of some of the important individual polymer systems that have been used is given below.

3
Poly(ethylene oxide) (PEO) and Related Systems

3.1
Poly (ethylene oxide) (PEO)

PEO which is a linear polymer containing donor oxygen atoms in the main backbone is prepared by the ring opening polymerization of ethylene oxide. High molecular weight polymers upto 5×10^6 are available commercially. PEO is a a semi-crystalline material; about 60% of the bulk is crystalline at room temperature and the rest is present in an amorphous phase. The melting point of the crystalline phase (T_m) is 65 °C and the glass transition temperature of the amorphous phase (T_g) is –60 °C. PEO forms metal-salt complexes with a wide range of metal salts such as alkali and alkaline earth metals, and also transition metal salts [37, 50-52]. These include a number of mono and divalent cations such as Li^+, Na^+, K^+, Cs^+, Ag^+, Mg^{2+}, Ca^{2+}, Zn^{2+}, Cu^{2+} etc. with different counter anions such as BPh_4^-, $CF_3SO_3^-$, BF_4^-, ClO_4^-, SCN^-, I^- etc.

In order to understand the nature of the PEO-metal salt complexes, attempts have been made to obtain and study phase diagrams. In contrast to simple inorganic systems, for polymers, obtaining phase diagrams is complicated as a result of slow crystallizations as well as the presence of chain ends and defects. These result in an apparent violation of the Gibbs rule. In spite of these complications phase diagram information is available for some PEO-metal salt compositions [53-56]. The techniques used for obtaining this information are mainly polarized light microscopy [57], differential scanning calorimetry [57-58], X-ray diffraction and NMR [59-62].

The maximum stoichiometry of the polymer-metal salt complexes (CH_2CH_2O: metal salt), as shown by NMR and DSC studies, is 3:1 for smaller metal ions such as Li^+ or Na^+. Ions such as K^+ or NH_4 tend to form 4:1 complexes. Bulky symmetrical anions such as ClO_4^-, AsF_6^- and $CF_3SO_3^-$ favor 6:1 complexes. A eutectic or a quasi eutectic exists between pure PEO and PEO-metal salt complexes. For example, the 3:1 complex between PEO and $LiCF_3SO_3$ shows a eutectic point which is very close to that of pure PEO itself in composition and melting temperature (30:1 and 60 °C). In contrast the 6:1 PEO-$LiClO_4$ complexes show a lower salt concentration (18:1 and 42 °C).

More recently, the phase diagram of a PEO system containing a divalent cation, PEO–$Ca[CF_3SO_3]_2$ has been elucidated by Bruce and coworkers by using variable temperature powder X-ray diffraction (VTXRD) and differential scanning calorimetry (DSC). A 6:1 complex has been shown to form with two polymorphic phases [63]. Figures 2, 3 and 4 show representative examples of the phase dia-

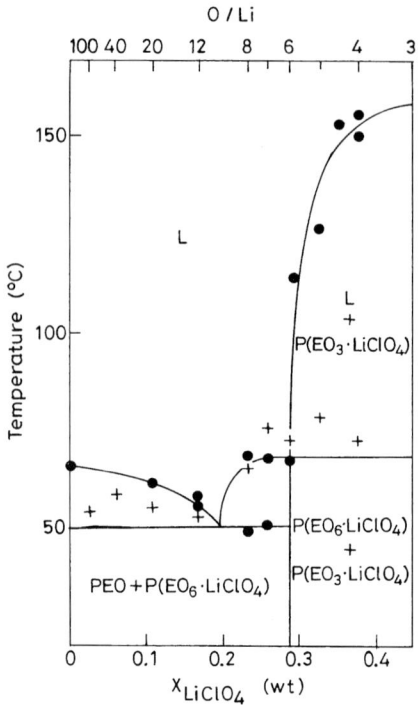

Fig. 2. Phase diagram of PEO-LiClO$_4$ complexes (taken from Ref.54)

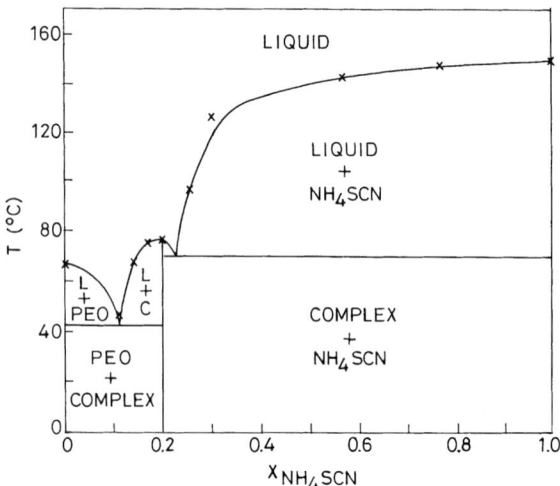

Fig. 3. Phase diagram of PEO-NH$_4$SCN complex (taken from Ref. 56)

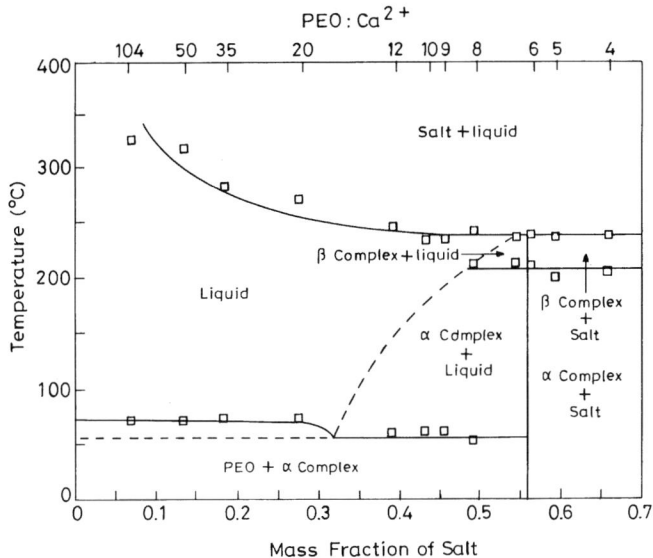

Fig. 4. Phase diagram of PEO-Ca(CF$_3$SO$_3$)$_2$ complex (taken from Ref. 63)

grams studies for the 3:1 (PEO-LiClO$_4$), 4:1 (PEO-NH$_4$SCN) and 6:1 (PEO-(Ca-CF$_3$SO$_3$)$_2$) systems.

Use of multinuclear NMR (^{1}H, ^{7}Li and ^{19}F) has corroborated evidence from DSC towards phase information. Thus faster relaxation times (T$_2$) have been associated with crystalline phases and while longer T$_2$, values have been attributed to amorphous solid solutions [59].

IR, Raman, NMR, X-ray and EXAFS studies on solid electrolytes (vide infra) give a substantial amount of information about the structure of the electrolytes and the dynamics of ion motion. Vibrational spectroscopy gives the direct evidence of the cation-polymer interaction. Far–IR studies confirm that the cation is coordinated to the ether oxygen atoms in PEO [64-67] and other comb like acrylate polymers (vide infra) having short oligo oxyethylene side chains. The mechanism of conduction of ions involves local, liquid like motions of the solvent (poly ether in the case of PEO) with the ions then moving in the amorphous, locally disordered (liquid) phase. This implies curved plots, (log σT vs 1/T) over the entire temperature range with no discontinuities, much weaker stoichiometry dependence, and an increase in conduction with a decreased glass transition temperature.

IR and Raman studies on PEO-NaBF$_4$ and PEO-NaBH$_4$ systems reported by Dupon and coworkers [65], indicate that extensive ion pairing occurs in NaBH$_4$ complexes leading to low conductivity values. Absence of such pairing in NaBF$_4$ complexes leads to enhanced conductivity.

NMR measurements and conductivity study correlations with phase diagrams have clearly established that the amorphous elastomeric phase in PEO is primarily responsible for the ionic conductivity [59].

Typical conductivity vs composition and temperature plots for PEO-LiClO$_4$ and PEO-LiCF$_3$SO$_3$ are shown in Fig. 5. Usually the conductivities are in the range of 10^{-3} to 10^{-4} S cm^{-1} at 100 °C and fall to 10^{-6} to 10^{-8} S cm^{-1} at room temperature. The molecular weight of the polymer changes the conductivity values. However, after a certain molecular weight the conductivities are invariant with an increase in molecular weight of PEO. The highest conductivities are seen for the Li$^+$ salts while the least conductive are given by the sodium salts. Free volume and configurational entropy models have been used to describe the temperature and concentration dependent behavior of the conductivity. These models have been described elsewhere [68-76].

In general the conductivity of the PEO polymer electrolyte varies with the concentration of the dissolved salt with the maximum conductivity being observed for an intermediate salt concentration. This is understood qualitatively by applying the VTF relation for conductivity (vide supra). As discussed earlier, the T$_0$ is a parameter closely related to the T$_g$ of the sample. In the absence of all other effects, increase of salt concentration should increase A and hence the conductivity. However, as the salt concentration is increased there is a simultaneous increase in T$_g$ and this leads to a decrease in conductivity. Therefore, optimum salt concentrations are required for maximum conductivities. Generally, where phase diagrams have been studied it is shown that for the eutectic composition a single amorphous phase is formed above the congruent melting point. This leads to an enhanced conductivity which is explained by the VTF equation. Above the eutectic point variation of conductivity appears linear on the Arrhenius plot [27, 37].

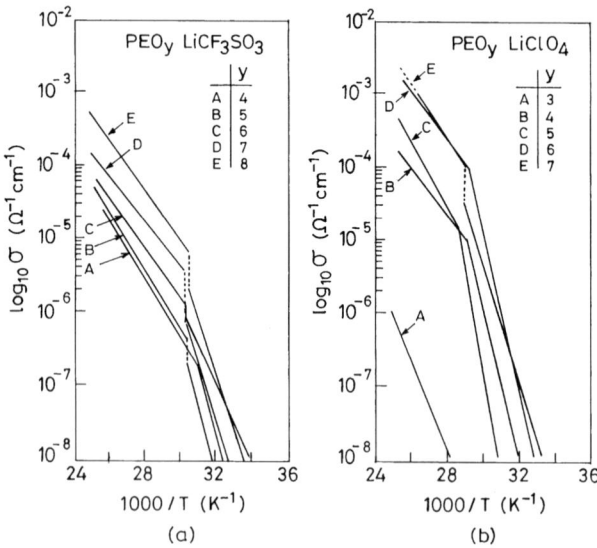

Fig. 5. Conductivity plots for (a) PEO$_4$-LiCF$_3$SO$_3$ and (b) PEO$_4$LiClO$_4$ polymer electrolytes (taken from Ref. 54)

Attempts to obtain transport number information by various methods such as pulsed field gradient NMR [62], radio tracer diffusion [77], and potentiostatic polarization technique [46] have suggested that both cation and anion mobilities are important for the total ionic conductivity seen. In general, however, the nature of charge carriers in polymer electrolytes is quite complex and ion aggregates such as triple ions have been implicated in conductivity [78-79].

3.2
Modification of PEO

Although PEO is an excellent solvent for the solvation of alkali metal ions, polymer electrolytes derived from pure PEO-metal salt complexes do not show high ionic conductivities at ambient temperatures, due to the partial crystalline nature of PEO [27, 29, 37, 59, 79] (vide supra).

There have been several attempts to reduce the crystallinity of PEO and enhance the ionic conductivity at ambient temperatures. These include (a) addition of plasticizers and other related additives to PEO, (b) use of plasticized salts, (c) cross linking of PEO by grafting and other methods. Some of these modifications are discussed below.

3.2.1
Addition of Plasticizers

Some of the plasticizers used in conjunction with PEO are listed in Table 1. These include poly THF [80] cyclic carbonates such as ethylene carbonate (EC) [81-87], propylene carbonate (PC) [81-87], modified propylene carbonate containing an oligo etheroxy side chain (MC-3) [88-89], poly ethylene glycol (PEG) [90], glycols [81], glycol ethers [81], di octyl sebacate (DOS) [82], diethyl phthalate (DEP) [82], and also crown ethers such as 12-crown-4-ether [91-92]. While many of these are simple organic compounds which are available commercially, some such as MC-3 were synthesized especially for their use as additives in polymer electrolytes, as shown in Scheme 1.

As is clearly evident most of the plasticizer additives are low molecular weight organic compounds. The main motivation for the addition of the plasticizer is to increase the free volume of the polymer and hence to lower the glass transition temperature.

Kelly and coworkers have studied PEG as a plasticizer for a PEO-LiCF$_3$SO$_3$ complex [90]. Although the conductivity increases from 3×10^{-7} S cm^{-1} to 10^{-4} S cm^{-1} at 40 °C when 65 mole % of PEG was added, the hydroxyl end groups of PEG react with lithium electrodes and so such a system would not be suitable for batteries.

Although addition of EC and PC individually or in a mixture have increased the conductivities of the PEO-Li salt polymer electrolytes it has been observed that with PC a conductivity of 10^{-4} S cm^{-1} cannot be reached (at room temperature) unless a high concentration of PC is used. However, at such high concentrations the mechanical properties of the polymer electrolyte film are adversely effected and the material loses dimensional integrity. A modified PC containing

Table 1. Some of the plasticizers studied with PEO

	Ref.
1. Ethylene carbonate (EC)	81-87
2. Propylene carbonate (PC)	81-87
3. 2-Keto-4-(2,5,8,11-tetra oxadodecyl) 1,3-dioxalane (Modified carbonate, MC-3)	88-89
4. Bis(2-ethyl hexyl sebacate (Di octyl sebacate, DOS)	82
5. Diethyl Phthalate (DEP)	82

Table 1. (continued)

	Ref.
6. Polyethylene glycol (PEG) HO$\mathrm{+CH_2CH_2O+_nCH_2CH_2OH}$	90
7. Tetra ethylene glycol (TEG) $HO(CH_2)_4OH$	81
8. Tetra ethylene glycol dimethyl ether (TEGDME) $MeO(CH_2)_4OMe$	81
9. 12-crown-4-ether 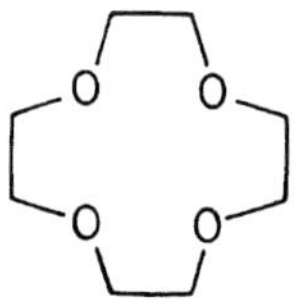	91-92

1. $H_3C(OCH_2CH_2)_3OH + CH_3SO_2Cl \xrightarrow{Et_3N} CH_3(OCH_2CH_2)_3OSO_2CH_3$ **A**

2. [2,2-dimethyl-1,3-dioxolane-4-methanol] $\xrightarrow[\text{'A'}]{NaH}$ [2,2-dimethyl-4-(CH$_2$(OCH$_2$CH$_2$)$_3$OCH$_3$)-1,3-dioxolane] **B**

3. **B** $\xrightarrow[H_2O]{HCl}$ HOCH$_2$CH(OH)CH$_2$(OCH$_2$CH$_2$)$_3$OCH$_3$ **C**

4. **C** $\xrightarrow[Na]{EtOCOOEt}$ [cyclic carbonate with CH$_2$(OCH$_2$CH$_2$)$_3$OCH$_3$] **MC-3**

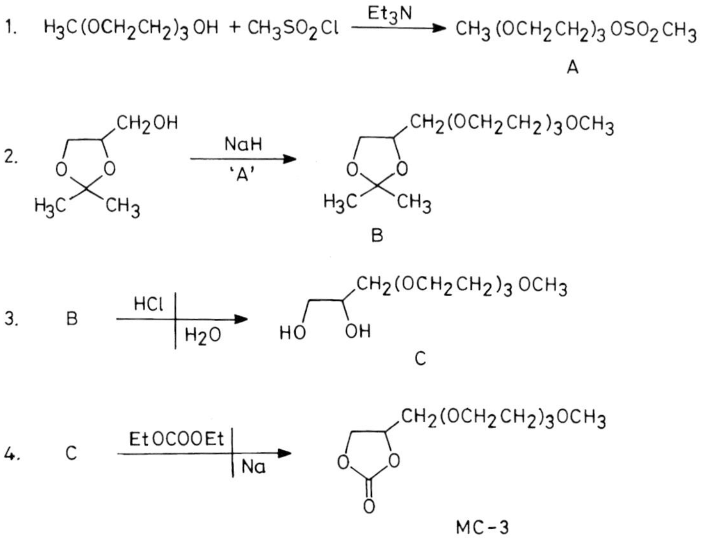

Scheme 1

an etheroxy side chain, MC-3 was synthesized (Scheme 1) to improve the plasticizer properties. With the addition of 50% of MC-3 by weight of PEO to the PEO-LiCF$_3$SO$_3$ complex a high ionic conductivity of 5x10^{-5} S cm^{-1} is achieved. Additionally in contrast to the loss of dimensional stability with the addition of PC, with the addition of MC-3 free standing films with good mechanical properties are achieved [88]. It was shown by Raman spectroscopy that MC-3 had a strong ion pair dissociation effect on LiCF$_3$SO$_3$ presumably because of the etheroxy side groups. This conclusion was also confirmed by near-EXAFS studies on the dissociation of a potassium salt of a long chain etheroxy sulfonate. It is believed that the etheroxy groups present in MC-3 make it more compatible with PEO resulting in a uniform blend when added to PEO [88-89].

Recently a number of studies have been carried out by Chintapalli and Frech on the effect of plasticizers such as EC, TEG and TEGDME on the conductivity and ionic association in a PEO-LiCF$_3$SO$_3$ polymer electrolyte system [81]. These studies clearly show that, with increasing amounts of the plasticizer, the conductivity behavior follows the VTF like curves typical of amorphous polymers (Fig. 6) and in each case the conductivity is enhanced with the progressive addition of plasticizer. Thus [(PEO)(TEG)$_3$]$_9$ LiCF$_3$SO$_3$ has a conductivity value of 6.5 x 10^{-5} S cm^{-1} at 30 °C which is more than three orders of magnitude higher than that of (PEO)$_9$ 9LiCF$_3$SO$_3$. The authors note that the plasticizers seem to preferentially interact with the crystalline phase of the (PEO) x-LiCF$_3$SO$_3$ complex and create new ion conducting pathways. Interestingly, the effect of the plasticizers on the ionic association seems to be varied. Thus EC and TEG tend to increase the concentration of less associated species while TEGDME increases the concentration of more associated species relative to PEO-LiCF$_3$SO$_3$. The reasons for this varied behavior are not clearly understood.

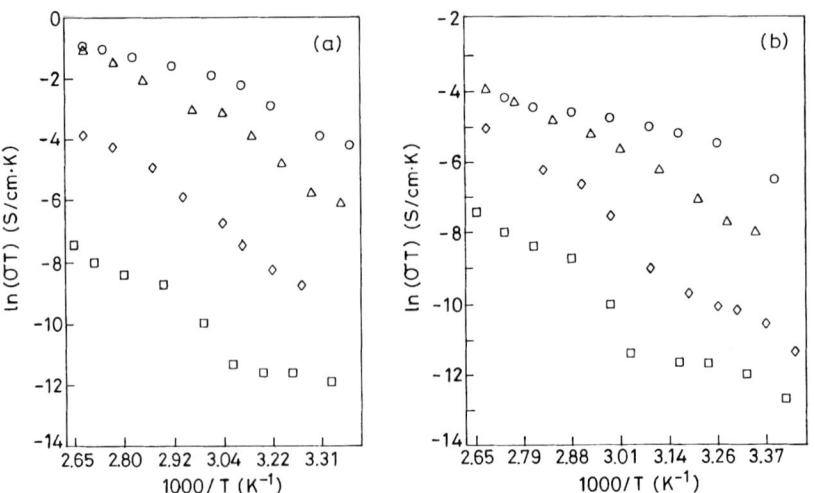

Fig. 6. VTF plots (ln(σT) vs 1000/T) for (a) [(PEO$_{1-x}$) (TEGDMEx)$_9$] LiCF$_3$SO$_3$. Values of x are 0.0□, 0.25 ◇, 0.50△, 0.75○ and (b) [(PEO1-x)-(TEGx)$_9$] LiCF$_3$SO$_3$. Values of x are 0.0□, 0.25 ◇, 0.50 △ and 0.75 ○ (taken from Ref. 81)

Table 2. Conductivity data for some modified PEO-polymer electrolytes[a]

Polymer Electrolyte	O/Li ratio	Conductivity	Ref
1. PEO-MC-3-LiCF$_3$SO$_3$[b]	9	5·10^{-5} S cm^{-1} (303)	88-89
2. PEO-TEG-LiCF$_3$SO$_3$[c]	9	6.5·10^{-5} S cm^{-1} (303)	81
3. PEO-Li TFSI	8	5·10^{-5} S cm^{-1} (298)	105
4. PEO-Li Tri TFSM	10	3·10^{-5} S cm^{-1} (298)	115
5. PEO-Li SO$_3$CF$_2$ SF$_5$	4	10^{-5} S cm^{-1} (303)	116
6. PEO-Li CH(SO$_2$CF$_3$)$_2$)	16	10^{-4} S cm^{-1} (303)	116
7. PEO-graft-polyester[d]-LiClO$_4$	50	2·10^{-5} S cm^{-1} (298)	119

[a] 'Pure' PEO-salt complexes show a low conductivity at room temperature, e.g., PEO-LiCF$_3$SO$_3$ (O/Li:20) shows a conductivity of 5x10^{-7} S cm^{-1} at 298 K.
[b] MC-3 is the plasticizer added.
[c] TEG is the plasticizer added.
[d] A PEO (400 or 600) reacted with maleic anhydride and phthalic anhydride and crosslinked.

Nagasubramanian and coworkers have studied the effect of 12-crown-4-ether on the conductivity behavior of PEO.LiBF$_4$ polymer electrolytes and on its performance in a battery consisting of Li as the anode and CoO$_2$ or TiS$_2$ as the cathodes. While it was found that the addition of the crown ether does bring about an increase in the ionic conductivity and seems to improve the cell performance in the battery, it has no effect on prolonging the active life of the cell [91-92].

In general the effect of the added organic plasticizer appears to increase the free volume of the polymer thereby decreasing the T$_g$ [93] and or reducing the content of the crystalline phase in PEO [83] and also to effect the ionic association in the polymer electrolytes [81]. Many of these effects have been studied by use of a variety of experimental methods such as IR spectroscopy, DSC, EXAFS, X-ray diffraction, NMR, conductivity studies, viscosity measurements etc. [81, 90, 93-103]. The effects of the plasticizers on the conductivity behavior of PEO polymer electrolytes along with the conductivity data of other PEO-polymer electrolytes discussed above are summarized in Table 2.

3.2.2
Plasticizing Salts

Although plasticizers have been quite effective in reducing the crystalline phase of PEO and in increasing the ambient temperature ionic conductivity in PEO-polymer salt complexes, they have certain disadvantages. Most of the plasticizers used are low molecular weight organic liquids and have the disadvantage of volatility. Slow evaporation of an improperly blended plasticizer can lead to the disintegration of polymer electrolyte films. Also, as discussed above, addition of some of the plasticizers such as EC or PC in large quantities leads to a loss of dimensional stability of the polymer electrolyte films. In view of some of these difficulties, there have been attempts at alternative methods of increasing the conductivities in PEO-metal salt complexes. These include the use of plasticizing salts, which are nonvolatile ionic compounds with bulky anions. Some of the commonly used plasticizing salts are shown in Table 3.

Table 3. Plasticizing salts used with PEO

		Ref
1. Lithium bis(trifluoromethane sulfonyl)imide (LiTFSI)	$LiN(SO_2CF_3)_2$	105-110, 113
2. Lithium tris(Trifluoromethane sulfonyl)methide (Li Tri TFSM)	$Li\,C(SO_2CF_3)_3$ [a]	112
3. Lithium pentafluoro sulfur difluoro methylene sulfonate	$LiSO_3CF_2SF_5$	115
4. Lithium bis(trifluoro methane sulfonyl) methane	$LiCH(SO_2CF_3)_2$	116

[a] Armand and coworkers have also reported a related plasticizing anion $[(CF_3SO_2)_2CR]_2$ $C_6H_4{}^{2-}$, R = CO or SO_2 [114].

$$LiOH + B(OH)_3 + 2\;\text{catechol} \longrightarrow [\text{spirocyclic borate}]^- \; Li^+$$

Scheme 2

In addition to the salts shown in Table 2, a new type of lithium salt containing a spirocyclic borate anion has recently been synthesized (Scheme 2) [102]. This salt was shown to possess good electrochemical and thermal stability. However, there are no reports on its use in polymer electrolytes.

Armand and coworkers first reported that lithium bis(trifluoromethane sulfonyl) imide $Li[(CF_3SO_2)_2N]$ (LiTFSI) forms polymer salt complexes with PEO [105-106]. They reported that the crystallinity of PEO is lowered and furthermore that the conductivities are enhanced. A higher dissociation of the cation, which was attributed to the dipolar solvation of the anion, was thought responsible in part to the high conductivities observed. A carbon-based salt, lithium tris(trifluoromethyl sulfonyl) methanide $Li(CF_3SO_2)_3C$ (Li Tri TFSM) [107-112] has also been used as a plasticizing salt with PEO. The synthesis of this salt is shown in Scheme 3. This salt also showed a plasticizing effect. Thus at a O:Li ratio of 5 no melting was detected (Table 4). The highest conductivity observed was 3×10^{-5} S cm^{-1} and 10^{-3} S cm^{-1} at 25 and 60 °C respectively. Further the Li Tri TFSM-PEO complex was shown to be electrochemically stable especially towards oxidation, even at E = +3.9 V. Borghini et al. have reported the electrochemical properties of PEO-Li $[(CF_3SO_2)_2N]$ salt which was further treated with γ-LiAlO$_2$ as a ceramic filler. This composite showed an exceptionally slow recrystallization rate and apart from an increase in the mechanical properties an enhancement of the lithium interface stability was also observed [113].

More recently new salts $[(CF_3SO_2)_2CR]_2C_6H_4{}^{2-}$, (R = CO or SO_2 [114], lithium pentafluoro sulfur difluoro methylene sulfonate $LiSO_3CF_2SF_5$ [115] and lithium bis(trifluoro methyl sulfonyl) methane, $LiCH(SO_2CF_3)_2$ [116] have been reported as plasticizing salts with PEO. While $[(CF_3SO_2)_2CR]_2C_6H_4{}^{2-}$ has been synthe-

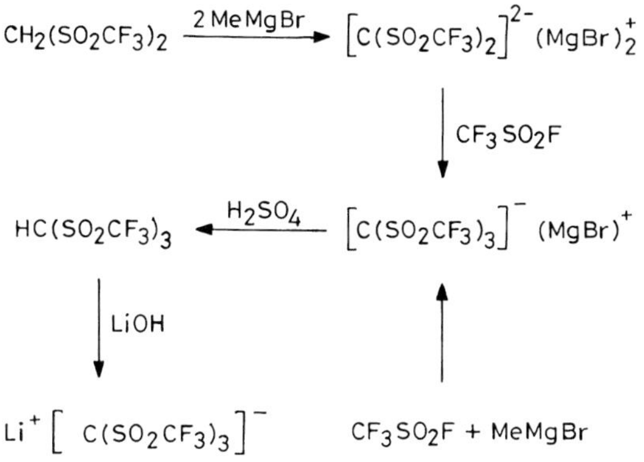

Polymer Solid Electrolytes: Synthesis and Structure

Scheme 3

Table 4. Glass transition temperatures and melting temperatures of PEO-Li tri TFSM samples as a function of composition (taken from Ref. 112)

O/Li	T_g °C	T_m °C
5	−16	–
6	−20	–
8	−29	–
10	−35	–
12	−41	35
20	−31	42

sized from the disubstitution of chlorines in terephthaloyl chloride or benzene-1,3-disulfonyl chloride with $CF_3SO_2CH^-$ [114], $LiSO_3CF_2SF_5$ has been prepared by the reaction perfluoro vinyl sulfur pentafluoride, $SF_5CF = CF_2$ with sulfur trioxide [115]. The preparation of $LiCH[SO_2CF_3]_2$ is carried out as follows [116],

$$2\ CH_2(CF_3SO_2)_2 + Li_2CO_3 \rightarrow 2\ LiCH(SO_2CF_3)_2 + CO_2 + H_2O \qquad (10)$$

The complexes of PEO with these plasticizing salts show much higher conductivities than 'normal' salt complexes. Thus, $(PEO)_{16}LiCH(SO_2CF_3)_2$ shows a σ of about 10^{-4} S cm^{-1} at 30 °C and 10^{-3} S cm^{-1} at 80 °C [116].

While the origin of the plasticizing effects of most of the salts described above are not clear, all of these have large stable anions with strong electron withdrawing substituents. The structural flexibility of these bulky anions with the associated possibility of a disordered arrangement might be responsible in some manner for the reduction of the crystalline phase in PEO. Further bulky anions, by not favoring ion pair formation, might also aid in the increased dissociation of the salt thereby increasing the Li$^+$ ion concentration. These effects are manifested in the increased conductivities (~ 10^{-5} S cm^{-1}) of PEO-salt complexes observed at room temperature [Table 2].

3.2.3
Other Modifications of PEO

Polymer properties can be modified by physical blending but this method has the inherent disadvantage of phase separation particularly when incompatible systems are mixed together. A better way of improving the properties of the polymer would be a modification at the molecular level by altering the structural features. Several strategies to modify the structural features of PEO so that the crystallinity is reduced and the ionic conductivities of the polymer electrolytes improved have been adopted. These include modifications of PEO by preparing copolymers of PEO by grafting and/or crosslinking. Table 2 contains ionic conductivity data for some of these modified PEO polymer electrolytes.

Polyester networks with various chain lengths of PEO have attracted attention [117-118]. In one approach, a PEO macromer was used as a cross-linking agent to link maleic anhydride and phthalic anhydride [119-120]. The resulting polyester resins were complexed with $LiClO_4$ and cured to obtain graft network polymer electrolytes which were flexible and amorphous. These showed good ambient temperature conductivity (σ_{298K} = 2x10^{-5} S cm^{-1}). Other approaches for the preparation of polyester linked PEO units include the reactions of an acrylate monomer such as methacryloyl chloride with mono methyl polyethylene glycol (MPEG) or polyethylene glycol (PEG). Scheme 4 summarizes the preparation of these monomers. The resulting monomers have been polymerized to obtain graft crosslinked networks [121-124]. Alternatively dicarboxylic acids have been used to link PEG (Scheme 5) [125-126]. Most of these polyester-PEO network polymers formed homogeneous lithium salt complexes and showed ambient temperature ionic conductivities of the order of 10^{-5} S cm^{-1}, concomitant with a reduction of the melting temperatures of the modified PEOs. The highest conductivity reported for these series is for a PMMA-PEO Li salt complex which showed a conductivity of 10^{-4} S cm^{-1} at room temperature [121-122]. This electrolyte was examined for its polarization characteristics. Also, the charge-discharge behavior of Li was investigated. It was found that the rate of charge transfer at the Li

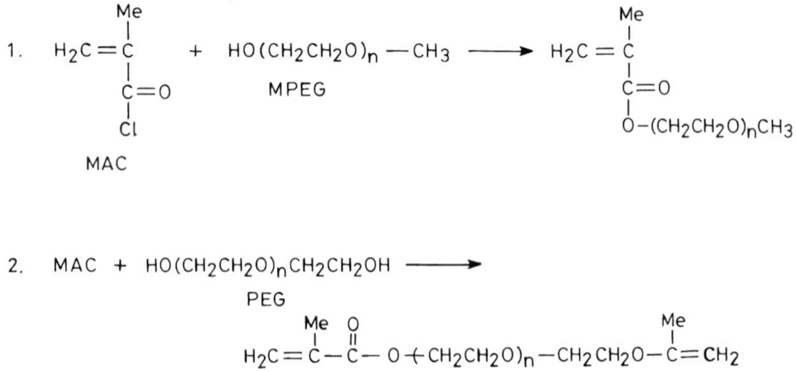

1. $H_2C=C(Me)(C=O)Cl$ (MAC) + $HO(CH_2CH_2O)_n-CH_3$ (MPEG) $\longrightarrow$ $H_2C=C(Me)(C=O)O-(CH_2CH_2O)_nCH_3$

2. MAC + $HO(CH_2CH_2O)_nCH_2CH_2OH$ (PEG) $\longrightarrow$

$H_2C=C(Me)-C(=O)-O+CH_2CH_2O)_n-CH_2CH_2O-C(Me)=CH_2$

Scheme 4

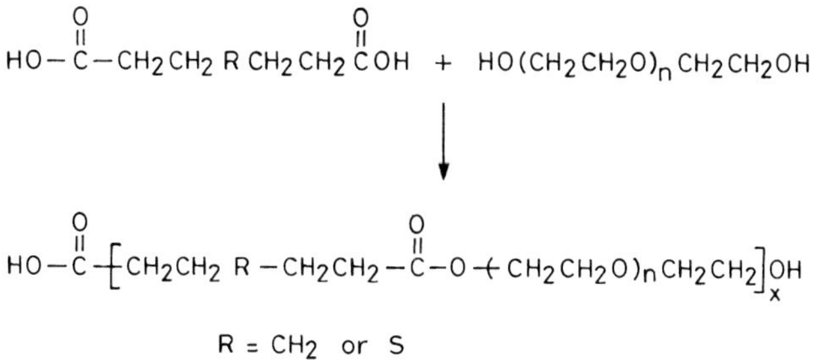

$$\text{HO}-\overset{\overset{\text{O}}{\|}}{\text{C}}-\text{CH}_2\text{CH}_2\ \text{R}\ \text{CH}_2\text{CH}_2\ \overset{\overset{\text{O}}{\|}}{\text{C}}\text{OH} \ +\ \text{HO}(\text{CH}_2\text{CH}_2\text{O})_n\text{CH}_2\text{CH}_2\text{OH}$$

$$\downarrow$$

$$\text{HO}-\overset{\overset{\text{O}}{\|}}{\text{C}}-\left[\text{CH}_2\text{CH}_2\ \text{R}-\text{CH}_2\text{CH}_2-\overset{\overset{\text{O}}{\|}}{\text{C}}-\text{O}-(\text{CH}_2\text{CH}_2\text{O})_n\text{CH}_2\text{CH}_2\right]_x\text{OH}$$

$$R = CH_2\ \text{or}\ S$$

Scheme 5

electrode/polymer interface was very high. A high coulombic efficiency (~ 88%) for lithium was observed during the charge discharge cycle at 50 $\mu A\ cm^{-2}$ cycling [121].

Recently Nagasubramanian and coworkers have prepared lithium polymer electrolytes which contained a blended mixture of photocured cycloaliphatic epoxide, PEO, LiASF$_6$, EC etc. Although the conductivity was reasonably good the prototype battery tested was poor in terms of reversibility and plating to stripping ratio [127]. Earlier, Peng and coworkers synthesized a completeley amorphous two component epoxy network polymer from diglycidyl ether of PEG and triglycidyl ether of glycerol. Added LiClO$_4$ functions both as the catalyst for ring opening of the epoxides as well as the source for the charge carriers [128].

Other types of modifications tried include preparation of ethylene oxide-propylene oxide copolymers [129] as well as polymers containing regular sequences of oxyethylene and oxymethylene units [130].

In a recent study, Park and coworkers [131] have studied comb shaped graft polymers based on acrylontrile-butadiene copolymer (NBR) having grafts of short chain PEO's on the butadiene unit, NBR-g-PEO. In this methodology first a partial epoxidation of NBR was carried out [131] followed by hydroxylation of some of the epoxides generated on the butadiene frame work, to produce a NBR diol. This is essentially a polymer with several reactive hydroxyl groups. The other reactant was a modified MPEG containing a reactive NCO group. Grafting is accomplished by reacting the MPEG-NCO with NBR diol [Scheme 6]. A large number of these grafted polymers have been found to be amorphous with their T_m's below room temperature. Further, complexes of these modified polymer with LiCF$_3$SO$_3$ showed a maximum conductivity of 3x10^{-5} S cm^{-1} at room temperature. In this context, it must be mentioned that diisocyanates have been used to modify PEO related systems and urethane grafts of α-ω dihydroxy polypropylene oxides have been prepared [132-135]. Studies carried out on poly urethane systems include positron annihilation to measure the free volume [136] and ^{7}Li NMR measurements [137]. These studies suggest that the urethane NH also participates in the solvation of Li$^+$ ions.

Another kind of modification of PEO involves the concept of composite polymer electrolytes [33, 37-38, 138-144]. These include the use of inorganic fillers such as NASICON [138], β-alumina [139], glassy fillers [139] γ-LiAlO$_2$ [140-143]

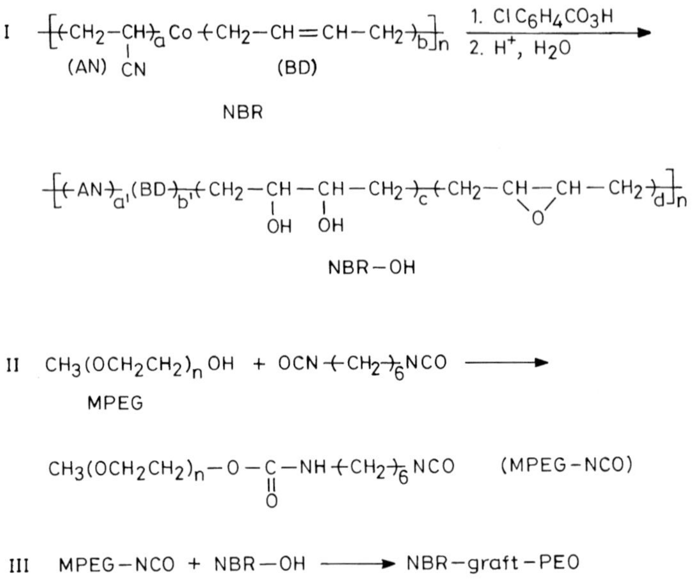

I $\ce{[(CH2-CH)_a Co (CH2-CH=CH-CH2)_b]_n}$ $\xrightarrow{\text{1. Cl C}_6\text{H}_4\text{CO}_3\text{H}}_{\text{2. H}^+,\text{ H}_2\text{O}}$
         (AN) CN   (BD)

NBR

$\ce{[(AN)_{a'}(BD)_{b'}(CH2-CH(OH)-CH(OH)-CH2)_c(CH2-CH-CH(O)-CH2)_d]_n}$

NBR–OH

II $CH_3(OCH_2CH_2)_n OH$ + $OCN(CH_2)_6 NCO$ ⟶
     MPEG

$CH_3(OCH_2CH_2)_n - O - \underset{\underset{O}{\|}}{C} - NH(CH_2)_6 NCO$ (MPEG–NCO)

III   MPEG–NCO + NBR–OH ⟶ NBR–graft–PEO

Scheme 6

etc. Significant improvement in the conductivity as well as in the electrochemical stability of the polymer electrolytes has been reported as a result of these modifications. Recently lithium salt polymer electrolytes containing blends of PEO or oxymethylene linked PEO with poly acrylamide (PAAM) have been studied [144]. It was found that the ionic conductivity of these electrolytes is in excess of 10^{-4} S cm^{-1} at room temperature. It has been shown that PAAM effectively inhibits the crystallization of PEO without impeding its segmental motion.

4
Other PEO Related Polymers

4.1
Some Linear Polymers

Poly(ethylene oxide) is a linear polymer containing the donor oxygen atoms in the main backbone. Some other similar systems known to function as polymer electrolytes include simple poly ethylene glycol (PEG) [145], end acetylated PEG [146], poly propylene oxide (PPO) [147-148], poly(β-propiolactone) [149], poly-(ethylene succinate) [150-151], poly (ethylene adipate) [152], poly (ethylene imine) [153] and poly (alkylene sulfide) [154]. Many of these form metal salt complexes. However, conductivities of the order of 10^{-5} S cm^{-1} are observed only at high temperatures. Table 5 summarizes this data.

Table 5. Conductivity data for polymer electrolytes containing linear polymers

Polymer	Metal Salt	O:Li ratio	Maximum Conductivity S cm^{-1} (K)	Ref.
1. Poly(propylene Oxide) $-[CH(CH_3)-CH_2-O]_n-$	LiBr, LiI NaB(C_6H_5)$_4$ LiCF$_3$SO$_3$ NaCF$_3$SO$_3$	9:1	~10^{-6} ~10^{-6} 2.2·10^{-5} (312) ~10^{-6}	147-148
2. Poly(β-propiolactone) $-[CH_2-CH_2-C(=O)-O]_n-$	LiClO$_4$	20:1	3.5·10^{-6}	149
3. Poly(ethylenesuccinate) $-[O-(CH_2)_2-O-C(=O)-(CH_2)_4-C(=O)]_n-$	LiClO$_4$ LiBF$_4$	33:1 12:1	~10^{-5} (363) 3.4·10^{-6} (288.2)	150-151
4. Poly(ethylene adipate) $-[CH_2-CH_2-C(=O)-(CH_2)_4-C(=O)]_n-$	LiCF$_3$SO$_3$	16:1	~10^{-6}	152
5. Poly(ethyleneimine) $-[CH_2CH_2-NH]_n-$	NaSO$_3$CF$_3$	6:1[a]	~10^{-7}	153
6. Poly(alkylene sulphide) $-[(CH_2)_x-S]_n-$ n=5	AgNO$_3$	4:1[b]	9·10^{-7} (318)	154

[a] O : Na ratio
[b] O : Ag ratio

4.2
Polymer 'Gel' Electrolytes

A variety of dimensionally stable solid electrolytes consisting of a mixture of organic plasticizers such as EC, PC etc., along with structurally stable polymers such as poly(acrylonitrile) (PAN) or poly(vinyl sulfone) (PVS), or poly vinyl pyrrolidine (PVP) or poly vinyl chloride (PVC) and several lithium salts have been tested and found to have excellent ionic conductivities at ambient temperatures [155-156]. In these 'gel' type electrolytes the primary role of the polymers PAN, PVS, PVP or PVC is to immobilize the lithium salt solvates of the organic plasticizer liquids. However, with polymers such as PAN a coordination interaction with Li^+ is also quite likely.

Watanabe prepared for the first time solid electrolytes comprising PC and $LiClO_4$ in PAN and reported a maximum conductivity of 2×10^{-4} S cm^{-1} [155]. Abraham and Alamgir prepared Li^+ conductive polymer electrolytes with extremely high ambient temperature conductivities of 4×10^{-3} S cm^{-1} [156-157]. These electrolytes are composed of Li salts such as $LiClO_4$ dissolved in organic solvents EC and PC and immobilized in a polymer network of PAN, poly(tetra ethylene glycol diacrylate) (PEGDA) or poly(vinyl pyrrolidone) PVP. Matsumuto, Rutt and Nishi described polymer electrolytes with high ionic conductivities (10^{-3} S cm^{-1}) and good mechanical strength [158-161]. A typical example of this type of polymer electrolyte is prepared by swelling poly(acrylonitrile-co-butadiene) (NBR)/poly(styrene-co-butadiene) (SBR)/$LiClO_4$ latex films with an organic solvent such as γ-butyrolactone [158]. The authors suggest that these systems have dual ion conductive channels, one which is the fused NBR-latex phase and the other is the $LiClO_4$ phase present at the interface of SBR/NBR latex particles. The pure SBR phase formed from SBR latex particles according to the authors is non polar and

Table 6. Conductivity data for polymer 'gel' electrolytes

Polymer Gel	Li Salt	Maximum Conductivity S cm^{-1} (K)	Ref.
1. NBR/SBR 1:1 +γ butyrolactone	$LiClO_4$[a]	$1.2 \cdot 10^{-3}$ (298)	158
2. PAN/EC/PC[b]	$LiClO_4$[b]	$1.7 \cdot 10^{-3}$ (293)	157
3. PAN/PC	$LiClO_4$	$2 \cdot 10^{-4}$ (293)	157
4. EC/PC/PAN/PEGDA[c]	$LiClO_4$[c]	$4.0 \cdot 10^{-4}$ (263); $1.2 \cdot 10^{-3}$ (293)	157
5. EC/PC/PAN[d]	$LiCF_3SO_3$	$4.0 \cdot 10^{-4}$ (263); $1.4 \cdot 10^{-3}$ (293)	157
6. EC/PC/PVP[e]	$LiCF_3SO_3$	$4.0 \cdot 10^{-5}$ (263); $5.0 \cdot 10^{-4}$ (293)	157

[a] Polymer gel saturated with 0.2 to 0.4 M $LiClO_4$ solution.
[b] The electrolyte comprises 38 mol % - 17x of EC, 33 mol % - 17x of PC containing 8 mol % - 17x of $LiClO_4$ immobilized in 21 mol % - 17x of PAN.
[c] The electrolyte comprises of 62 mol % - 17x of EC, 13 mol % - 17x PC and 8 mol % - 17x $LiClO_4$ immobilized in 16 mol % - 17x PAN and 1 mol % - 17x PEGDA.
[d] The electrolyte comprises of 42 mol % - 17x EC, 36 mol % - 17x PC, and 7 mole per cent $LiCF_3SO_3$ immobilized in 15 mol % - 17x PAN.
[e] The electrolyte comprises of 35 mol % - 17x EC and 31 mol % - 17x PC and 10 mol % - 17x $LiCF_3SO_3$ immobilized in 24 mol % - 17x of poly(vinyl pyrrolidine) PVP.

Polymer Solid Electrolytes: Synthesis and Structure

therefore is not impregnated and merely provides mechanical support. Table 6 summarizes the conductivity data for some of these nonconventional polymer electrolytes.

Scrosati and coworkers [165] have fabricated an all solid lithium battery by combining a PAN based polymer electrolyte (containing EC and PC) with a lithium metal anode and a poly pyrrole (pPy) film cathode. Although the Coulombic efficiency was found to be high, near 90%, the battery has a poor shelf life.

4.3
'Polymer-in-Salt' Electrolytes

In the preceding discussion on polymer electrolytes it was shown that ionic conductivity is realised as a result of doping metal salts in the host polymer which contains suitable donor sites. Generally, however, high conductivities are observed only at low Li(M)/O ratios. Increase in lithium (metal) salt concentration leads to a rapid increase in the crystallinity of the materials and/or an increase in the glass transition temperatures [27,29]. Both of these effects are detrimental to ionic conductivities and hence the advantage of higher charge carrier concentrations are lost.

In an alternative approach adopted by Angell and coworkers [44] instead of doping the metal salts into a polymer, small amounts of polymer are combined with salt mixtures. Thus, when a mixture of LiI, LiOAc and $LiClO_4$ in mole percentages of 50, 30 and 20 are doped with a small amount of polypropylene oxide, a "polymer (in salt) electrolyte" is obtained with an ambient temperature conductivity of $\sim 10^{-4}$ S cm^{-1}. These materials have glass transition temperatures low enough to remain rubbery at room temperature while preserving good lithium ion conductivities and a high electrochemical stability.

5
Pendant Polymer Systems

All the polymers such as PEO and related systems discussed above have, in their backbone, coordinating sites which interact with the added metal ions. Other types of polymers have also been investigated as hosts for polymer electrolyte formation. These polymers contain short side chains of oligoetheroxy groups capable of complexing with alkali metal salts. These "comb" like polymers include both organic and inorganic polymers. Among the former the most well studied systems are itaconates and methacrylates, and among the latter polyphosphazenes and polysiloxanes.

5.1
Poly Acrylates and Itaconates

There have been reports on the use of polyacrylates and itaconates [166-173] as hosts for polymer electrolytes. Table 7 summarizes the conductivity data observed for these systems. Most of these show conductivities of about 10^{-5} S cm^{-1} only at high temperatures. Tsuchida and coworkers have investigated poly [(ω-carboxy)] oligo (oxyethylene) methacrylate with an immobilized anion (COO^-) as a host

Table 7. Conductivity data for some acrylate, itaconate and crown ether polymers

Polymer	Metal salt	Maximum conductivity S cm^{-1} (K)	Ref.
1. Poly(methoxy poly ethylene-glycolmono methacrylate)			167-169

$$\left[-CH_2-\underset{\underset{\underset{O-(CH_2CH_2O)_xCH_3}{|}}{C=O}}{\overset{CH_3}{\underset{|}{C}}}-\right]_n$$

x = 22	LiCF$_3$SO$_3$	6·10^{-5} (373), 2x10^{-5} (293)	
x = 9	LiCF$_3$SO$_3$	3·10^{-4}(373), ~10^{-6} (293)	
x = 3, 7, 12 & 17	LiClO$_4$	~10^{-5} (298)	
x = 7	LiPF$_6$	~10^{-5} (298)	
2. Poly[diethoxy(3) methyl itaconate]	LiClO$_4$	5.6·10^{-5} (333)	170

$$\left[-CH_2-\underset{\underset{CH_2COO(CH_2CH_2O)_xCH_3}{|}}{\overset{CH_2COO(CH_2CH_2O)_xCH_3}{\underset{|}{C}}}-\right]_n$$

3. Poly(di-poly(propylene glycol) itaconate)	LiClO$_4$ NaClO$_4$ ZnCl$_2$ LiCl	~10^{-6} (303) 3.4·10^{-7} (333) 1.2·10^{-7}(333)	171
4. Poly[(w-carboxy) oligooxyethylene methacrylate]	Li Na K	8·10^{-11} 7·10^{-8} 2·10^{-8}	172

$$\left[-CH_2-\underset{\underset{O-(CH_2CH_2O)_2CH_2COO^-}{|}}{\overset{CH_3}{\underset{|}{C}}}-\right]_n$$

Table 7. (continued)

Polymer	Metal salt	Maximum conductivity S cm^{-1} (K)	Ref.
5. Poly(crown ether) $\left[-CH_2-\underset{\underset{\underset{CH_2-Crown}{\mid}}{\underset{O}{\mid}}}{\overset{\overset{CH_3}{\mid}}{C}}-\right]_n$ $C=O$	Li Na K LiClO4	3·10^{-9} 3·10^{-8} 1·10^{-7} 2·10^{-8} (333)	168

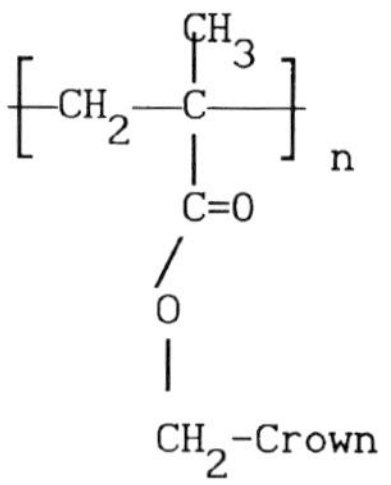

polymer with various metal salts such as Li, Na or K. However, very low conductivities (1.1x10^{-7} S cm^{-1}) were observed [172].

More recently a methacrylate polymer with a short etheroxy side group, poly (methoxy ethoxy ethyl methacrylate) poly (MEEMA) was synthesized (Scheme 7) and investigated as a polymer host for polymer electrolytes [174-177]. This polymer readily forms polymer salt complexes with LiX (X = ClO$_4^-$, BF$_4^-$, CF$_3$SO$_3^-$) and MI (M = Na,K). The ionic conductivities observed with these polymers are quite high with ambient temperature conductivities of the order of 10^{-5} S cm^{-1}. The conductivity plot (log σ vs 10^3/T) for poly(MEEMA)-4LiBF is shown in Fig. 7. Table 8 summarizes the conductivity data. Figure 8 shows the conductivity isotherm plots. These studies suggest that whereas in the poly (MEEMA)-LiClO$_4$ system an enhancement in conductivity of about 20 times is seen at 328 K, with respect to pure poly MEEMA in the analogous poly(MEEMA)-LiBF$_4$ system the enhancement is 2240 times at 324 K [175].

Further, the conductivity behavior in these systems follows the VTF equation. Thus [ln($\sigma \sqrt{T}$)A]$^{-1}$ vs. T plots afford straight lines [Fig. 9]. The values of A and B obtained from these plots [Table 9] show that the apparent activation energy B is quite low.

The conductivities of these systems are comparable to the polyphosphazene polymers discussed below.

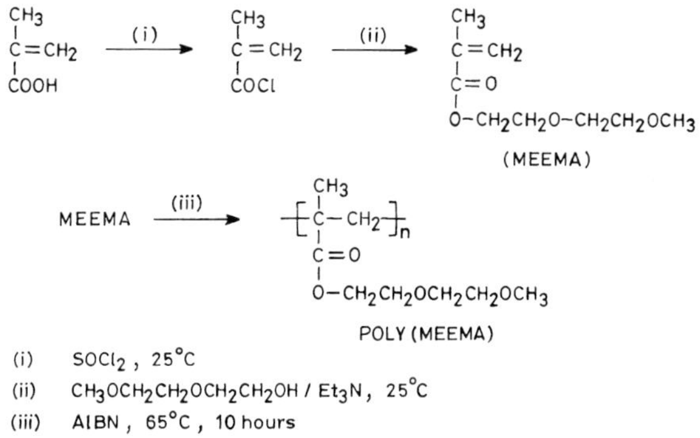

$$\underset{\text{COOH}}{\overset{CH_3}{\underset{|}{C}}=CH_2} \xrightarrow{(i)} \underset{\text{COCl}}{\overset{CH_3}{\underset{|}{C}}=CH_2} \xrightarrow{(ii)} \overset{CH_3}{\underset{\underset{O-CH_2CH_2O-CH_2CH_2OCH_3}{\underset{|}{C=O}}}{\underset{|}{C}}=CH_2}$$

(MEEMA)

$$\text{MEEMA} \xrightarrow{(iii)} \left[\underset{\underset{O-CH_2CH_2OCH_2CH_2OCH_3}{\underset{|}{C=O}}}{\overset{\overset{CH_3}{|}}{C}}-CH_2\right]_n$$

POLY (MEEMA)

(i) SOCl$_2$, 25°C
(ii) CH$_3$OCH$_2$CH$_2$OCH$_2$CH$_2$OH / Et$_3$N, 25°C
(iii) AIBN, 65°C, 10 hours

Scheme 7

Table 8. Conductivity data of poly(MEEMA)-Li salt complexes

Metal salt	Li/O Ratio	σ S cm^{-1} (K)	Ref.
1. LiCF$_3$SO$_3$	1:20	2.3·10^{-6} (270)	174-177
		6.0·10^{-5} (303)	
2. LiClO$_4$	1:10	3.3·10^{-6} (305)	174-177
		1.0·10^{-5} (323)	
3. LiBF$_4$	1:5	5.8·10^{-5} (300)	174-177
		2.0·10^{-4} (318)	
	1:6	4.0·10^{-5} (300)	

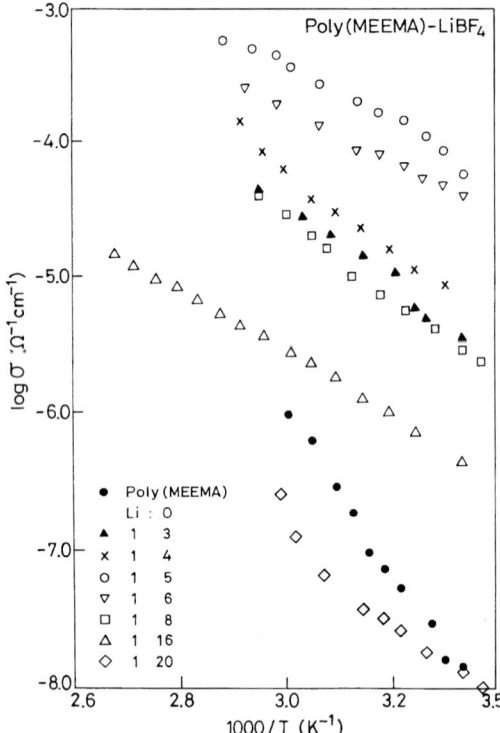

Fig. 7. Conductivity plots (log σ vs 10^3/T) for poly(MEEMA-LiBF$_4$) complexes (taken from Ref. 175)

Polymer Solid Electrolytes: Synthesis and Structure

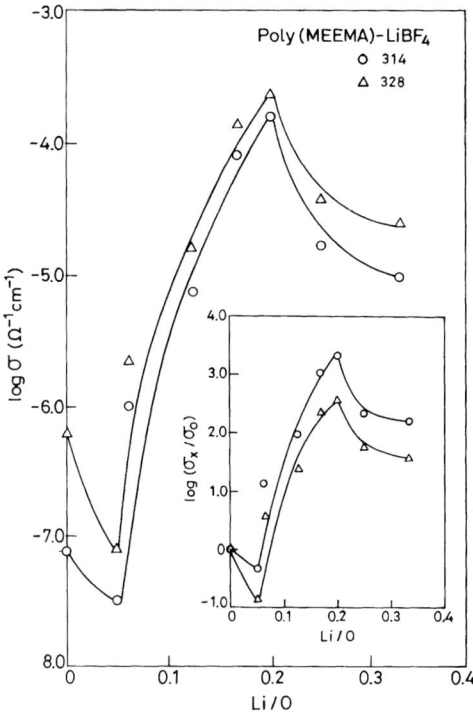

Fig. 8. Conductivity isotherm plots at 314 and 328 K for different compositions of poly(MEEMA)-LiBF$_4$ complexes. Insets are the log [σx/σo] vs Li/O ratio plots (taken from Ref. 175)

Table 9. Activation energy parameters obtained for poly(MEEMA)-Li salt complexes from fitting VTF equation (taken from Ref. 175)

Composition (Li:O ratio)	A	B (eV)	T (K°)
Poly(MEEMA)-:LiClO$_4$			
1:8	0.333	4.24·10^{-2}	263
1:10	0.286	8.66·10^{-2}	187
1:12	0.250	4.64·10^{-2}	254
1:20	0.167	7.38·10^{-2}	224
1:30	0.118	1.15·10^{-1}	196
Poly(MEEMA)-LiBF$_4$			
1:3	0.571	7.69·10^{-2}	204
1:4	0.500	5.44·10^{-2}	229
1:5	0.444	4.08·10^{-2}	220
1:6	0.400	5.19·10^{-2}	208
1:8	0.333	9.31·10^{-2}	177
1:16	0.200	6.06·10^{-2}	252

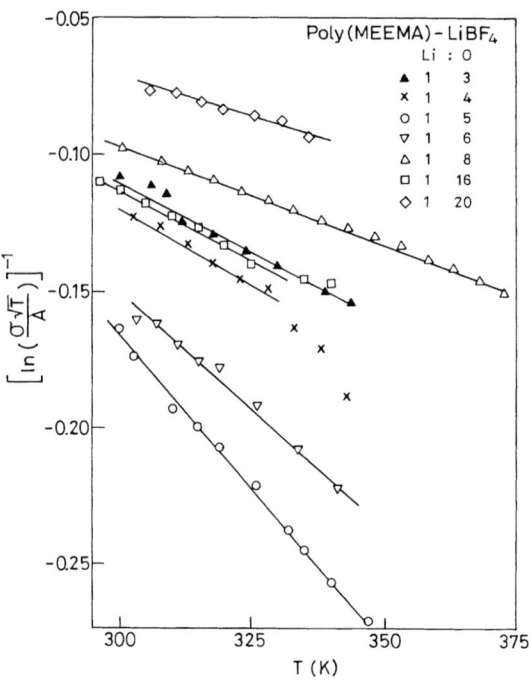

Fig. 9. VTF plots for poly(MEEMA)-LiBF$_4$ complexes (taken from Ref. 175)

5.2
Poly(crown-ether)

Acrylate polymers containing crown-ether pendants have been synthesised as shown in Scheme 8 [178]. This polymer forms complexes with LiClO4. However, the conductivity observed is quite low.

5.3
Poly(phosphazenes)

Poly(phosphazenes) are inorganic polymers with a $\begin{smallmatrix} R & R \\ \diagdown & \diagup \\ [\,P & = & N\,]_n \end{smallmatrix}$ repeating unit in the backbone [179-183]. The precursor poly(dichlorophosphazene)

$\begin{smallmatrix} Cl & Cl \\ \diagdown & \diagup \\ [\,P & = & N\,]_n \end{smallmatrix}$ is prepared by a ring opening polymerization of the cyclic six membered inorganic compound, N$_3$P$_3$Cl$_6$:

$$N_3P_3Cl_6 \xrightarrow[\text{vacuum}]{250\,°C} [\,P = N\,]_n \begin{smallmatrix} Cl & Cl \\ \diagdown & \diagup \\ \, & \, \end{smallmatrix} \qquad (11)$$

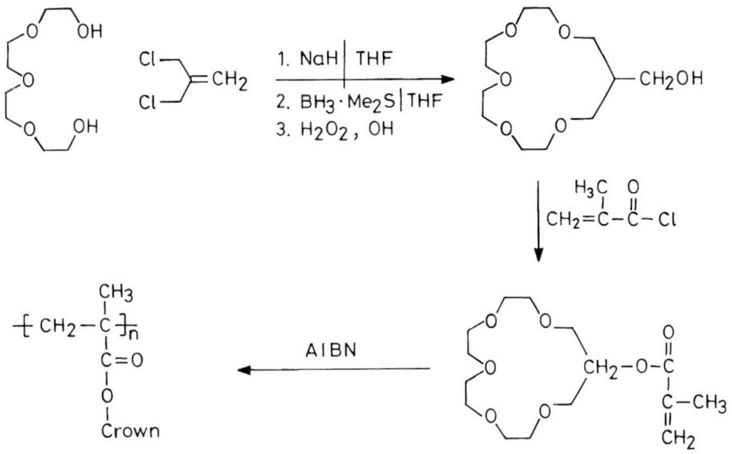

Scheme 8

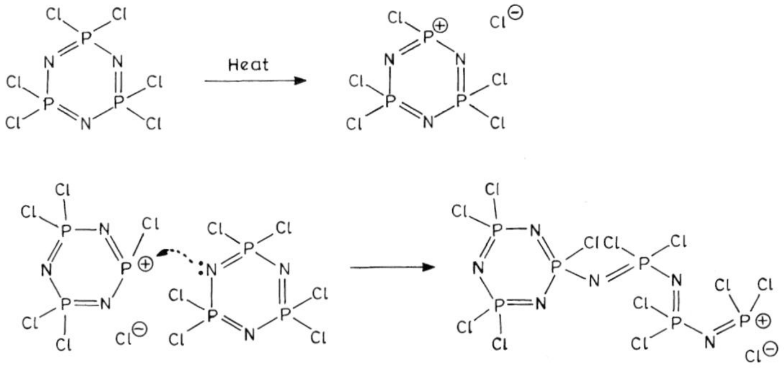

Scheme 9

Although there are now several methods of preparing the linear poly(dichlorophosphazene), the most widely used route involves a thermal ring opening of the inorganic heterocyclic ring, $N_3P_3Cl_6$ [181-183]. The polymerization is catalyzed by small amounts of Lewis acids or even moisture [184]. The mechanism of polymerization is believed to proceed through an initial heterolytic cleavage of a P-Cl bond leading to the formation of a phosphorus centered cation. Subsequent intermolecular nucleophilic attack by another ring nitrogen leads to a ring opening. Chain growth continues to afford eventually a high molecular weight polymer. This mechanism is summarized in Scheme 9 [185-188].

Poly(dichlorophosphazene) by itself is not useful since it is hydrolytically unstable. This however implies a high reactivity of the P-Cl bonds, a characteristic which implies the possibility of facile nucleophilic substitution reactions of chlorines by hydrolytically stable organic groups (Scheme 10). Also, the macromolecular

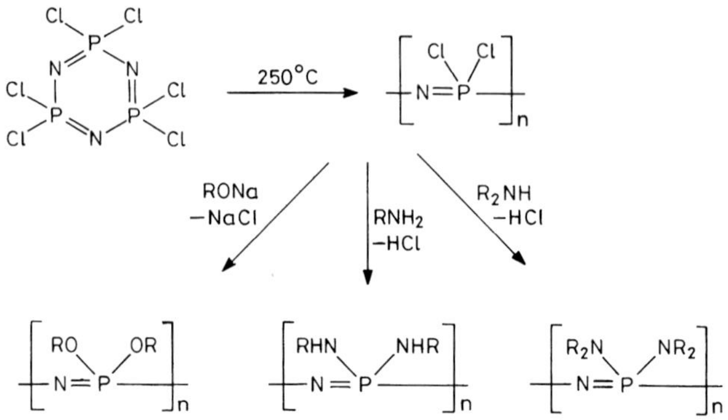

Scheme 10

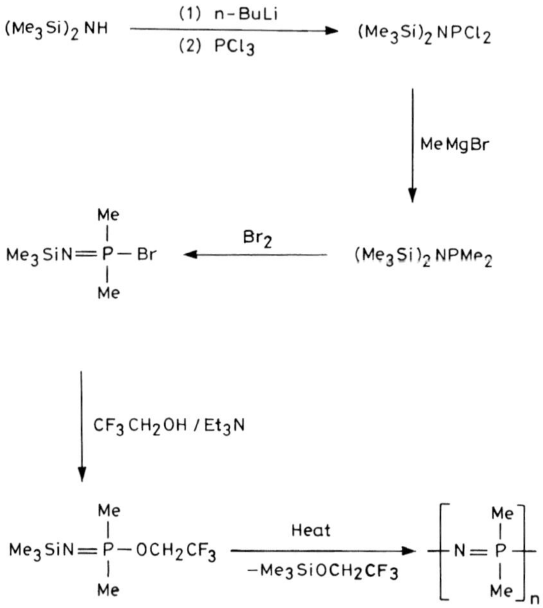

Scheme 11

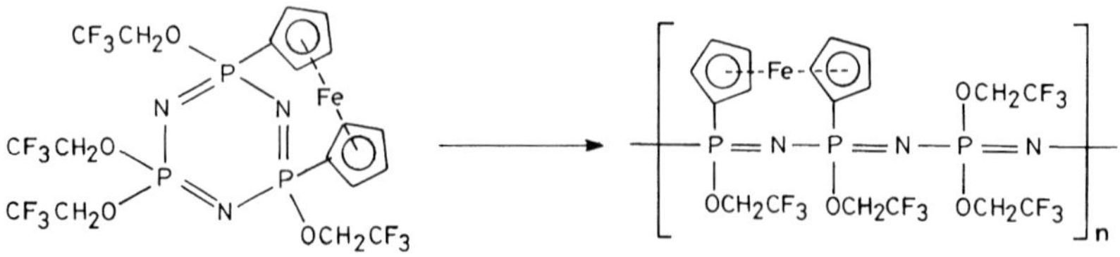

Scheme 12

Polymer Solid Electrolytes: Synthesis and Structure

substitution route apart from providing stability, also provides a way to fine tune the polymer properties [181, 188-192]. Other important methods of preparing polyphosphazenes with P = N backbones are summarized in Schemes 11 and 12 respectively [193-198]. Scheme 11 shows a condensation polymerization method [193-196]. This method which takes advantage of the facile cleavage of a N-Si bond with a simultaneous Si-O bond formation in the leaving group is used for exclusive preparation of alkyl and aryl substituted poly(phosphazenes). Scheme 12 shows the strain induced ring opening polymerization of ferrocenyl cyclophosphazenes [197-198].

5.3.1
Structural Features of Poly(phosphazenes)

In order to appreciate the utility of appropriately substituted poly(phosphazenes) as hosts in polymer electrolytes it is essential to obtain a preliminary idea about the nature of bonding in these polymers [179, 181, 185, 186].

Each phosphorus has five valence electrons and each nitrogen also has five. Each phosphorus atom utilizes four electrons for the formation of σ-bonds (two in the polymeric frame work and two for exo substituent bonds). This leaves one odd electron on phosphorus. Nitrogen utilizes two electrons for frame work σ-bonding, two are present in a lone pair and again one odd electron remains on nitrogen. These odd electrons present in a $2p_z$ orbital in nitrogen and in a $3d$ orbital in phosphorus can be involved in multiple bonding. Since each phosphorus can use as many as five $3d$-orbitals, torsion of a P-N bond can bring the nitrogen p-orbital into an overlapping position with a d-orbital at almost any torsion angle. [Scheme 13]. Because of this the torsional mobility of poly(phosphazenes) is inherently much higher than that present in corresponding organic polymers. It has been suggested that the inherent torsional barrier for the backbone bonds may

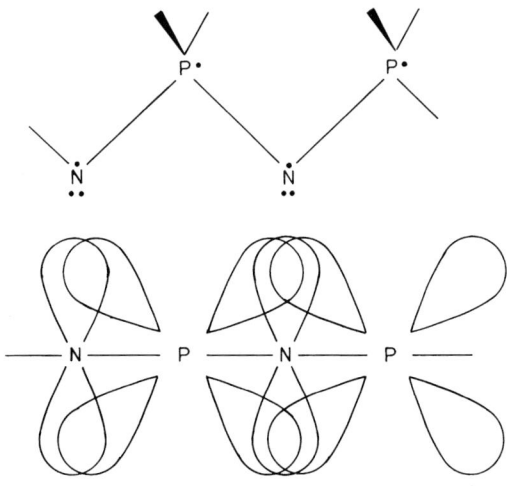

Scheme 13

be quite low (0.1 kcal per bond). The torsional flexibility in poly(phosphazenes) is also accentuated by the presence of side groups in every other skeleton atom rather than in every skeleton atom in the backbone. This situation is different from other organic polymers.

The backbone conformational flexibility of poly(phosphazenes) is important for ionic mobility and hence suitably substituted poly(phosphazenes) have attracted considerable interest as hosts in polymer electrolytes.

5.3.2
Oligo Etheroxy Substituted Poly(phosphazenes)

5.3.2.1
Methoxy Ethoxy Ethoxy Poly(phosphazene) (MEEP)

MEEP is an etheroxy side chain containing poly(phosphazene) and has been synthesized as follows [199-200]:

$$[\text{P(Cl)(Cl)}=N]_n + 2n\,\text{RONa} \xrightarrow[\text{THF}]{n\text{-Bu}_4\text{NBr}} [\text{P(OR)(OR)}=N]_n + 2\text{NaCl} \quad (12)$$

$$R = CH_2CH_2O\text{-}CH_2CH_2OCH_3$$

The sodium salt of 2-(2-methoxy ethoxy) ethanol was allowed to react with poly(dichlorophosphazene) to afford the fully substituted MEEP. However, the solubility of MEEP in water requires an extensive dialysis procedure for the separation of the pure polymer from sodium chloride.

MEEP is a completely amorphous polymer with a glass transition temperature of $-83.5\,°C$. MEEP forms salt complexes with a wide range of metal salts. The glass transition temperatures of the polymer-salt complexes increase with increasing salt concentration. The conductivity values first increase with an increase of salt concentration and then after attaining an optimum value begin to drop. Some data are summarized in Table 10. A typical conducivity plot of MEEP-metal salt complexes is shown in Fig. 10. The conductivity of MEEP $(\text{LiCF}_3\text{SO}_3)_{0.25}$ is as high as 10^{-5} S cm^{-1} at room temperature. This is at least three orders of magnitude higher than that observed for the analogous PEO-metal salt complexes. It was reasoned by Shriver and coworkers that the high conductivity is largely due to the highly flexible backbone of the polymer leading to a large segmental motion [200]. The high reorientational mobility of the backbone and the side groups as reflected by the glass transition temperature of MEEP allows for the observed conductivity. It has been suggested that the mode of transport of ions obeys the VTF equation (vide supra). Blonsky and coworkers attempted to identify the nature of the carrier species and the transport numbers of the ions in the MEEP-metal salt complexes [201]. Since virgin MEEP is an insulator it was assumed that in the MEEP-metal salt complexes the observed conductivity has no electronic component. Using the potentiostatic polarization method the transference number

Table 10. Conductivity values of some polymer electrolytes derived from MEEP and modified MEEP

Polymer	Metal Salt	O:Li ratio	Conductivity S cm^{-1}	Ref.
1. [NP(OCH$_2$CH$_2$OCH$_2$CH$_2$OCH$_3$)$_2$]$_n$, MEEP	—	—	8.1·10^{-8} (303); 2.1·10^{-7} (363)	200
	AgCF$_3$SO$_3$	48:1	2.6·10^{-4} (303); 1.4·10^{-3} (363)	200
	LiCF$_3$SO$_3$	24:1	2.7·10^{-5} (303); 2.2·10^{-4} (363)	200
	LiBF$_4$	36:1	5.2·10^{-5} (303); 8.7·10^{-4} (363)	213
	LiBr	36:1	8.4·10^{-6} (303)	213
	LiNO$_3$	36:1	1.1·10^{-5} (303)	213
	LiSCN	36:1	9.1·10^{-6} (303)	213
	NaCF$_3$SO$_3$	21:1[a]	6.1·10^{-5} (303)	213
	I$_2$	2.5:1	<10^{-3} (303)	203
	LiCF$_3$SO$_3$	16:1	6.0·10^{-5} (303); 6.0·10^{-5} (363)	213
2. [NP(OCH$_2$CH$_2$OCH$_3$)$_2$]$_n$				
3. [NP{(OCH$_2$CH$_2$)$_x$-OCH$_3$}]$_n$				
x = 7, ME7P	LiCF$_3$SO$_3$	95:1	2.4·10^{-5} (303); 1.1·10^{-4} (363)	213,24
x = 12, ME12P	LiCF$_3$SO$_3$	24:1	1.4·10^{-5} (303); 1.5·10^{-4} (354)	213
x = 17, ME17P	LiCF$_3$SO$_3$	222:1	1.2·10^{-6} (305); 6.2·10^{-5} (343)	213,24
4. MEEP-PEG[b] crosslinked	LiCF$_3$SO$_3$	6.4[c]	4.1·10^{-5} (303); 1.7·10^{-4} (343)	204
5. MEEP-PEG[d] crosslinked	LiCF$_3$SO$_3$	8.9[c]	3.0·10^{-5} (303); 1.0·10^{-4} (343)	204
6. MEEP[e] crosslinked	LiCF$_3$SO$_3$	24:1	7.0·10^{-5} (295); 7.0·10^{-4} (373)	205
7. MEEP-PEO Blend	LiBF$_4$	114:0.033	4.0·10^{-6} (298)	206
	LiBF$_4$	20.83:0.18	1.3·10^{-6} (298)	206
	LiClO$_4$	53.57:0.07	1.2·10^{-5} (298)	206
	LiCF$_3$SO$_3$	28.85:0.13	<10^{-6} (298); <10^{-4} (373)	206
	LiAlCl$_4$	28.85:0.13	~10^{-7} (298)	206
	LiAsF$_6$	28.85:0.13	<10^{-6} (298) <10^{-4} (373)	206
	LiN(CF$_3$SO$_2$)$_2$	28.85:0.13	6.7·10^{-3} (298)	207

[a] O:Na ratio [b] 1 mole% PEG [c] wt% of the salt [d] 10 mole % PEG [e] crosslinking by γ-radiation.

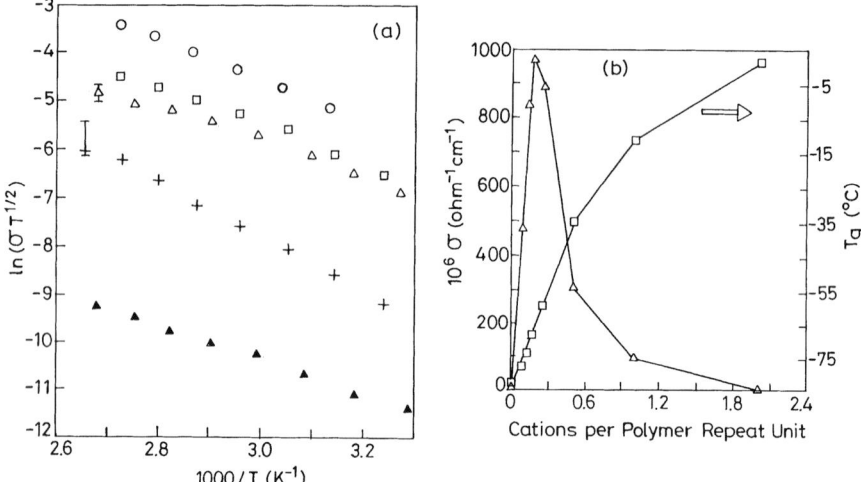

Fig. 10. (a) Conductivity plots (ln (σ $T^{1/2}$) vs 1000/T) for MEEP-metal salt complex. ▲, pure MEEP; +, MEEP . [Sr(SO$_3$CF$_3$)$_{0.25}$]; △, MEEP . [NaSO$_3$CF$_3$]$_{0.25}$; □, MEEP . [LiSO$_3$CF$_3$]; 0.25 O, MEEP . [Ag SO$_3$CF$_3$]
(b) Left axis: △, electrical conductivity at 70 °C composition of MEEP. (AgSO$_3$CF$_3$) complexes. Right axis : , T$_g$ (extrapolated to 0 °C/min heating rate) vs composition of MEEP.(AgSO$_3$CF$_3$)$_x$ complexes (taken from Ref. 200)

for Li ions was found to be in the range of 0.34 to 0.42 for MEEP-LiCF$_3$SO$_3$ complexes. However, Blonsky and coworkers conclude that the potentiostatic polarization technique does not provide reliable and straightforward results. The overall picture was that both the cations and anions contribute to ionic conductivity [199-203].

5.3.2.2
Modified MEEP Systems

Although MEEP-metal salt complexes show high ionic conductivities, these materials have undesirable creep properties at ambient temperatures, thus precluding their use in practical devices. To overcome this problem, one approach has been to synthesize dimensionally stable poly(phosphazenes) by crosslinking poly-[(chloro)(methoxy ethoxy ethoxy) phosphazene] with poly(ethylene glycol). If some of the chlorines in poly(dichlorophosphazene) are left unsubstituted after the initial reaction with the sodium salt of methoxy ethoxy ethanol, these can be reacted further with the difunctional poly(ethylene glycol) to afford crosslinked materials (Eq. 13 and Scheme 14) [204]

$$[\text{Cl}_2\text{P}=\text{N}]_n \xrightarrow{\text{RONa}} [(\text{RO})_2\text{P}=\text{N})_x (\text{(OR)(Cl)P}=\text{N})_y]_n \xrightarrow{\text{PEG}} \text{"crosslinked MEEP"} \quad (13)$$

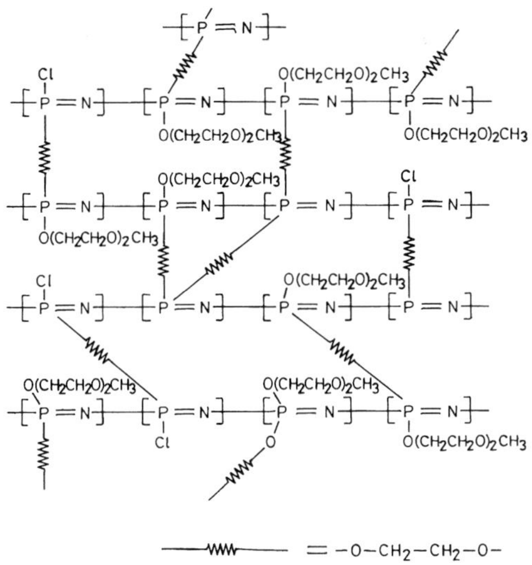

Scheme 14

A marked increase in the stability of the crosslinked materials was noted and there was no flow of the polymer even at 140 °C. The overall conductivity of the (PEG crosslinked MEEP) -LiCF$_3$SO$_3$ complexes compares favorably with parent MEEP (Table 10).

Alternative methods of increasing the mechanical properties of MEEP have included crosslinking by irradiation of pure MEEP or MEEP-(LiX)$_{0.25}$ complexes with ^{60}Co γ-rays and by the use of porous fiber glass support matrices [205]. Also, physical blends of MEEP with various polymers such as poly(ethylene oxide), poly(propylene oxide), poly(ethylene glycol diacrylate) and poly(vinyl pyrrolidine) have been made [206-208]. These blends were shown to possess better mechanical properties than pure MEEP, while showing comparable conductivities (Table 10). Thus, for example, a mixed (MEEP)(PEO)$_5$ – (LiBF$_4$)$_{0.033}$ complex shows a conductivity of 4.0x10^{-6} S cm^{-1} at room temperature. Comparable magnitude of ionic conductivity is shown by PEO-(LiBF$_4$)$_{0.125}$ only at 57 °C.

Similarly, (MEEP)(PEO)$_5$-[LiN(CF$_3$SO$_2$)$_2$]$_{0.013}$ showed a high conductivity of 6.7x10^{-5} S cm^{-1} at 25 °C (Table 10). Another modification of MEEP involved preparing a metal salt complex with LiAlCl$_4$. The MEEP-LiAlCl$_4$ system itself is mechanically stable presumably owing to the AlCl$_4^-$ anions serving as crosslinks between the polymer chains involving the P=N bonds [206-208].

DSC studies on the MEEP-PEO composite system clearly suggests a multiphase character in the composite with both amorphous MEEP-like and crystalline PEO-like phases being present. ^{7}Li-NMR studies based on T$_1$ measurements and line widths have been helpful in distinguishing lithium ions present in the crystalline and amorphous phases. Thus Li$^+$ ion in a crystalline phase is associated with a longer T$_1$ than that in an amorphous phase [207]. Further, it is expected that enhanced ionic mobility (Li$^+$) in polymer electrolytes leads to a line narrowing in the ^{7}Li NMR spectrum of the corresponding ion. Arrhenius plots of ^{7}Li NMR

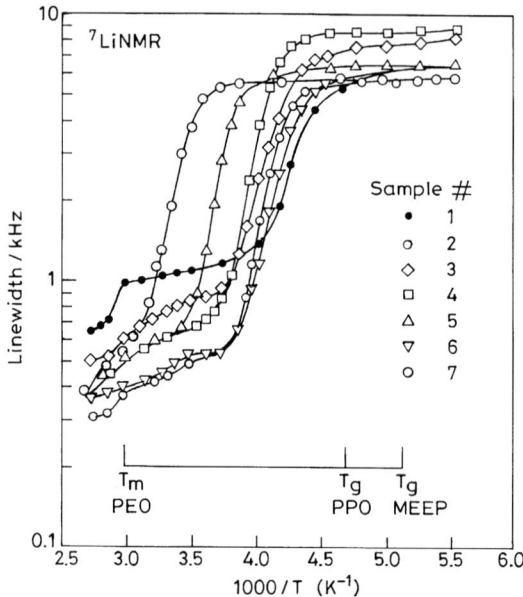

Fig. 11. Arrhenius plots of ^{7}Li-NMR linewidths : Sample *1*, MEEP : PEO : LiClO$_4$, 46:38:16; Sample *2*, MEEP:PEO:LiClO$_4$, 60:26:14; Sample *3*, MEEP:PEO:LiBF$_4$, 47:38:15, Sample *4*, MEEP:PEO:LiBF$_4$, 62:26:12; Sample *5*, MEEP:PPO:LiClO$_4$, 48:39:13; Sample *6*, MEEP:LiClO$_4$, 92.8; Sample *7*, PPO:LiClO$_4$, 81:19 (taken from Ref. 209)

linewidths (full widths at half maximum, fwhm) for several MEEP-PEO composites (Fig. 11) clearly shows that the onset of line narrowing corresponding to onset of ionic mobility correlates with the T_g of the material.

Attempts have been made to use MEEP-PEO-LiBF$_4$ polymers as polymer electrolytes in Li/TiS$_2$ batteries. It was shown that the polymer electrolyte was stable over a 200 cycle period, suggesting the potential uses of these materials [208].

5.3.2.3
Surfactant Substituted Polyphosphazenes

The synthesis of MEEP involves the reaction of poly(dichlorophosphazene) with the sodium salt of methoxy ethoxy ethanol. The byproduct in this reaction is sodium chloride which has to be separated from the polymer completely, since even traces of the ionic impurities would lead to spurious results. However, unfortunately MEEP is also soluble in water and therefore separation from sodium chloride is rendered extremely difficult. A cumbersome and lengthy dialysis procedure is required to effect the separation and purification of the polymer. Further MEEP is also hydrophilic and residual water in the polymer is an undesirable feature for a solid electrolyte particularly when involved with alkali metal salt complexes. Additionally the dimensional stability of MEEP is poor and has been commented upon above.

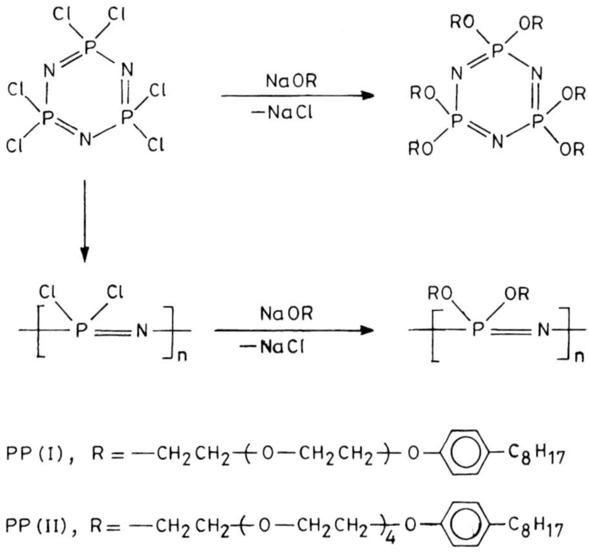

Polymer Solid Electrolytes: Synthesis and Structure

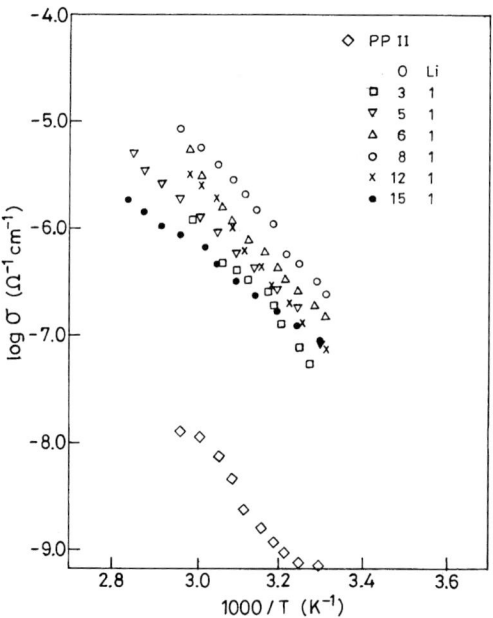

PP (I), R = —CH$_2$CH$_2$—(O—CH$_2$CH$_2$)—O—⟨◯⟩—C$_8$H$_{17}$

PP (II), R = —CH$_2$CH$_2$—(O—CH$_2$CH$_2$)$_4$—O—⟨◯⟩—C$_8$H$_{17}$

Scheme 15

Fig. 12. The conductivity plot (log σ vs 10^3/T) of PP(II)-LiBF$_4$ complexes of different O:Li ratios (taken from Ref. 211)

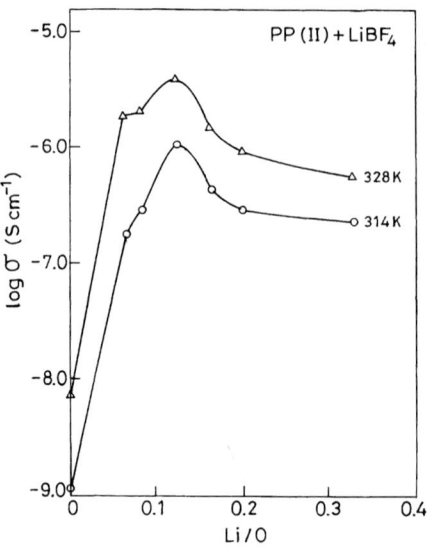

Fig. 13. Plot of log σ vs Li/O ratios for PP(II)-LiBF$_4$ complexes at 314 and 328 K (taken from Ref. 211)

Table 11. Conductivity values of polymer electrolytes derived from surfactant substituted poly(phosphazenes) PP(I) and PP(II)

	PP(I)-LiClO$_4$			Ref.
Composition O:Li	σ (S cm^{-1}) 310 K	σ (S cm^{-1}) 333 K	σ (S cm^{-1}) 348 K	
6:1	4.0·10^{-8}	4.4·10^{-7}	1.0·10^{-6}	212
12:1	2.4·10^{-8}	8.7·10^{-8}	1.3·10^{-7}	212
18:1	1.9·10^{-8}	2.6·10^{-8}	4.2·10^{-8}	212

	PP(II)-LiBF		
	σ (S cm^{-1}) 303 K	σ (S cm^{-1}) 333 K	
5:1	7.6·10^{-7}	1.3·10^{-6}	211, 212
8:1	2.3·10^{-7}	5.8·10^{-6}	211, 212
12:1	6.9·10^{-8}	2.5·10^{-6}	211, 212

To obviate these difficulties, poly(phosphazenes) containing surfactant side groups have been synthesized (Scheme 15) [210–212]. These surfactant substituted poly(phosphazenes), PP(I) and PP(II) show remarkable solubility properties. Unlike most polymers they are soluble in a wide range of organic solvents including hydrocarbons such as hexane or heptane. Also PP(I) and PP(II) are com-

Table 12. Activation energies obtained for PP(II) – LiBF$_4$ from fitting the Arrhenius equation (taken from Ref. 211)

O:Li	Activation Energy (eV)	A
3:1	0.47	0.800
5:1	0.65	0.674
6:1	0.84	0.666
8:1	0.69	0.600
12:1	0.63	0.500
15:1	0.49	0.444

pletely hydrophobic and insoluble in water. Thus separation of sodium chloride from the reaction mixture is easily accomplished. PP(I) and PP(II) show upfield signals in the ^{31}P-NMR (–7.4 and –7.2 ppm) characteristic of most alkoxy substituted poly(phosphazenes) and are completely amorphous as shown by DSC and XRD studies. The T$_g$'s of the virgin polymers are –13.2 and –28.8 °C respectively.

Both PP(I) and PP(II) form complexes with lithium salts. The conductivity plot of the lithium salt complex of PP(II) is shown in Fig. 12. The plot of log σ vs. Li/O ratios for PP(II)-LiBF$_4$ complexes is shown in Fig. 13. Table 11 summarizes the conductivity data.

These studies indicate that while the conductivities observed for the polymer metal salt complexes involving the short etheroxy containing PP-I are not very significant, those observed with PP-II are quite good. The highest conductivity observed for PP(II)-LiBF$_4$ system is for an O:Li ratio of 8:1 which shows a conductivity of 5.8×10^{-6} S cm^{-1} (333 K). Also the conductivity behavior in these polymer electrolytes is explained by an Arrhenius behavior. In this context it may be noted that the conductivity behavior of MEEP-alkali metal salt complexes fits both Arrhenius and VTF equations. Table 12 summarizes the activation energy parameters for PP(II)-LiBF$_4$ complexes.

5.3.2.4
Other Etheroxy Side Chain Containing Poly(phosphazenes)

The effect of (side) chain length in a series of poly(etheroxy phosphazenes), NP(O(CH$_2$CH$_2$O)$_x$CH$_3$)$_2$]$_n$ (where x = 1, 2, 7, 12 and 17), on the conductivity of their complexes with lithium trifluoromethane sulfonate has been investigated [213-214]. The maximum conductivity at 30 °C (5.3×10^{-5} S cm^{-1}) was observed when x = 7. It was also noted that while the glass transition temperatures of the pure polymers increased with increasing chain length, the T$_g$'s of the polymer-metal salt complexes showing an optimum conductivity of 2×10^{-4} S cm^{-1} at 70 °C was nearly constant at 220 K. Some of this data have been summarized in Table 10.

5.3.2.5
Mixed Substituent Poly(phosphazenes)

Allcock and coworkers have recently reported the synthesis of mixed substituent containing poly(phosphazenes) (Scheme 16) [215]. In a given set of polymers one

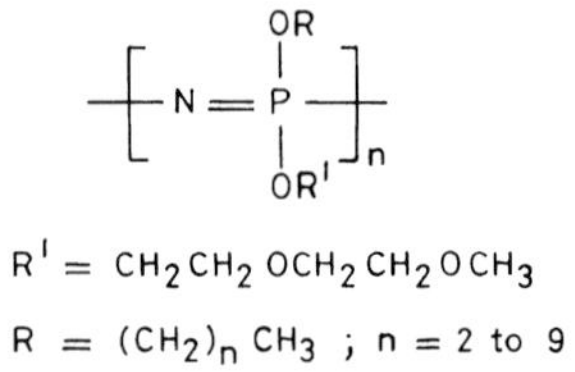

Scheme 16

$R' = CH_2CH_2OCH_2CH_2OCH_3$

$R = (CH_2)_n CH_3$; $n = 2$ to 9

Table 13. Thermal transition data for the mixed substituent poly(phosphazenes), $\{NP(OCH_2CH_2OCH_2CH_2OCH_3)_x(O(CH_2)_yCH_3)_{2-x}\}_n$ (taken from Ref. 215)

Polymer	T_g °C (mol % of lithium triflate added)				
2	−91(0);	−86(7.0);	−73(20.2);	−60(26.3);	−54(37.9)
3	−95(0);	−88(7.4);	−73(21.1);	−71(27.5);	−55(39.4)
4	−95(0);	−86(7.8);	−72(22.1);	−71(28.6);	−67(40.8)
5	−95(0);	−90(8.2);	−75(23.1);	−71(29.8);	−58(42.1)
6	−94(0);	−87(8.6);	−76(24.0);	−69(30.8);	−47(43.3)
7	−90(0);	−84(9.0);	−68(24.9);	−62(31.9);	−59(44.6)
8	−85(0);	−77(9.4);	−63(25.8);	−60(32.9);	−64(45.7)
T_m^1	−68 ;	−51 ;			
T_m^2	−60 ;				
9	–	–	−64(18.6);	−40(34.0);	−42(46.8)
T_m^1	−43 ;	−40(9.8);	−36 ;	– ;	
T_m^2	−37 ;	−35 ;	−29 ;	– ;	

Table 14. Maximum conductivity data for $\{NP(OCH_2CH_2OCH_2CH_2OCH_3)_x(O(CH_2)_yCH_3)_{2-x}\}_n$ at 25°C (taken from Ref. 215)

Polymer (y)	Max. conductivity S cm^{-1}	mol. % LiCF$_3$SO$_3$	Mole ratio Polymer:salt
2	1.3·10^{-5}	3.5	27.3:1
3	1.1·10^{-5}	14.4	5.9:1
4	9.1·10^{-6}	22.1	3.5:1
5	7.2·10^{-6}	15.9	5.3:1
6	5.4·10^{-6}	16.5	5.0:1
7	3.8·10^{-6}	17.3	4.8:1
8	2.1·10^{-6}	25.8	2.9:1
9	2.0·10^{-6}	18.6	4.4:1

of the substituents was kept constant and was the etheroxy group −OCH$_2$CH$_2$OCH$_2$CH$_2$OCH$_3$ and the other substiuent was an alkoxy substituent −OR. The length of R was varied from (CH$_2$)$_2$−CH$_3$ to (CH$_2$)$_9$CH$_3$. It was felt that increased chain lengths in the alkoxy group would increase the free volume of polymer and would assist ion mobility. It was observed that all polymers (salt free and salt complexed) were amorphous over a wide temperature range (−120 to +80 °C). Crystallinity occurred only when the chain length of R increased beyond nine carbons. In general, as with other systems discussed in this account so far, the T'_gs of the polymer salt complexes increased with an increase in the amount of dissolved salt in the system. Some of these trends are summarized in Table 13. Conductivity studies on the polymer-Li salt complexes revealed that initial addition of a lithium salt such as $LiSO_3CF_3$ leads to an increase in conductivity. At low salt concentrations these polymeric systems showed an excellent ability to solvate and coordinate to Li^+ ions. The maximum conductivity data are summarized in Table 14 which shows that the best conductivity is seen for R = $(CH_2)_2CH_3$: $\sigma = 1.3 \times 10^{-5}$ S cm^{-1} for a polymer/salt ratio of 27.3:1.

Addition of a higher salt reduced the conductivity for each system. This is believed to be due to at least three reasons. One, as the salt concentration is increased there are less sites available for coordination. Two, the polymers become less flexible owing perhaps to ionic cross linking which also raises the glass transition temperatures. Three, salt aggregates and ion pair formations also increase with increased salt concentration and these factors are also detrimental to ionic conductivity.

Another effect observed was that the maximum conductivities for each polymer system decreased with an increase in the length of the alkyl group side chain, R. Thus, although the free volume of the polymer perhaps is increased as the chain length in R increases, this also leads to the availability of fewer coordination sites per volume of polymer. Clearly this latter feature is quite important in fine tuning the ionic conductivities.

Another series of cosubstituent poly(phosphazenes) containing various side groups such as 2-(2-methoxy ethoxy) ethoxy, PMEG, poly(oxy ethylene (4))lauryl ether units have been investigated as hosts for lithium salts [216]. Conductivity studies carried out on the lithium salt complexes of these polymers also reveal that polymers with long chain alkyl containing side groups have poorer ionic conductivities than systems containing methyl end groups.

5.3.3
Polymers containing Pendant Oligo(Oxyethylene) Cyclotriphosphazenes

MEEP and other related poly(phosphazenes) studied (vide supra) as hosts for polymer salt complexes are flexible polymers, whose flexibility is attributed mainly to the torsional mobility of the P-N backbone. In contrast to these linear poly-(phosphazenes), there have also been studies on polymers which contained, cyclophosphazenes as pendant groups [217-219]. In these systems the backbone of the polymer is made out of an entirely carbon framework. Inoue and coworkers have reported the synthesis of styrene and substituted styrene polymers containing oligo oxyethylene substituted cyclophosphazenes (Scheme 17). These polymers known as poly (VBDEP), poly (VBTEP), poly(SDEP) and poly(STEP) are also amorphous polymers with glass transition temperatures around -60 °C. It is believed that the introduction of many short oxy ethylene chains into the phosphazene rings brings about the low T'_gs. All of these polymers form complexes with lithiums salts. As in other instances addition of lithium salt increases the T_g of the polymers. The conductivity studies on the lithium salt complexes reveal that high conductivities in the order of 1.8×10^{-5} S cm^{-1} are observed.

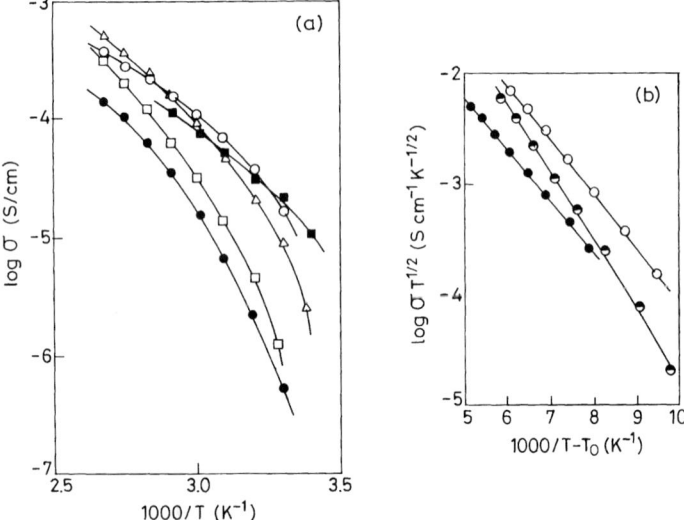

Scheme 17

Fig. 14. (a) Conductivity plot (log σ vs 1000/T) for poly(SDEP) - (■, ●) and poly STEP-LiClO$_4$ complexes (○, △, ◻). Li/O = 0.01 (●), 0.03 (■), 0.050 (○), 0.075 (△) and 0.10 (◻) and (b) VTF plots for Poly(STEP)-LiClO$_4$ complexes. Li/O=0.035 (●), 0.075 (◐), and 0.1 (○) (taken from Ref. 218).

Table 15 summarizes the ionic conductivity data. The temperature dependance of conductivity is explained by the VTF equation (Fig. 14).

In view of the high conductivities observed with these "rigid" backbone polymers, it is interpreted that flexibility of the backbone need not be a critical fac-

Table 15. Ionic conductivities of polymer electrolytes derived from cyclophosphazene pendant polymers

Polymer	Metal	O/Li ratio	Conductivity S cm^{-1} (K)	Ref.
1. Poly(VBDEP)	LiClO$_4$	30:1	7.6·10^{-6} (313), 6.5·10^{-5} (363)	217
	LiClO$_4$	15:1	1.2·10^{-4} (363)	
2. Poly(VBTEP)	LiClO$_4$	30:1	3.6·10^{-5} (313), 2.9·10^{-4} (363)	217
	LiClO$_4$	15:1	1.1·10^{-4} (333)	
3. Poly SDEP	LiClO$_4$	21:1	1.8·10^{-5} (303), 2.6·10^{-4} (373)	218
4. Poly STEP	LiClO$_4$	20:1	1.8·10^{-5} (303), 3.9·10^{-4} (373)	218

tor to achieve high ionic conductivity. It is believed that in these systems ion transport is mediated by the segmental motion of the side oxy ethylene chains without any assistance from the mobility of the main chain.

5.3.4
Ionic Poly(phosphazenes)

In order to increase the fraction of charge carried by cations or anions, Shriver and coworkers have synthesized a new class of poly(phosphazene) polyelectrolytes [220-222]. In these systems an anionic or cationic structural unit is covalently linked to the polyphosphazene leaving them immobilized. The counter ions, only, are "free" in principle for ionic mobility. Schemes 18 and 19 summarize the synthesis of these class of poly(phosphazenes). Thus the synthesis of the cation conductors CC-1 and CC-2 (Scheme 18) [220-221] involves the partial replacement of chlorines on poly(dichlorophosphazene) with the sodium salt of hydroxy ethane sulfonic acid. The remainder of the chlorines are replaced by the etheroxy side group. Several polymers with varying substituent ratios were prepared. In another modification the sodium ions were completely replaced by an ion exchange with magnesium ions [221]. The anionic conductors AC-I and AC-II were prepared by first an initial partial substitution of chlorines in poly(dichloro phosphazene) with the methoxy ethoxy ethoxy substituent, followed by reaction with the sodium salt of an amino alcohol. Quaternization of the nitrogen by a proton or an R group is easily accomplished [222]. These polymers have an immobile cation unit and a mobile halide anion.

Although these polymers are interesting their ionic conductivities at ambient temperatures were quite low (< 10^{-6} S cm^{-1}) presumably due to significant ion pair formation.

5.4
Poly(siloxanes)

In view of the low T_g values normally exhibited by poly(siloxanes), several modified forms of these polymers have been tried as hosts in polymer electrolytes [223-228]. The polymers investigated include block copolymers consisting of

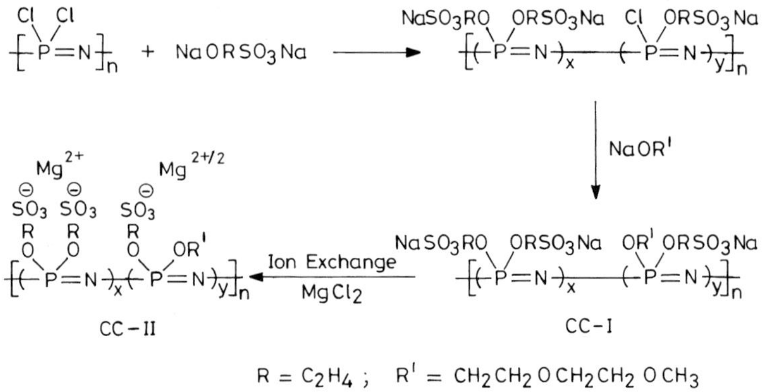

Scheme 18

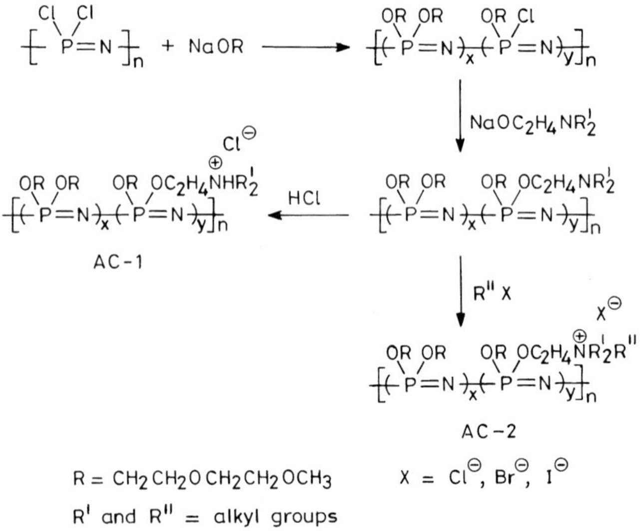

Scheme 19

dimethyl siloxane and ethylene oxide units [223-224], urethane crosslinked net works of poly(dimethyl siloxane-graft-ethylene oxide) polymers [225], and other polymers based on poly(methyl hydrosiloxane), poly(ethylene glycol) monomethyl ether and poly(ethylene glycol) [226-227]. More recently polyelectrolytes based on poly(siloxanes) have also been described [228]. The ionic conductivity data for polymer electrolytes derived from some of these polymers are summarized in Table 16.

Table 16. Conductivity data for some poly(siloxanes)

Polymer	Metal salt	Ionic Conductivity S cm^{-1} (K)	Ref.
1. Poly(dimethyl siloxane) [a] - Poly(ethylene oxide)-urethane Net work	LiClO$_4$	10^{-5} (303)	225
2. Crosslinked siloxane [b]	LiSO$_3$CF$_3$	7.3·10^{-5} (313)	226
3. Graft siloxane Poly electrolyte [c]	Na$^+$	10^{-7} (303) (1.7·10^{-5})	228

[a] (H$_3$C)$_3$SiO[(Si(CH$_3$)$_2$-O)$_x$-Si(CH$_3$)(PEO)-O]$_n$SiCH$_3$ crosslinked with hexamethylene diisocyanate. The conductivity data is for an O/Li ratio of 17.
[b] Crosslinked siloxane polymer prepared from the reaction of poly(methyl hydrosiloxane), poly(ethylene glycol)monomethyl ether and poly (ethylene glycol). The conductivity data is for a 15 wt% complex with LiSO$_3$CF$_3$
[c] Poly(methyl hydro siloxane) grafted with vinyl terminated poly (ethylene glycol) monomethyl ether and sodium salt of vinyl end group containing disubstituted phenol. Conductivity value in parenthesis refers to conductivity upon addition of crypt [2.2.2].

6
Structure of Polymer Electrolytes

Various methods have been employed to find out about the structure of polymer electrolytes. These include thermal methods such as differential scanning calorimetry (DSC), differential thermal analysis (DTA), X-ray methods such as X-ray diffraction and X-ray absorption fine structure (XAFS), solid state NMR methods particularly using ^{7}Li NMR, and vibrational spectroscopic methods such as infrared and Raman [27]. The objective of these various studies is to establish the structural identity of the polymer electrolyte at the macroscopic as well as the molecular levels. Thus the points of interest are the crystallinity or the amorphous nature of materials, the glass transition temperatures, the nature and extent of interaction between the added metal ion and the polymer, the formation of ion pairs etc. Ultimately the objective is to understand how the structure (macroscopic and molecular) of the polymer electrolyte is related to its behavior particularly in terms of ionic conductivity. Most of the studies have been carried out, quite understandably, on PEO-metal salt complexes. In comparison, there has been no attention on the structural aspects of the other polymers particularly at the molecular level.

As discussed in the sections on individual polymers, PEO is a semicrystalline polymer with about 60% of the bulk being crystalline at room temperature with the rest being present in an amorphous phase. In spite of the difficulties involved in obtaining phase diagrams with polymer electrolytes such as the slow kinetics of crystallization and randomness and therefore to determine exact boundaries between the various phases involved, phase diagram studies on PEO system suggests 3-4 repeat units of PEO per metal salt (vide supra) [53-56].

Other polymers such as MEEP, poly MEEMA and polysiloxanes are completely amorphous polymers and remain amorphous even in the polymer-metal salt complexes at least upto certain concentrations of the metal salt. The question of the amorphous nature of the polymer electrolytes is important in view of Berthier's demonstration in the PEO-polymer electrolyte system that the ionic conductivity occurs mostly in the amorphous phase [59]. Thus the room temperature ionic conductivities of the completely amorphous MEEP-LiX complexes is at least three orders of magnitude higher than the corresponding PEO-LiX complexes. This has led to the use of plasticizers and other modifications to suppress crystallinity and increase free volume as discussed above.

The question of the molecular level structure of polymer electrolytes and its relevance to understanding ionic conductivity has captured the attention of many researchers. This aspect has been investigated mainly by X-ray methods, NMR and vibrational spectroscopy. In the following a brief summary of these investigations is presented.

6.1
EXAFS of Polymer Electrolytes

Extended X-ray Absorption Fine Structure (EXAFS) and X-ray Absorption Near Edge Structure (XANES) have been used to probe the structure of polymer electrolytes. Unlike X-ray diffraction methods which yield information on crystalline materials EXAFS can also be used to obtain structural information on amorphous materials [229-230]. This is particularly relevant for polymer electrolytes since Berthier and coworkers have proved conclusively that ionic conductivity occurs predominantly in the amorphous phase of the polymer (PEO) [59]. The EXAFS experiment consists of irradiation of the sample with a high energy monochromatic x-radiation usually produced from a synchotran source where the energy of the incident radiation is varied from about 100 eV below to about 1000 eV above a characteristic core electron energy level of a chosen element within the sample. The EXAFS data can be interpreted to obtain information about the nearest neighbour atoms around the target element and the distances from the target element to these neighbours [207], i.e. the immediate local environment around the target element can be determined. However, due to the limitations of the method, while EXAFS can determine distances within a precision of 0.01 Å the number of the neighbours determined suffer from an imprecision of ~10%. A further limitation of EXAFS is that study of light elements is rendered difficult because these require a large intensity of soft X-rays. Because of these experimental difficulties most polymer electrolytes studied involve the heavy metal salts. Also, if more than one local structure is present in the polymer electrolyte the EXAFS technique usually produces an average structure. Nevertheless since the technique gives an idea about the local environment around the chosen metal ion, this information can be extrapolated in conjunction with other techniques to throw light on polymer-metal salt complex structures involving light atom metals such as Li^+. Table 17 summarizes the data obtained on a few PEO-metal salt complexes. Thus work on $PEO_n.RbI$ showed that each Rb was surrounded by 4 neighboring oxygens from various PEO chains [231-232]. From studies on

Table 17. Coordination around metal ions in polymer electrolytes: data from EXAFS (taken from Ref. 229)

A. Cation K-edge
Polymer electrolyte: $PEO_n \cdot MBr_2$ System

n	M	CA[a]	CN[b]	M-CA[c]
4	Co	O	3.8	2.12
		Br	2.2	2.38
8	Co	O	4.0	2.16
		Br	3.0	2.39
30	Co	O	1.9	2.06
		Br	1.4	2.39
8	Ni	O	5.8	2.12
		Br	4.5	2.40
100	Ni	O	5.7	2.03
		Br	---	----
8	Cu	O	1.1	2.02
		Br	3.2	2.36
8	Zn	O	2.2	2.03
		Br	2.4	2.34
4	Ca	O	7.3	2.40
		Br	1.7	2.83
		Ca	1.9	3.38
12	Ca	O	9.1	2.38
		Br	1.0	2.83
		Ca	2.8	3.36

B. Zn K Edge
Polymer Electrolyte $PEO_8 \cdot ZnX_2$

	CA	CN	M-CA
X=I	O	3.6	2.11
	I	1.0	2.51
X=Br	O	2.2	2.03
	Br	2.4	2.34
X=Cl	O	1.4	2.05
	Cl	0.9	2.20

[a] Coordination atoms
[b] Coordination number around the metal ion
[c] Distance in Å

PEO-divalent metal salt complexes it is found that the M–O distance is nearly invariant at about 2.1 Å for Co, Ni, Cu and Zn, but is slightly larger for Ca (2.4 Å) [229, 233–236]. Another general trend that is seen is that for high ratios of PEO/metal salt large coordination numbers for the metal ion are seen. Thus Co^{2+} has a coordination number of six (4 oxygens and 2 Br^-) in the $PEO_4 \cdot CoBr_2$ complex, whereas in the analogous $PEO_{15} \cdot CoBr_2$ the coordination reduces to ~3. It can be said that increase in tight coordination by neighbours will lead to low ionic conductivity. Thus, the PEO$_n$.CaBr$_2$ complexes which shows high coordination numbers has a low ionic conductivity. It may be pointed out that even for most PEO-LiX complexes high ionic conductivities are observed only at high O/Li ratios. In the complexes PEO$_n$.ZnX$_2$, when X = I$^-$ or Br$^-$ the coordination to Zn^{2+} is through four oxygens while when X = Cl$^-$ the coordination number is reduced, showing the effect of the anion on the coordination. Thus EXAFS can provide useful information on the local structure around the metal ion even in the amorphous region. It would have been more useful if it can be applied to PEO-alkali metal salts directly.

6.2
X-Ray Structures of PEO-metal Salt Complexes

Early attempts at determining the X-ray structures of some PEO-metal salt complexes have been hampered because of the difficulty of determining these structures by single crystal methods. However, these efforts have results in the elucidation of X-ray structures, albeit of poor resolution, of PEO$_4$.HgCl$_2$ [237] and PEO.HgCl$_2$ [238] and later of PEO-NaI [239] PEO$_3$.NaSCN and PEO.NaSCN [240] complexes. More recently better quality X-ray structures of PEO$_3$.NaClO$_4$ [241], PEO$_3$.LiCF$_3$SO$_3$ [242], PEO$_4$.KSCN and PEO$_4$.NH$_4$SCN [243] have become available from high resolution powder X-ray diffraction data. Table 18 summarizes the crystallographic parameters for some of the PEO-metal salt complexes. As can be seen from the table all these complexes belong to the monoclinic space group.

Contrary to the expectations of Armand [34] effective anion-cation separation is *not* achieved in any of the PEO-metal salt complexes whose X-ray structures have been solved. This includes structures with bulky anions such as I$^-$ as in PEO-NaI [239] or structures with bulky and "non-coordinating" anions such as CF$_3$SO$_3^-$ as in PEO$_3$-LiCF$_3$SO$_3$ [242] or ClO$_4^-$ as in PEO$_3$-NaClO$_4$ [241]. In the PEO-NaI structure infact a tight zig-zag chain of the cation and the anion is seen with a coordination number of two for each ionic species. This chain is wrapped

Table 18. Crystallographic parameters for some PEO-metal salt complexes

PEO-Metal salt	Space group	Cell dimensions (a,b,c in Å, β in °)		Ref.
1. (PEO)$_3$·LiCF$_3$SO$_3$	P 2$_1$/a	a = 16.768(2); c = 10.070(1);	b = 8.613(1); β = 121.02(1)	242
2. (PEO)$_4$·NH$_4$SCN	C 2/c	a = 25.512(3); c = 16.097(1);	b = 8.0813(1); β = 125.98(1)	243
3. (PEO)$_4$·KSCN	C 2/c	a = 25.663(2); c = 15.801(1);	b = 8.231(7); β = 125.26(1)	243
4. (PEO)$_3$·NaI	P 2$_1$/a	a = 18.15; c = 8.41;	b = 7.98; β = 122.3	239
5. (PEO)$_3$·NaSCN	P 2$_1$/a	a = 16.83; c = 10.64;	b = 7.19; β = 125.5	240
6. (PEO)·NaSCN	P 2$_1$/c	a = 7.55; c = 5.83;	b = 12.10; β = 97.5°	240

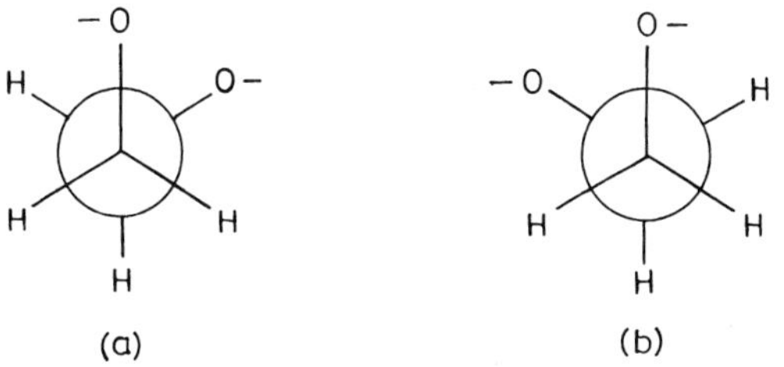

Polymer Solid Electrolytes: Synthesis and Structure

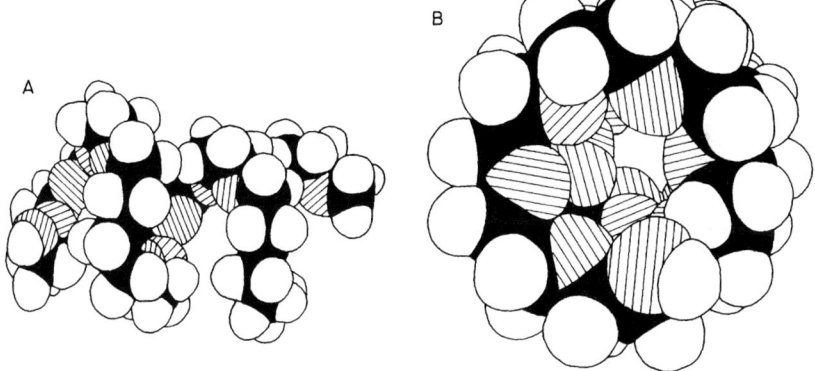

Scheme 20

Fig. 15. Molecular Models of poly(ethylene oxide) in a $T_2GT_2\bar{G}$ conformation as proposed for complexation to sodium and lithium cations, (A) side view, (B) end view. Hydrogen atoms are not shaded, carbon atoms are black and oxygen atoms are *crosshatched* (taken from Ref. 67).

around by the helical PEO structure [239]. Infact in all the PEO-metal salt structures the conformation of the PEO chains can be described as helical as was found for PEO itself [244]. Although the fine structural details of the PEO helix varies from structure to structure, in all of them the conformation was found to be *trans* (T) (CC-OC), *trans* (T) (CO-CC) and *gauche* (G) or minus *gauche* (G), The *gauche* (G) and minus *gauche* (G) conformations for the O-CH_2-CH_2-O group are shown in Scheme 20. The oxygen atoms are located in the inner surface of the tunnel cavity so as to provide appropriate coordination environment for the cation *inside* the PEO helix. Figure 15 shows a molecular model of a PEO in a $T_2G\,T_2G$ conformation. Because of this invariably the metal ion is present *inside* the PEO helix and not outside, at least in all the X-ray structures described. Furthermore, in all cases there is negligible interchain interactions between PEO chains nor is there the presence of any "ionic-crosslinking". The minimum coordination number is 5 seen for Li^+ in the complex $(PEO)_3.LiCF_3SO_3$ and the maximum 7 for the $PEO_4.MSCN$ (M = K or NH_4^+) complexes. In all cases, the anion also participates in coordination to the metal ion. Some of these structural features are summarized in Table 19.

Table 19. Coordination environment around the cation and polymer conformation in PEO-metal salt complexes as determined from X-ray diffraction studies

PEO-Metal salt complex	CN[a]	CA[b]	Polymer[c] conformation	Ref.
1. PEO·NaI	5	3 oxygens from PEO; 2I$^-$	(TTGTTGTT$\bar{\text{G}}$)	239
2. (PEO)$_3$·NaSCN	6	4 oxygens from PEO; 2 nitrogens from SCN$^-$	(TTGTTGTT$\bar{\text{G}}$)	240
3. PEO·NaSCN	6	2 oxygens from PEO; 2 nitrogens from SCN$^-$; 2 sulfurs from SCN$^-$	(TGGT$\bar{\text{G}}\bar{\text{G}}$)	240
4. (PEO)$_3$·LiCF$_3$SO$_3$	5	3 oxygens from PEO; 2 oxygens from CF$_3$SO$_3$	TTGTTGTT$\bar{\text{G}}$	242
5. (PEO)$_4$·MSCN[d]	7	5 oxygens from PEO; 2 nitrogens from SCN$^-$	TTGTTGTT$\bar{\text{G}}$TTG	243

[a] Coordination number around the cation
[b] The coordinating atoms around the cation
[c] C-O bonds trans (T); C-C bonds gauche (G) or gauche minus ($\bar{\text{G}}$)
[d] M = K or NH$_4^+$.

The X-ray structures of (PEO)$_3$LiCF$_3$SO$_3$ (Figs. 16 and 17) and (PEO)$_4$.MSCN (M = K$^+$ or NH$_4^+$) have been determined to a very good degree of accuracy and consequently allow a closer inspection of the bond parameters around the cation. These data are summarized in Tables 20 and 21 respectively. In the (PEO)$_3$.LiCF$_3$SO$_3$ complex the lithium ion is coordinated to five oxygens in an approximate trigonal bipyramidal arrangement. The two axial positions and one equatorial position are taken by oxygens from the PEO chain, while the two other equatorial positions are taken by oxygens derived from the anion, CF$_3$SO$_3^-$. In fact, each anion uses two oxygens to bridge adjacent lithium ions, reminiscent of many bridging ligands such as the carboxylate moiety. The bond lengths show that as expected the axial Li-O bond lengths (derived from PEO) are longer than the much shorter equatorial Li-O bond length (Table 20). In fact, the inequality of all the Li-O bond lengths reflects the non symmetric coordination environment present. Another point of interest is the fact that replacement of the CF$_3$ substituent by other bulkier asymmetric substituents such as found in the plasticizing salts LiN(SO$_2$CF$_3$)$_2$ and LiC(SO$_2$CF$_3$)$_3$ can lead to an inhibition of crystallization. This feature partly explains the origin of the plasticizing effect of these salts.

Polymer Solid Electrolytes: Synthesis and Structure

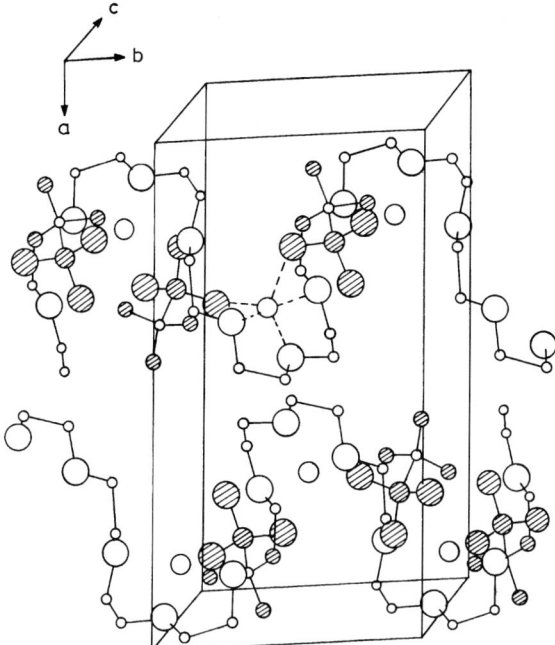

Fig. 16. View of the X-ray structure of (PEO)$_3$·LiCF$_3$SO$_3$ along c axis. CF$_3$SO$_3$-groups are shaded. Coordination around one Li$^+$ is shown in *dashed lines*.

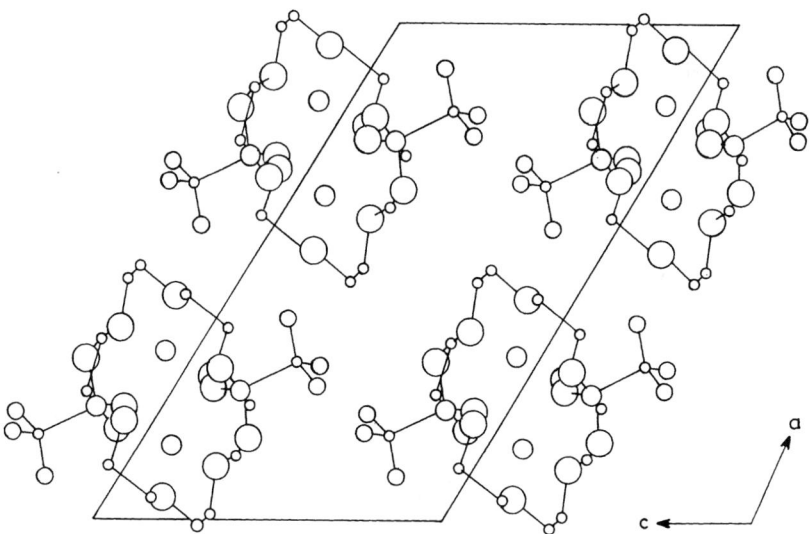

Fig. 17. View of the X-ray structure of (PEO)$_3$·LiCF$_3$SO$_3$ along the chain axis b (taken from Ref. 242).

Table 20. Bond length and angle data around Li in the X-ray structure of $PEO_3 \cdot LiCF_3SO_3$ (taken from Ref. 242)

Bond distances (Å)	Bond Angles (°)
Li - O(1) : 2.38(9)	O(1) - Li - O(2) : 174(5)
Li - O(2) : 2.01(8)	O(3) - Li - O(4) : 134(5)
Li - O(3) : 1.72(7)	O(3) - Li - O(5) : 107(4)
Li - O(4) : 2.21(8)	O(4) - Li - O(5) : 112(4)
Li - O(5) : 2.14(8)	

Table 21. Bond distance data around the cation in the structure $PEO_4 \cdot MSCN$ (M=K or NH_4) (taken from Ref. 243)

$(PEO)_4 \cdot KSCN$	$PEO_4 \cdot NH_4 SCN$[a]
K - O(1) : 2.79(2)	N(1) - O(1) : 3.11(2)
K - O(2) : 2.63(2)	N(1) - O(2) : 2.88(2)
K - O(3) : 2.83(3)	N(1) - O(3) : 3.04(2)
K - O(4) : 2.96(2)	N(1) - O(4) : 3.16(2)
K - O(4') : 3.09(2)	N(1) - O(4'): 2.91(2)
K - N : 3.05(2)	N(1) - N : 2.96(2)
K - N' : 2.74(2)	N(1) - N' : 2.68(2)

[a] N(1) is the nitrogen of the ammonium ion.

The thiocyanide metal salt complexes, $(PEO)_4.KSCN$ and $(PEO)_4.NH_4SCN$ are isostructural [243]. The coordination around the cation is seven (Table 21). In these complexes the PEO repeat distance along the axis of helix is shortened leading to eight ethylene oxy units within a shorter fiber axis compared to six ethylene oxy units found for $(PEO)_3.LiCF_3SO_3$. This probably accounts for the availability of five ether oxygens for coordination to the metal ion. Two other coordination positions are taken by the nitrogens of the anion. In fact, one nitrogen and two ether oxygens are involved in simultaneous coordination to two cations and act as bridging units. Contrary to an earlier structural report, the X-ray structure reported by Lightfoot and coworkers conclusively has shown that the cation is present *inside* the PEO helix, putting rest to the belief that cations larger than Na^+ cannot be accommodated *within* the PEO helix and therefore have to reside *outside* [245].

It is of course likely that complexation of lithium (or other metal) ions in the amorphous region of PEO is outside the helix and therefore random. However, this assumption does not have concrete structural proof although Shriver and coworkers have presented evidence by vibrational spectroscopic methods that in $(PEO)_n.RbX$ complexes probably Rb^+ resides outside the helix [246].

6.3
Infrared and Raman Spectroscopic Studies

Infrared and Raman spectroscopic studies have been extensively carried out on PEO-metal salt complexes and have aided our understanding of the polymer structure, interaction of the ions with the polymer, as well as ion-ion interactions [67, 246].

Based on the vibrational spectroscopic study on the *gauche* and *trans* forms of ethylene dichloride and its application to PEO by Davidson [247], Shriver and coworkers were able coassign a gauche configuration for the $O-CH_2CH_2-O$ group in PEO-alkali metal salt complexes [246]. This was arrived at by the assignment of bands in the 800–1000 cm^{-1} region. There was no evidence even upto 120 °C, for a *trans* conformer. Further, evidence for cation-ether oxygen coordination was obtained by the broadening of the C-O-C stretch (1150 cm^{-1}) observed in PEO and its movement to lower frequencies upon complexation [Fig.18]. Also, a Raman vibrational band at 865 cm^{-1} for the PEO-NaSCN complex (and related compounds) in view of its relationship with Raman spectra for a number of crown-ether complexes [248] was taken as evidence for a metal-oxygen stretch and it was suggested that the PEO helix wraps around the metal ion. Although Shriver and coworkers discounted the possibility of strong ion pair formations in the complexes PEO-MX (M = Na, X = SCN, I; M = Na, X = CF_3SO_3), later X-ray evidence [239–243] does show that the anions are invariably involved in coordination to the cation. However, work done by the Evanston group does show strong ion pair formation in $PEO-LiNO_3$, [67] $PEO.NaBH_4$ and $PEO.NaBF_4$ complexes [65].

Raman scattering spectroscopic methods have been used in combination with deconvolution techniques to analyze dissociation and association of salts dissolved in PEO and PEO related polymers [249–256]. Thus the splitting of the sym-

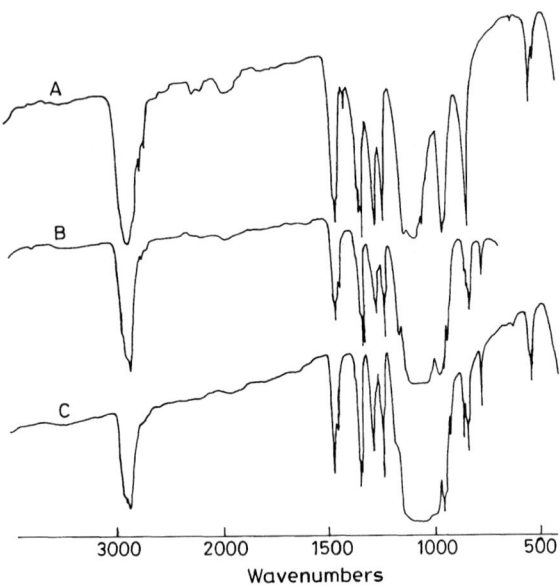

Fig. 18. Infrared spectra for (A) pure PEO, 600,000 (B) PEO.LiBF$_4$ and (C) PEO.NaBF$_4$ complexes (taken from Ref. 67).

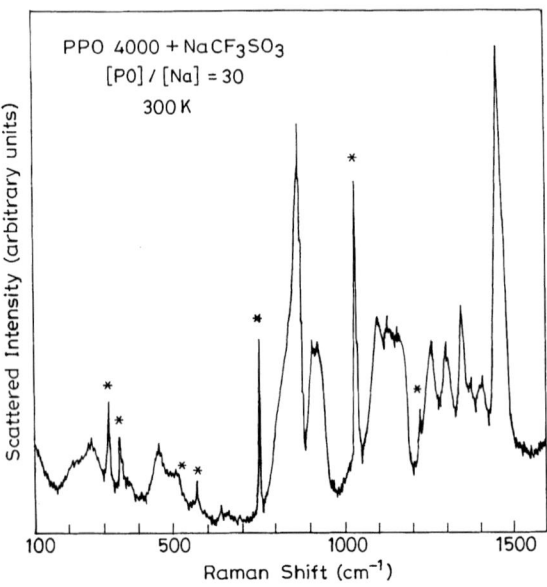

Fig. 19. Raman spectrum for PPO(4000)-NaCF$_3$SO$_3$. The internal vibrational modes of CF$_3$SO$_3$ are marked with an asterisk (taken from Ref. 252).

metric stretching mode of the $CF_3SO_3^-$ anion into a double band in the PPO-$NaCF_3SO_3$ complex (O/Na,30:1) was taken as evidence for the presence of free and paired ions [252]. The Raman spectrum of this complex is shown in Figure 19. Figure 20 shows the SO_3 symmetric stretching mode at different temperatures. An interesting finding was that although the amount of dissociated free ions decreased rapidly above the T_g of the polymer (i.e. ion pair formation was higher at higher temperatures) this does not lead to a decrease in conductivity, but rather to an increase. This underscores the probability that the temperature dependance of ionic conductivity (higher conductivities at higher temperatures) is due to an increased mobility of the charge carriers rather than an increase in the number of charge carriers [242]. Similarly a Raman scattering study on a poly(propylene

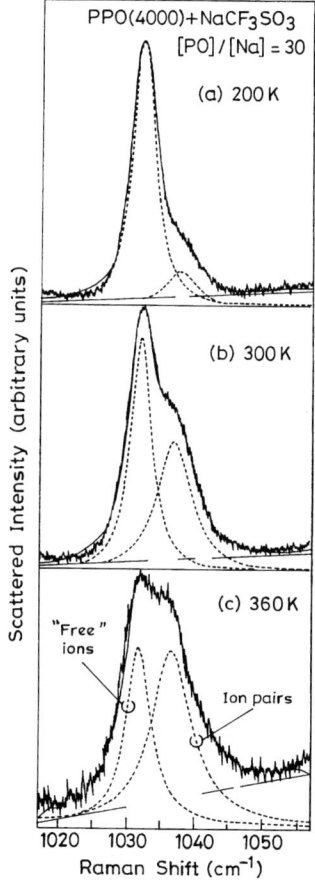

Fig. 20. The SO_3 symmetric stretching mode for PPO(4000).$NaCF_3SO_3$ complex at different temperatures.
Dashed lines correspond to the two components of the Lorentzian envelope fitted to the spectrum (smoothed solid line). Linear solid lines represent fluorescence scattering and are taken as ground levels (taken from Ref. 252).

glycol) (PPG)-NaCF$_3$SO$_3$ system revealed that the number of ion pairs increases with increasing molecular weight of the polymer as well as with increasing temperature [255]. Independent studies using Near Edge X-ray Absorption Fine Structure (NEXAFS) spectroscopy on the effect of temperature on ion pairing of a potassium salt (di potassium -NN' Jeffamine dipropane sulfonate) JDPS, in a PEO-K salt complex containing MC-3 as a plasticizer also showed that increasing ion-pair formation occurs with increased temperatures [257].

A recent Raman and infrared vibrational spectroscopic study of ethylene carbonate containing solutions of lithium perchlorate indicated that a strong interaction is present between the Li$^+$ ions and all the three oxygens of the plasticizer including the carbonyl group [258]. This result confirms the role of EC and PC as plasticizing solvents and has implications in understanding the conductivities of polymer gel electrolytes.

6.4
NMR of Polymer Electrolytes

Solid-state NMR spectroscopy has been used to study polymer electrolytes [27, 259-266]. Among the various nuclei that have been used as probes ^{7}Li and ^{23}Na have received maximum attention. However, the quadrupolar nature of these nuclei compounded with the solid state of the sample result in considerable line broadening of the NMR signal. Thus, a lot of valuable information pertaining to the structure of polymer electrolytes is not accessible. However, an analysis of the linewidths and the spin lattice relaxation times of the nuclei affords considerable information on the nature of the ions present in a polymer electrolyte.

Although both lithium isotopes are NMR active (^{6}Li and ^{7}Li) most studies have been centered on ^{7}Li owing to its higher receptivity [267] (Table 22). ^{7}Li-NMR studies based on T$_1$ measurements and linewidths have been helpful in distinguishing between lithium ions present in a crystalline or an amorphous phase [59]. Thus a Li$^+$ ion in a crystalline phase is associated with a longer T$_1$ than that in an amorphous phase. Secondly it is expected that enhanced ionic mobility (Li$^+$) in polymer electrolytes would lead to a line narrowing in the ^{7}Li NMR spectrum. Based on the measurement of T$_1$'s and the onset of line narrowing of ^{7}Li signal with reference to temperature and its correlation to glass transition temperature (T$_g$) of the polymer electrolyte, it has been possible to establish that ionic conduction in PEO-LiX complexes occurs mainly in the amorphous phase [59].

In a NMR study of PPO-NaCF$_3$SO$_3$ polymer electrolyte, Schantz and coworkers have observed that the ^{23}Na-NMR signal is composed of a two component band with a narrow signal (1 KHz, 273 K) being superimposed on a broad signal (10 KHz, 273 K) [263]. Increase of temperature leads to the broading of the whole signal (Fig. 21). From inversion recovery experiments it was found that the narrow line was associated with a short T$_1$ of 9 μs (297.5 K) whereas the broad line was associated with a T$_1$ of 13 ms. It was proposed that the narrow ^{23}Na-NMR signal is associated with dissociated cations while the broad component was assigned as due to cation-anion pairs. From the linewidth analysis a lifetime of 10^{-4} s was computed for the ion pairs. Greenbaum and coworkers who have studied a similar system by ^{23}Na-NMR are of the opinion that the broad component

Table 22. Nuclear properties of some important nuclei used as probes in polymer electrolytes and their comparison with proton[a] (taken from Ref. 267)

Nucleus	NA(%)	I	Q $(10^{-28} m^2)$	γ $(10^7 \text{ rad } S^{-1}T^{-1})$	R^c (MHz)	Frequency
1. ^{6}Li	7.43	1	$-6.4 \cdot 10^{-4}$	3.9371	3.58	14.71
2. ^{7}Li	92.57	3/2	$-3.7 \cdot 10^{-2}$	10.3976	$1.54 \cdot 10^3$	38.866
3. ^{23}Na	100	3/2	0.10	7.0704	$5.24 \cdot 10^2$	26.429
4. ^{13}C	1.108	1/2	–	6.7283	1.00	25.145
5. ^{19}F	100	1/2	–	25.1815	$4.73 \cdot 10^{-3}$	94.094
6. ^{1}H	99.985	1/2	–	26.7522	$5.67 \cdot 10^3$	100.000

[a] NA is the natural abundance of the nucleus; I is the nuclear spin quantum number, R^c is the receptivity relative to ^{13}C; Q is the electric quadrupole moment; γ the magnetogyric ratio and frequency refers to the resonance frequency relative to protons (100 MHz)

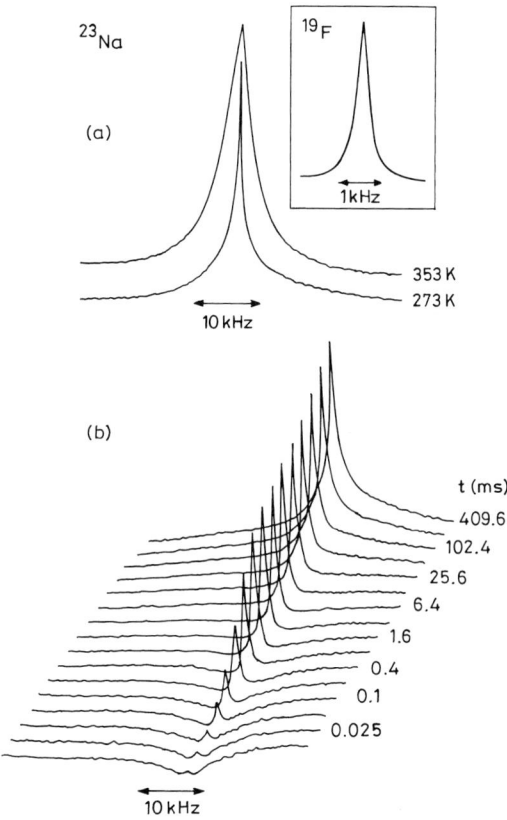

Fig. 21. (a) ^{23}Na NMR absorption spectra for PPO-NaCF$_3$SO$_3$ for O/Na ratio 30 at 273 and 353 K. Inset shows ^{19}F absorption spectra at 273 K (b) ^{23}Na spin lattice relaxation spectra for PPO-NaCF$_3$SO$_3$ complex (O/Na:30) at 297.5 K using the inversion recovery pulse sequence. The delay time t is indicated for every second spectrum (taken from Ref. 263).

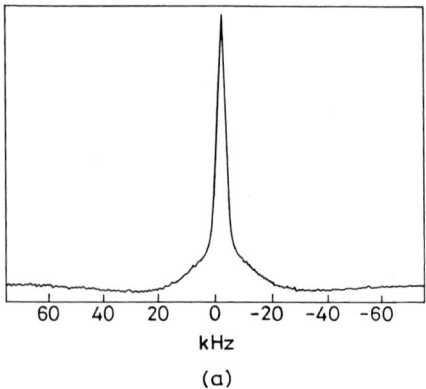

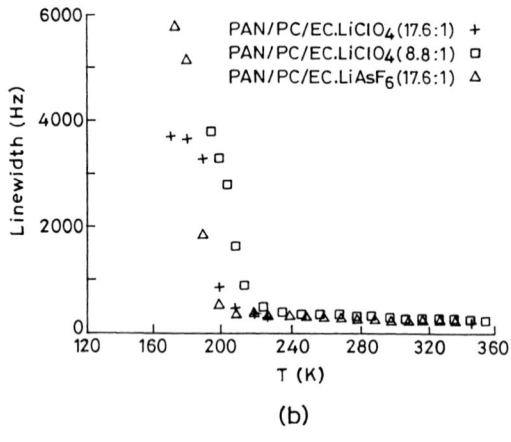

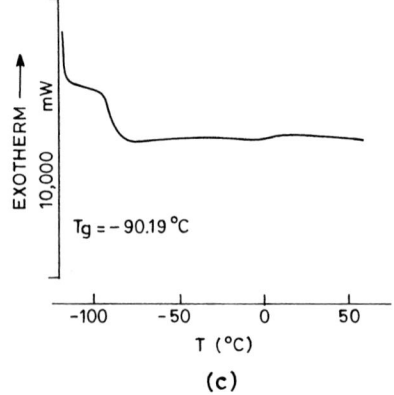

Fig. 22. (a) Li NMR spectrum of $PAN_{0.21}$ $EC_{0.38}$ $PC_{0.33}$ (LiClO4) at 195 K (b) Temperature dependance of ^{7}Li NMR central line width (full width at half maximum) for three gel samples (c) DSC thernogram for $PAN_{0.21}$ $EC_{0.40}$ $PC_{0.35}$ ($LiClO_4$) (taken from Ref. 266).

of the NMR signal arises due to the formation of ionic clusters [264]. Narrow linewidths (5 Hz, 353 K) observed in the ^{19}F-NMR spectrum of PPO-NaCF$_3$SO$_3$ indicates little or no interaction of the fluorine in the anion with the cation. As discussed above, the X-ray structure of PEO-LiCF$_3$SO$_3$ [242] also shows quite clearly that the coordination to the Li$^+$ occurs through the oxygen of the anion and not through the fluorines.

Recent ^{7}Li-NMR studies on gel electrolytes based on PAN/EC/PC/LiClO$_4$ confirms that the onset of the narrowing of linewidth of the ^{7}Li-NMR signal is closely correlated with the T$_g$ of the material [Fig. 22] as is found for PEO-LiX electrolytes [266]. Further from the similarity of the ^{7}Li-NMR linewidths (~ 300 Hz at 300 K) observed in gel electrolytes with that of PEO-LiX electrolytes in comparison with liquid electrolytes containing EC-PC and LiX (15 Hz at 295K), it is concluded that the interpretation of the high conductivity of gel electrolytes in terms of microscopic "packets" of liquid electrolyte separated by regions of inert PAN matrix may be erroneous [266].

7
Summary

Since the initial discovery of polymer electrolytes by P. Wright and M. Armand there has been a phenomenal growth in this area with research encompassing multi-disciplinary areas such as materials science, chemistry, and physics. Although the initial focus has been on PEO and on its modifications, with progressive understanding on the structural aspects of polymer electrolytes several new polymers have been synthesized and investigated. There has also been considerable focus on developing new types of ionic salts. Most of these modifications have resulted in the attainment of routine ambient temperature conductivities of the order of 10^{-4} to 10^{-5} S cm^{-1} particularly for lithium electrolytes. It is clear that dramatically new approaches are required to enhance this conductivity to about 10^{-2} to 10^{-1} S cm^{-1}. Also, enhancing the dimensional and electrochemical stability of the polymer electrolytes would be an important goal. The practical application of polymer electrolytes at least on a commercial scale awaits the emergence of this new generation of materials, although there have been reports based on existing materials on the development of polymer-electrolyte-based batteries for portable devices such as camcorders, and also on the development of systems suitable for commercial traction battery purposes.

Acknowledgement. We are thankful to Department of Atomic Energy, India for financial support.

References

1. Aono H, Imanaka N, Adachi GY (1994) Acc Chem Res 27:265
2. Reuter B, Hardel K (1961) Naturwissenshaften 48:161
3. Owens B B, Argue AG (1967) Science 157:308
4. Takahashi T, Ikeda S, Yamamoto O (1972) J Electrochem Soc 119:477

5. Tatsumisago M, Shinkuma Y, Minami T (1991) Nature 354:217
6. Nicholson PS, Whittingham MS, Farrington GC, Smeltzer W W, Thomas J (eds) (1992) Solid State Ionics-91, North-Holland, Amsterdam
7. Chowdari BVR, Chandra S, Singh S, Srivastava PC (eds) (1992) Solid state ionics, materials and applications, World Scientific, Singapore
8. Bange K, Gambke T (1990) Adv Material 2:10
9. Visco SJ, Liu M, Doeff MM, Ma YP, Lampert C, De Jonghe C (1993) Solid State Ionics 60:175
10. Hagenmuller P, Van Gool W (eds) (1978) Solid electrolytes, general principles, characterization, materials, applications. Academic Press, New York
11. Vincent CA, Bonino F, Lazzari M, Scrosati B (1983) Modern batteries. Edward Arnold, London
12. Gabano F (ed) (1983) Lithium batteries. Academic Press, London
13. Rickert H (1978) Angew Chem Int Ed Engl 17:37
14. Bonino F, Ottaviani M, Scrosati B, Pistoia G (1988) J Electrochem Soc 135:12
15. Weber N, Kummer JT (1967) Proc Annu Power Sources Conf 21:37
16. Whittingham MS, Huggins RA (1971) J Chem Phys 54:414
17. Goodenough JB, Hong HYP, Kafalas JA (1976) Mater Res Bull 11:203
18. Hong HYP (1976) *ibid* 11:173
19. Alpen UV, Rabenau A, Talat GH (1977) Appl Phys Let 30:621
20. Boukamp BA, Huggins RA (1978) Mater Res Bull 13:23
21. Mercier R, Malugnani JP, Fahys B, Saida A (1981) Solid State Ionics 5:663
22. Wada H, Menetrier M, Levasseur A. Hagenmuller P (1983) Mater Res Bull 18:189
23. Kennedy JH, Sahami S, Shea SW, Zhang Z (1985) Solid State Ionics 18/19:368
24. Bates JB, Farrington CG (eds) (1981) Fast-Ion Transport in Solids, North-Holland, Amsterdam
25. Kleitz M, Sapoval B, Ravaine D (eds) (1983) Solid state ionics, North Holland, Amsterdam
26. Kanehori K, Matsumoto K, Miyauchi K, Kudo T (1983) *ibid* 9/10:1445
27. Ratner MA, Shriver DF (1988) Chem Rev 88:109
28. Scrosati B (1993) Applications of Electroactive Polymers Chapman and Hall, London
29. Gray FM (1991) Solid Polymer Electrolytes : Fundamentals and Technological Applications VCH, New York
30. Linford RG (Ed) (1990) Electrochemical Science and Technology of Polymers Vol. II. Elsevier Applied Science, London
31. Liplowski J, Ross PN (eds) (1994) Polymeric materials for lithium batteries in the electrochemistry of novel materials. VCH, New York
32. Abraham KM, Alamgir M (1993) J Power Sources 43:195
33. Scrosati B (Ed) (1993). Applications of electroactive polymers. Chapman and Hall, London
34. Vashishta P, Mundy JN, Shenoy GK (eds) (1979). Fast ion transport in solids. Elsevier Applied Science, North Holland, New York
35. Le Nest JP, Gandini A (1990) Proc. 2nd International Symposium on Polymer Electrolytes. Scrosati B, (ed), Elsevier, Amsterdam, p 129
36. Armand MB (1986) Annu Rev Mater Sci 16:245
37. MacCallum JR, Vincent CA (eds) (1987) Polymer electrolyte reviews-I, Elsevier Applied Science, New York
38. MacCallum JR, Vincent CA (eds) (1989) Polymer electrolyte reviews-II, Elsevier Applied Science, New York
39. Fenton DE, Parker JM, Wright PV (1973) Polymer 14:589
40. Wright PV (1975) Br Polym J 7:319
41. Wright PV (1976) J Poly Sci Polym Phys Ed 14:955
42. Armand MB, Chabagno JM, Duclot M (1978) Extended Abstracts, Second International Conference on Solid Electrolytes. St. Andrews, Scotland
43. Duclot MJ, Vashishta P, Mundy N, Shenoy GK (eds) (1979) Fast ion transport in solids. North-Holland, Amsterdam

44. Angell CA, Liu C, Sanchez E (1993) Nature 362:137
45. Cole KS, Cole RH (1941) J Chem Phys 9:341
46. Macdonald JR (1973) *ibid* 58:4982
47. Mahan GD, Roth WL (eds) (1976) Super ionic conductors. Plenum, New York
48. Fulcher J (1925) J Am Chem Soc 8:339
49. Vogel H (1921) Phys Z 22:645
50. Jones GK, Mc Ghie AR, Farrington GC (1991) Macromolecules 24:3285
51. Wintersgill MC, Fontanella JJ, Greenbaum SG, Adamic KJ (1988) Brit Polym J 20:195
52. Smith MJ, Silva CJR (1992) Solid State Ionics 98:269
53. Chiang CK, Davis GT, Harding CA, Takahashi T (1985) Macromolecules 18:825
54. Robitaille CP, Fanteux D (1983) J Electrochem Soc 133:315
55. Lee YL, Crist B (1986) J Appl ,Phys 60:2683
56. Stainer M, Hardy LC, Whitmore DH, Shriver DF (1984) J Electrochem Soc 131:784
57. Hibma T (1983) Solid State Ionics 9/10:1101
58. Parker JM, Wright PV, Lee CC (1981) Polymer 22:1305
59. Berthier C, Gorecki W, Minier M, Armand MB, Chabagno JM, Rigand P (1983) Solid State Ionics 11:91
60. Minier M, Berthier C, Gorecki W (1983) *ibid* 9/10:1125
61. Gorecki W, Andreani Z, Berthier C, Armand MB, Mali M, Roos J, Brinkmann D (1986) *ibid* 18/19:295
62. Bhattacharya S, Smoot SW, Whitmore DH (1986) *ibid* 18/19:306
63. Mehta MA, Lightfoot P, Bruce PG (1993) Chem Mater 5:1338
64. Papke, BL, Ratner MA, Shriver DF (1982) J Electrochem Soc 129:1694
65. Dupon R, Papke BL, Ratner MA, Whitmore DH, Shriver DF (1982) J Am Chem Soc 104:6247
66. Teeters D, Frech R (1986) Solid State Ionics 18/19:271
67. Papke B, Ratner MA, Shriver DF (1982) J Electrochem Soc 129:1434
68. Cohen MH, Turnbull D (1959) J Chem Phys 31:1164
69. Cohen MH, Grest GS (1980) Phys Rev B B21:4113
70. Gibbs JH, Dimarzio EA (1958) J Chem Phys 28:373
71. Adam G, Gibbs JH (1965) *ibid* 43:139
72. Goldstein M (1973) J Phys Chem 77:667
73. Angell CA, Sichina W (1976) Ann NY Acad Sci 279:53
74. Angell CA (1983) Solid State Ionics 9/10:3
75. Angell CA (1986) *ibid* 18/19:72
76. Angell CA, Williams E (1973) J Polym Sci Poly Lett 11:383
77. Chadwick AV, Strange JA, Worboys MR (1984) Solid State Ionics 9/10:1155
78. Clancy S, Shriver DF, Ochrymomycz LA (1986) Macromolecules 19:606
79. Fauteux D, Prud'homme J, Harvey PE (1988) Solid State Ionics 28-30:923
80. Cameron GG, Ingram MD, Sarmouk K (1990) Eur Polym J 26:1097
81. Chintapalli S, Frech R (1996) Macromolecules 29:3499
82. Walker Jr CW, Salomon M (1993) J Electrochem Soc 140:3409
83. Abraham KM, Alamgir M (1990) *ibid* 137:1657
84. Hong H, Liquan C, Xuefie H, Rongjian X (1992), Electrochim Acta 37:1671
85. Huq R, Farrington GC, Koksbang R, Tonder, PE (1992) Solid State Ionics 57:277
86. Abraham KM, Alamgir M (1993) J Power Sources 43/44:195
87. Koksbang R, Olsen II, Tonder PE, Knudsen N, Lundsgaard J S, Yde-Andersen S (1990) *ibid* 32:175
88. Lee HS, Yang XQ, McBreen J, Xu ZS, Skotheim TA, Okamoto Y (1994) J Electrochem Soc 141:886
89. Yang XQ, Lee HS, Hansan LK, McBreen J, Xu ZS, Skotheim TA, Okamota Y, Lu F (1995) *ibid* 142:46
90. Kelly IE, Owen JR, Steele BCH (1984) J Electroanal Chem 168:467
91. Nagasubramanian G, Attia AI, Halpert G, (1992) J Electrochem Soc 139:3043

92. Nagasubramanian G, Distefano S (1990) J Electrochem Soc 137:3830
93. Sheldon MH, Glasse MD, Latham RJ, Linford RG (1989) Solid State Ionics 34:135
94. Wang C, Liu Q, Cao Q, Meng Q, Yang L (1992) *ibid* 53/56:1102
95. Kelly I, Owen JR, Steele BCH (1985) J Power Sources 14:13
96. Watanabe M, Kanba M, Matsuda H, Tsunemi K, Mizoguchi K, Tsuchida E, Shinohara I (1981) Makromol Chem Rapid Commun 2:741
97. Paulmer RDA, Kulkarni AR (1994) Solid State Ionics 68:243
98. Huq R, Koksbang R, Tonder PE, Farrington GC (1992) Electrochim Acta 37:1681
99. Yang L, Lin J, Wang Z, Wang C, Zhou R, Liu Q (1990) Solid State Ionics 40/41:616
100. Fish D, Smid J (1992) Electrochim Acta 37:2043
101. Cameron GG, Ingram MD, Sarmouk K (1990) Eur Polym J 26:1097
102. Hyodo S, Okabayashi K (1989) Electrochim Acta 34:1557
103. Frech R, Huang W, Johansson P, Lindgren J (1995) J Electrochem Act. 40:2147
104. Barthel J, Wuhr M, Buestrich R, Gores HJ (1995) J Electrochem Soc 142:2527
105. Scrosati B (Ed) (1989). Proceedings of 2nd International Symposium on Polymer Electrolytes. Elsevier Applied Science, London
106. Goulart G, Sylla S, Sanchez JY, Armand M (1989). In: Scrosati B (ed) Proceedings of 2nd International Symposium on Polymer Electrolytes. Elsevier Applied Science, London, p 99
107. Turowsky L, Seppelt K (1988) Inorg Chem 27:3135
108. Dominey LA, Koch VR, Blakley TJ (1992) Electrochim Acta 37:1551
109. Dominey L A (1993) US Pat 5,273,840
110. Walker Jr CW, Cox DJ, Salomon M (1996) J Electrochem Soc 143:L80
111. Croce F, D'Aprano A, Nanjundiah C, Koch VR, Walker Jr CW, Salomon M (1996) *ibid*, 143:154
112. Benrabah D, Baril D, Sanchez JY, Armand M, Gard MG (1993) J Chem Soc Faraday Trans 89: 355
113. Borghini MC, Mastragostino M, Passerini S, Scrosati B (1995) J Electrochem Soc 42:2118
114. Benrabah D, Sanchez JY, Deroo D Armand M, (1994) Solid State Ionics 70/71:157
115. Nafshun RL, Lerner MM, Hamel NN, Nixon PG, Gard GL (1995) J Electrochem Soc 142: L153
116. Holcomb NR, Nixon PG, Gard GL, Nafshun RL, Lerner MM (1996) *ibid*, 143:1297
117. Payne DR, Wright PV (1982) Polymer 23:690
118. Hu CP, Wright PV (1989) Br Polym J 21:421
119. Fang B, Hu C P, Xu HB, Ying SK (1991) Polymer Commun, 32:382
120. Shi J, Hu CP, Chen R, Ying SK (1990). In: Chowdary BV R, Liu Q, Chen L (eds) Recent advances in fast ion conducting materials and devices. World Scientific Pub Co., Singapore, p 267
121. Morita M, Fukumasa T, Motoda M, Tsutsumi TY, Matsuda Y, Takahashi T, Ashitaka H (1990) J Electrochem Soc 137:3401
122. Lobitz P, Fullbier H, Reiche A, Illner JC, Reuter H, Horing S (1992) Solid State Ionics 58:41
123. Bannister DJ, Davies GR, Ward IM, McIntyre JI (1984) Polymer 25:1600
124. Tada H, Kawahara H (1988) J Electrochem Soc 135:1728
125. Manaresi, P, Bignozzi MC, Pilati F, Munari A, Mastragostino M, Meneghello L, Chille A (1993) Polymer 334:2422
126. Borghini C, Mastragostino M, Menezhello L, Manaresi P, Munari A (1994) Solid State Ionics 67:263
127. Nagasubramanian G, Surampudi S, Harpert G (1994) J Electrochem Soc 141:1414
128. Peng X, Ba H, Chen D, Wang F (1992) Electrochim Acta 37:1556
129. Przyluski J, Wieczorek (1992) Solid State Ionics 53-56:1071
130. Nekoomanesh M, Nagae SH, Booth C, Owen JR (1992) J Electrochem Soc 139:3046
131. Kim DW, Park JK, Gong MS, Song HY (1994) Polym Engg Sci 34:1305
132. Gray FM, MacCallum JR, Vincent CA, Giles JR (1988) Macromolecules 21:392
133. Ichikawa K, Macknight WJ (1992) Polymer 33:4693

134. Ichikawa K, Dickinson LC, Macknight WJ, Watanabe M, Ogata N (1992) Polymer 33:4699
135. Watanabe M, Ogata N (1988) Brit Polym J 20:181
136. Matsumoto M, Ichino T, Rutt JS (1994) J Electrochem Soc 141:1989
137. Rutt M, Nishi S (1995) J Electrochem Soc 142:3052
138. Plocharski J, Wieczorek W (1988) Solid State Ionics 28-30:979
139. Wieczorek W (1992) Mater Sci Eng B 15:108
140. Panero S, Scrosati B, Greenbaum SG (1992) Electrochim Acta 37:1533
141. Gang W, Ross J, Brinkmann D, Capuano F, Croce F, Scrosati B (1992) Solid State Ionics 53-56:1102
142. Croce F, Capuano F, Selvaggi A, Scrosati B (1990) J Power Sources 32:389
143. Capuano F, Croce F, Scrosati B (1991) J Electrochem Soc 138:1918
144. Wieczorek W, Such K, Florjanczyk Z, Stevens JR (1994) J Phys Chem 98:6840
145. Manoravi P, Selvaraj II, Chandrasekhar V, Shahi K (1993) Polymer 34:1339
146. MacCallum JR, Tomlia AS, Vincent CA (1986) Eur Polym J 22:787
147. Armand MB (1983) Solid State Ionics 9/10:745
148. Watanabe M, Rikukawa M, Sanui K, Ogata N, Kato H, Kobayashi T, Ohtaki Z (1983) Polymer J 15:65
149. Watanabe M, Togo M, Sanui K, Ogata N, Kobayashi T, Ohtaki Z (1984) Macromolecules 17:2908
150. Watanabe M, Ritukawa M, Sanui K, Ogata N, Kato H, Kobayashi T, Ohtaki Z (1984) Macromolecules 17:2902
151. Dupon R, Papke BL, Ratner MA, Shriver DF (1987) J Electrochem Soc 131:586
152. Armstrong RD, Clarke MD (1984) Electrochim Acta 29:1443
153. Harris CS, Shriver DF. Ratner MA (1986) Macromolecules 19:987
154. Clancy S, Shriver DF, Ochrymomycz LA (1986) Macromolecules 19:606
155. Watanabe M, Kanba M, Nagaoka K, Shinohara I (1983) J Polym Sci Polym Phys., 21:939
156. Abraham KM, Alamgir M (1990) J Electrochem Soc, 137:1657
157. Abraham M, Alamgir M (1994) Solid State Ionics 70/71:20
158. Matsumoto M, Rutt SJ, Nishi S (1995) J Electrochem Soc 142:3052
159. Matsumoto M, Ichino T, Rutt JS, Nishi S (1993) *ibid*, 140:2151
160. Matsumoto M, Ichino T, Rutt JS (1994) *ibid*, 141:1989
161. Matsumoto M, Ichino T, Rutt JS (1994) J Polym Sci Polym Chem Ed. 32:2551
162. Ferillade G, Perche Ph (1975) J Appl Electrochem 25:63
163. Nagasubramanian G, Attia AI, Harpert G (1994), J Appl Electrochem 24:298
165. Kakuda S, Momma T, Osaka T, Appetecchi BG, Scrosati B (1995) J Electrochem Soc 142:L1
166. Cowie JMG, Cree SH (1989) Rev Phys Chem 40:85
167. Xia DW, Solt Z, Smid J (1984) Solid State Ionics 14:221
168. Bannister DJ, Davies GR, Ward IM, McIntyre JE (1984) Polymer 25:1600
169. Kobayashi N, Uchigama M, Shizehara T, Tsuchida E (1985), J Phys Chem 89:987
170. Cowie JMG, Martin ACS (1985) Polymer Commun 25:298
171. Cowie JMG, Martin ACS (1987) Polymer 28:627
172. Tsuchida E, Ohno H, Kobayashi N, Zshizaka H, (1989) Macromolecules 22:1771
173. Cowie JMG, Ferguson R (1985) J Polym Sci Polym Phy 23:2181
174. Selvaraj II, Chaklanobis S, Chandrasekhar V (1993) J Polym Sci Polym Chem 31:2643
175. Selvaraj II, Chaklanobis S, Chandrasekhar V (1995) J Electrochem Soc 142:366
176. Selvaraj II, Manoravi P, Chandrasekhar V (1992) In: Chowdari BVR, Chandra S, Singh S, Srivastava PC (eds) Solid state ionics, materials and applications. World Scientific, Singapore, p 591
177. Selvaraj II, Chaklanobis S, Chandrasekhar V (1995) Polymer 36:2693
178. Deramunage D, Fernandez JE, Garciz Rubie H (1989) Macromolecules 22:2845
179. Chandrasekhar V, Thomas KRJ (1993) Struct Bond 81:41
180. Allcock HR (1972) Phosphorus-Nitrogen Compounds, Academic Press, New York
181. Allcock HR (1985) Chem Eng News 63:22
182. Allcock HR, Kugel RL (1965) J Am Chem Soc 87:4216

183. Allcock HR, Kugel RL (1966) Inorg Chem 5:1716
184. Ganapathiappan S, Dhathathreyan KS, Krishnamurthy SS (1987) Macromolecules 20:1501
185. Allcock HR, Lampe FW (1990) Contemporary polymer chemistry 2nd edn. Prentice-Hall, New Jersey, USA
186. Mark JE, Allcock HR, West R (1992) Inorganic Polymers Prentice Hall, New Jersey, USA
187. Allcock HR (1980) Polymer 21:673
188. Allcock HR (1979) Acc Chem Res 12:351
189. Singler RE, Schneider NS, Hagnauer G (1975) Polym Eng Sci 15:321
190. Allcock HR, Fuller TJ (1980) Macromolecules 13:1338
191. Allcock HR, Fuller TJ (1981) J Am Chem Soc, 103:2250
192. Allcock HR, Kwon S (1988) Macromolecules 21:1980
193. Neilson RH, Wisian-Neilson P (1988) Chem Rev 88:541
194. Neilson RH, Hani R, Wisian-Neilson P (1987) Macromolecules 20:910
195. Wisian-Neilson P, Ford RR, Neilson RH Ray AK (1986) ibid 19:2089
196. Wisian-Neilson P, Neilson RH(1980) J Am Chem Soc 102:2848
197. Allcock HR, Riding, GH, Lavin KD (1985) Macromolecules 18:1340
198. Allcock HR, Riding GH, Lavin KD (1987) ibid 20:6
199. Austin PE, Riding GH, Allcock HR (1983) Macromolecules 16:719
200. Blonsky PM, Shriver DF, Austin PE, Allcock HR (1984) J Am Chem Soc 106:6854
201. Blonsky PM, Shriver DF, Austin PE, Allcock HR (1986) Solid State Ionics 18/19:258
202. Allcock HR, Austin PE, Neenan TX, Sisko JT, Blonsky PM, Shriver DF (1986) Macromolecules 19:1508
203. Lerner MM, Tipton AL, Shriver DF, Denbek AA, Allcock HR (1991) Chem Mater 3:1119
204. Tonge JS, Shriver DF (1987) J Electrochem Soc 134:269
205. Nazri GA, Meibuhr SG, (1989) ibid 136:2450
206. Abraham KM, Alamgir M, Perrotti SS (1988) ibid 105:535
207. Abraham KM (1992) In: Chowdari BVR, Chandra S, Singh S, Srivastva PC (eds) Solid state ionics: materials and applications. World Scientific, Singapore, p 277
208. Abraham KM, Alamgir M (1991) Chem Mater 3:339
209. Adamic KJ, Greenbaum SG, Abraham KM, Alamgir M, Winstersgill MC, Fontanella JJ (1991) Chem Mater 3:534
210. Selvaraj II, Chaklanobis S Chandrasekhar V (1997) Manuscript submitted to Polymer
211. Selvaraj II, Chaklanobis S, Chandrasekhar V (1995) J Electrochem Soc 142:3434
212. Selvaraj II (1993) Ph.D. thesis, Indian Institute of Technology, Kanpur, India
213. Blonsky PM (1986) Ph.D. thesis, North Western Univ., USA
214. Tonge JS, Blonsky PM, Shriver DF, Allcock HR, Austin PE, Neenan TX, Sisko JT (1987) Proc. Symp. on Lithium Batteries 87-1:533
215. Allcock HR, Napierala ME, Cameron CG, O'Connor SJM (1996) Macromolecules 29:1951
216. Allcock HR, Kuharcik SE, Reed CS, Napierala ME (1996) Macromolecules 29:3384
217. Inoue K, Nishikawa Y, Tanigaki T (1991) *ibid* 24:3464
218. Inoue K, Nishikawa Y, Tanigaki T (1991), J Am Chem Soc 113:7609
219. Inoue K, Nishikawa Y, Tanigaki T (1992) Solid State Ionics 58:217
220. Ganapathiappan S, Chen K, Shriver DF (1988) Macromolecules 21:2299
221. Chen K, Shriver DF (1991) Chem Mater 3:771
222. Ganapathiappan S, Chen K, Shriver DF (1989) J Am Chem Soc 111:4091
223. Nagaoka K, Naruse H, Shinohara I, Watanabe MJ (1984) J Polym Sci Polym Lett 22:659
224. Adamic KJ, Greenbaum SG, Wintersgill M, Fontanella JJ (1986) J Appl Phys 60:1342
225. Bouridah A, Dalard F, Deroo D, Cheradame H, Le Nest JF (1985) Solid State Ionics 15:223
226. Spindler R, Shriver DF (1988) Macromolecules 21:648
227. Hall PG, Davies GR, McIntyre JE, Ward IE, Bannister DJ, LeBrocq KMF (1986) Polym Commun 27:98
228. Lonergan MC, Ratner MA, Shriver DF (1995) J Am Chem Soc 117:2344
229. Linford RG, (1995) Chem Soc Rev, 267

230. Teo BK (1986) 'EXAFS : Basic Principles and Data Analysis' Springer-Verlag, Heidelberg
231. Catlow CRA, Chadwick AV, Greaves GN, Moroney LM, Worboys MR (1983) Solid State Ionics 9/10:1107
232. McBreen J, Yand XQ, Lee HS, Okamoto Y (1995) J Electrochem Soc 142:348
233. Schlindwein WS, Pynenburg RAJ, Latham RJ, Linford RG (1995) Nucl Instrum Methods Phys Res B 97:292
234. Latham RJ, Linford RG, Pynenburg RAJ, Schlindwein WS, Farrington GC (1993) J Chem Soc Faraday Trans 89:349
235. Badara HMN, Schlindwein WS, Latham RJ, Linford RG (1994) J Chem Soc Faraday Trans 90:3549
236. Latham RJ, Linford RG, Schlindwein WS (1989) Faraday Discuss Chem Soc 88:103
237. Iwamoto R, Saito Y, Ishihara H, Tadokoro H (1968), J Polym Sci (A-2) 6:1509
238. Yokoyama M, Ishihara H, Iwamoto R, Tadokoro H (1969) Macromolecules 2:184
239. Chatani Y, Okamura S (1987) Polymer 28:1815
240. Chatani Y, Fujii Y, Takayanagi T, Honma A (1990) *ibid*, 31:2238
241. Lightfoot P, Mehta MA, Bruce PG (1992) J Mater Chem 2:379
242. Lightfoot P, Mehta MA, Bruce PG (1993) Science 262:883
243. Lightfoot P, Nowinski JL, Bruce PG (1994) J Am Chem Soc 116:7469
244. Takahashi Y, Tadokoro H (1973) Macromolecules 6:672
245. Hibna T (1983) Solid State Ionics 9/10:1101
246. Papke BL, Ratner MA, Shriver DF (1981) J Phys Chem Solids 42:493
247. Davidson WHT (1955) J Chem Soc p 3270
248. Sato H, Kusumoto Y (1978) Chem Lett p 635
249. Akashi H, Hsu SL, Macknight WJ, Watanabe M, Ogata N (1995) J Electrochem Soc 142:205
250. Schantz S, Sandahl J, Borjesson L, Torel LM, Stevens J (1988) Soid State Ionics 28-30:1047
251. Schantz S, Torrel LM, Stevens JR (1988) J Appl Phys 64:2038
252. Kakihana M, Schantz S, Torrel LM (1990) J Chem Phys 92:6271
253. Kakihana M, Schantz S, Torrel LM, Stevens JR (1990) Solid State Ionics 40-41:641
254. Schantz S (1991) J Chem Phys 94:6296
255. Schantz S, Torrel LM, Stevens JR (1991) *ibid*, 94:6862
256. Torrel LM, Jacobsson P, Sidebottom D, Peterson G (1992) Solid State Ionics 53-56:1037
257. Yang XQ, Lee HS, McBreen J, Xu ZS, Skotheim TA, Okamoto Y, Lu F (1994) J Chem Phys 101:3230
258. Wang Z, Huang B, Huang H, Xue R, Chen L, Wang F (1996) J Electrochem Soc 143:1510
259. Gupta S, Shahi K, Binesh N, Bhat SV (1993) Solid State Ionics 67:97
260. Chung SH, Jeffrey KR, Stevens JR (1991) J Chem Phys 94:1803
261. Panero S, Scrosati B, Greenbaum SG (1992) Electrochim Acta 37:1533
262. O'Gara JF, Nazri G, MacArthur DM (1991) Solid State Ionics, 47:87
263. Schantz S, Kakihana M, Sandberg M (1990) ***ibid***, 40/41:645
264. Greenbaum SG, Pak YS, Wintersgill MC, Fontanella JJ (1988) *ibid* 31:241
265. Lemmon JP, Kohnert RL, Lerner MM (1993) Macromolecules 26:2767
266. Croce F, Brown SD, Greenbaum SG, Slane SM Salomon M (1993) Chem Mater 5:1268
267. Mason J (1987) Multinuclear NMR, Plenum Press, New York

Editor: Prof. Sir S. Edwards
Received: December 1996

Standard Methods for Testing the Aerobic Biodegradation of Polymeric Materials. Review and Perspectives

Anne Calmon-Decriaud[1,2], Véronique Bellon-Maurel[1], Françoise Silvestre[2]

[1] Cemagref, 361 .rue J.F Breton, BP 5095, 34033 Montpellier Cedex 01, France.
[2] Laboratoire de Chimie Agro-Industrielle, Ecole Nationale Supérieure de Chimie, 118 Route de Narbonne, 31077 Toulouse Cedex, France.
E-mail: anne.decriaud@cemagref.fr; E-mail: veronique.bellon@cemagref.fr

There is an on-going worldwide research effort to develop biodegradable polymers for packaging from renewable sources. This development has caused a need to evaluate the biodegradation of these polymers in different environments, e.g. dumping in marine, freshwater, compost or landfill sites. Therefore, many organizations such as ASTM, OECD, the European Committee of Normalization, the Japan Biodegradable Plastics Society, etc., have developed accelerated laboratory test procedures for evaluating potentially biodegradable materials. This report gives an overview of the standardization activities for biodegradability assessment of polymers and a comparison of the methods used for biodegradability tests on solid polymers and packaging materials.

Keywords: biodegradation, standard methods, review, agromaterials, renewable source.

1	Introduction	208
2	Defining Biodegradability	209
2.1	Degradation	209
2.2	Biodegradation	209
3	Standards Organizations and Their Standard Tests	211
3.1	American Normative Reference (ASTM)	211
3.2	Japanese Normative Reference (JIS)	211
3.3	European Normative Reference (ECN)	213
3.4	Other Committees	213
4	Critical Presentation of Standard Test Methods in Liquid Conditions	213
4.1	Principle of Biodegradation Liquid Tests	213
4.2	Individual Measurement Units	214
4.2.1	For O_2 Measurement	214
4.2.2	For CO_2 Measurement	214
4.3	Overall System Structure	215
4.4	Sample Preparation	218
4.5	Inoculum Choice	219
4.5.1	Inoculum Quality	219
4.5.2	Inoculum Quantity	219

Advances in Polymer Science, Vol. 135
© Springer-Verlag Berlin Heidelberg 1998

| 4.6 | Experimental Conditions | 221 |
| 4.7 | Conclusions | 221 |

5	**Standard Test Methods in Solid Media**	222
5.1	Petri Disk Screen	222
5.2	Compost Chamber Method	223
5.3	Conclusions	223

| 6 | **Discussion and Conclusion** | 224 |

References ... 224

1
Introduction

The durable properties of plastics (e.g., polyethylene, PVC) make them the ideal material for a large number of applications. A consequence of this phenomenal use of plastic materials is their increasing presence in municipal solid waste throw-away products. Plastics constitute approximately 10% by weight of the total solid wastes in Europe, and more than 25% by volume [1, 2]. Furthermore, plastic wastes may represent an undesired pollutant in many ecosystems (e.g., soil, freshwater, marine habitats). Because of their resistance to microbial attack, they tend to accumulate. Current limitation of waste deposit sites in Europe [3, 4], discussion about renewable versus fossil resources, new environmental regulations [5-7] and introduction of composting infrastructures in waste management have led to enhanced R&D on materials which are claimed to be biodegradable. Many companies throughout the world are developing biodegradable polymers [8, 9] which belong to three categories: 1-*natural polymers* such as cellulose, starch and proteins, 2-*modified natural polymers*: prepared by biological and/or chemical modification, such as cellulose acetate, lignocellulose esters [10], polylactidacid, polyalkanoates, of which the most promising is the copolymer polyhydroxybutyrate/valerate (PHB/HV) [11-12], and 3-*composite materials* which combine biodegradable particles (e.g., starch, regenerated cellulose or natural gums) and a synthetic biodegradable polymer (e.g., starch-vinylalcohol-copolymer) [13]. Degradable materials may be a viable solution for some plastic items, especially those that are litter-prone, not easily collected or not easily recycled, such as plastics used in agriculture or in packaging [14].

The development of biodegradable polymers for packaging purposes has generated a need for evaluating these biodegradable polymers in solid waste treatment and aquatic environments. In order to successfully establish biodegradability, standard test methods are required. Therefore, standard tests for degradation of polymers under various conditions have been the object of many organizations such as ASTM, OECD, the European Committee of Normalization, the Japan Biodegradable Plastics Society, etc. This paper gives an overview of standardization activities for biodegradable polymers and a comparison of the biodegradability tests for solid polymers and packaging materials.

2
Defining Biodegradability

Although the term "biodegradable polymers" is becoming more popular, it has different meanings. The definitions are not always clear and they are open to a large diversity of interpretations [15-19]. On the one hand, in the biomedical area, "biodegradation" refers to hydrolyses and oxidations that are the primary polymer degradation processes[20]. On the other hand, for materials exposed to a natural environment, this term means fragmentation, loss of mechanical properties or chemical modifications through the action of microorganisms. Thus many different degradation modes, either abiotic degradation (e.g., degradation due to agents such as oxygen, water and sunlight) or biotic degradation (e.g., degradation due to microorganisms), can synergistically combine in natural conditions to degrade polymers, leading to different degrees of degradability (e.g., fragilisation, fragmentation, solubilisation) and each specific action is difficult to isolate [21-27].

In addition, a number of standard authorities have sought to produce definitions for biodegradable plastics.

2.1
Degradation

Degradation can be defined as "a change in the chemical structure of a plastic involving a deleterious change in properties" [28]. The material is degraded under environmental conditions (e.g., microorganisms, temperature, light, water) and in a reasonable period of time in one or more steps. It has been suggested that the degradation of plastics could be divided into "deterioration" and "decomposition". Deterioration corresponds to physical and chemical degradation with a permanent change in the physical properties of the plastics (ISO 472:1988). Decomposition always induces physical and chemical degradation which results in a decrease in weight, while deterioration, such as fragmentation by UV light, does not necessarily imply a decrease in weight [29].

Many studies show that plastics can be degraded by five interacting mechanisms:

- photodegradation by natural daylight
- oxidation by chemical additives (e.g., catalysts)
- thermal degradation by heat
- mechanical degradation by mechanical effects
- biodegradation by microorganisms (e.g., bacteria, fungi). In the definition of biodegradable plastics, the biodegradation process only is pointed out.

2.2
Biodegradation

Biodegradation is the natural and complex process of decomposition facilitated by biochemical mechanisms, and each standard organization gives its own defin-

Table 1. General definitions of a biodegradable polymer (or plastic) proposed by Standard Authorities

Standard Authorities	Biodegradable plastics
ISO 472-1988	A plastic designed to undergo a significant change in its chemical structure under specific environmental conditions resulting in a loss of some properties that may vary as measured by standard test methods appropriate to the plastic and the application in a period of time that determines its classification. The change in the chemical structure results from the action of naturally occurring microorganisms.
ASTM sub-committee D20-96	A degradable plastic in which the degradation results from the action of naturally occurring microorganisms such as bacteria, fungi and algae.
DIN 103.2-1993 German working group	A plastic material is called biodegradable if all its organic compounds undergo a complete biodegradation process. Environmental conditions and rates of biodegradation are to be determined by standardized test methods.
E.C.N (May 1993)	A degradable material in which the degradation results from the action of microorganisms and ultimately the material is converted to water, carbon dioxide and/or methane and a new cell biomass.
Japanese Biodegradable Plastics Society (1994)	Polymeric materials which are changed into lower molecular weight compounds where at least one step in the degradation process is through metabolism in the presence of naturally occurring organisms.

ition of a biodegradable polymer (Table 1). In summary, biodegradability can be defined as the intrinsic capacity of a material to be degraded by the action of microorganisms, i.e., to progressively obtain a simpler structure. More precisely, there are two definitions according to the fate of polymers [30-31]:
– **Primary biodegradability** (or partial biodegradability) is the alteration in the chemical structure. It results in a loss of specific properties of the polymer. It is important to measure it for polymers used in durable applications (e.g., building, vehicle sets).
– **Ultimate biodegradability** (or total biodegradability) deals with total mineralization: the material is totally degraded by microorganisms with production of carbon dioxide (under aerobic conditions) or methane (under anaerobic environment), water, mineral salts and new microbial cellular constituents (biomass). However, in some cases, this mineralization is not complete because some end-products are resistant to degradation and remain as oligomers. It is interesting to measure it for polymers intended for one-use applications (e.g., packaging, many plastics).

Browsing through various standards dealing with biodegradation two other definitions come to light for organic compounds [32]:

– **Ready biodegradability** is assessed in "stringent tests which provide limited opportunity for biodegradation and acclimatisation to occur".
– **Inherent biodegradability** assessed in tests "based on a prolonged exposure of the test compound to the microorganisms, a more favorable ratio of biomass/chemical or other conditions favoring biodegradation".

Recently, G. Swift introduced the concept of "environmentally acceptable biodegradable polymer" [33]. A biodegradable polymer may be partially or totally degraded whereas an "environmentally acceptable biodegradable polymer" must be totally mineralized or -if only partially mineralized- must not produce any environmentally harmful residues. Consequently, this definition includes not only the degree of biodegradation but also, the impact of the polymer on the environment. It is also interesting to introduce the property of "compostability" which is more restrictive than biodegradability [34]. Compostability is the ability of a material to be degraded by microorganisms in a compost made of green waste, water and wood chips, residues are non-toxic neither for the compost nor for human beings.

In relation to various definitions of biodegradability, several standard tests have been developed by Standards Organizations all around the world. We will present the most dynamic ones and the standard tests that they propose.

3
Standards Organizations and Their Standard Tests

Standards Organizations have shown an interest in biodegradation tests only for a short time (five years) [35-36]. This explains why standard tests are rather simple, similar to one another and still very few. Table 2 gives a rapid review of these tests and the biodegradability definition they refer to.

3.1
American Normative Reference (ASTM) [37]

The D20.96 Committee of the American Society of Testing and Materials (ASTM) worked intensively on test methods for water-insoluble polymers and plastic materials [38, 39]. Since 1993, five standards for biodegradation of plastic materials in various conditions have been published (Table 2), and other test methods with different environments are nearing completion [39], i.e., anaerobic high-solid digestor and anaerobic accelerated landfill to simulate the fate of a material during solid waste management. Other experimental methods need to be developed for freshwater and simulated marine conditions.

3.2
Japanese Normative Reference (JIS)

Since 1990, the government and the Ministry of International Trade and Industry (MITI) have tried to encourage the development of biodegradable plastics; the Japanese industries have created a research group called "Biodegradable Plastics Society" to coordinate work on the development of biodegradable polymers

Table 2. Title and general presentation of existing standard methods for assessing the biodegradability of plastic materials

	Test N°	Biodegra-dability	Title	Conditions	Measured parameters
ASTM	D5210-92	Inherent and ultimate	Test method for determining the *anaerobic* biodegradation of plastic materials in the presence of *municipal sewage sludge*.	Anaerobic	CO_2/CH_4
	D5209-92		Test method for determining the *aerobic* biodegradation of plastic materials in the presence of *municipal sewage sludge*	Aerobic	CO_2
	D5338-92		Test method for determining aerobic biodegradation of plastic materials under *controlled composting conditions*		CO_2
	D5271-93		Test method for determining the aerobic biodegradation of plastic materials *an activated sludge wastewater treatment system*.		O_2
	D5247-92	Primary	Test method for determining the aerobic biodegradability of degradable plastics *by specific microorganisms*		Mw + MP*
JIS	K 6950-94	Ready and Ultimate	Determination of the biodegradability of polymeric materials. *Aerobic* biodegradation in the presence of *activated sludge*.		O_2
ECN	Draft 085-1995		Evaluation of the "ultimate" *aerobic* biodegradability and disintegration of packaging materials under *controlled composting conditions* – Method by analysis of released carbon dioxide		CO_2
OECD	301B 301C 301D 301F		Modified Sturm test (evolved CO_2) Modified MITI test (I) Closed bottle test Manometric respirometric test		O_2
	302B 302C 304A	Inherent Ultimate	Modified Zahn-Wellens/EMPA test Modified MITI test (II) Inherent biodegradability in soil		$^{14}CO_2$

* Mw: molecular weight, MP: mechanical properties.

Standard Methods for Testing the Aerobic Biodegradation of Polymeric Materials 213

and assessment of polymer biodegradability [40]. The method based on the "Modified MITI Test" (OCDE 302C-1981) is recognized as a test method for measuring the biodegradability of plastics [36] (Table 2).

3.3
European Normative Reference (ECN)

In Europe, there is currently no standard test method [41-43]. The technical committee 261, which has been operating since 1991, is setting up the standardization of biodegradability testing methods in the field of packaging and the environment. The European Committee of Normalization (ECN) has written a draft about the evaluation of the ultimate biodegradability under controlled composting conditions (Table 2). Other tests are in preparation in different environmental conditions: aerobic and anaerobic environment, freshwater, etc.

3.4
Other Committees

The OECD (Organization for Economic Cooperation and Development), an intergovernmental organization consisting of 24 industrialized countries, worked on biodegradability testing methods [44]. Even if the OECD has not worked specifically on materials, the methods developed for insoluble chemicals can be adapted (Table 2). The test strategy of the OECD is based on a three-tiered approach, namely: ready biodegradability (Level I), inherent biodegradability (Level II) and simulation of environmental compartments (Level III).

A more detailed analysis of these various procedures shows their common points and their relative advantages and disadvantages. These tests can be simply classified into "tests in liquid conditions" and "tests in solid conditions".

4
Critical Presentation of Standard Test Methods in Liquid Conditions

A liquid environment is currently used to assess the biodegradability of materials despite the fact that the fate of most films is soil, compost or landfill.

4.1
Principle of Biodegradation Liquid Tests

The general principle of these tests is the same, i.e., mainly aerobic fermentation of polymers in an aquatic environment, regardless of whether they differ in terms of the parameters used to estimate the biodegradability, measure the inoculum, and to determine the mode of operation (continuous or static). Generally, the source of microorganisms is activated sludge freshly sampled from a well-operated municipal sewage treatment plant. The material to be tested is the sole organic carbon source in the medium. Through the action of microorganisms and in the presence of oxygen, this carbon is converted into biomass, CO_2, residual polymer or oligomers, and inorganic dissolved carbon. Biodegradation is generally assessed, either by oxygen consumption or by carbon dioxide production. More precisely, the criterion is the amount of O_2 consumed or CO_2 produced divided by the maximal theoretical quantity of oxygen, or carbon dioxide, that should have been consumed, or produced, by microorganisms, knowing the amount of carbon in the material.

The respiration test can be either static (closed bottle) or continuous (air supplying). The advantage of static systems is that there are few risks of leakage and the set-up is fairly simple, but these tests use a destructive process which implies problems of reproducibility and of dimensions. For instance, in the OECD 301F test, 20 bottles are necessary to assess the biodegradability of only one material. On the other hand, the advantage of continuous systems is that they are non-destructive, a kinetic study is possible and it is possible to automate the measurement. However, the set-up is complex and leakage is possible.

Whatever the operating mode, each respiration test requires several individual "measurement units" working in parallel. Each unit always includes a bioreactor and a measurement or trapping system for the criterion of interest, either CO_2 or O_2. Several parallel units are required to test not only the material sample, but also to control the viability of the microorganisms, inherent carbon dioxide production, inhibition, etc.

In conclusion, we have to outline the limits of respiratory methods. Indeed, in the metabolic reaction, polymer carbon is used by microorganisms to produce not only CO_2, but also oligomers and new biomass. Thus, a single measurement of CO_2 (or, what is worse, consumed O_2) is not sufficient to clearly indicate the amount of transformed polymer. Dissolved organic carbon (DOC) is sometimes measured at the end of the test and indicates the amount of oligomer remaining in the medium. This procedure should be obligatory in order to make the balance of carbon, and to check if there are no unexpected losses.

4.2
Individual Measurement Units

4.2.1
For O_2 Measurement

In a closed respirometer, consumed oxygen is measured in a non-destructive way by using an oxygen electrode or a BOD meter[45-46]. The reactor contains a liquid medium inoculated with activated sludge. To avoid CO_2 concentration in the headspace, KOH (pellets or solution) is placed in the headspace to absorb carbon dioxide produced by conversion and endogenous respiration (Fig. 1a). The biodegradation is calculated as the ratio of biological oxygen demand (BOD) to the theoretical oxygen demand (TOD).

4.2.2
For CO_2 Measurement

Carbon dioxide production is measured in a system that is basically the same as the Sturm test [47]. A measurement line is divided in three parts: air treatment (production of CO_2-free air), degradation reactor and carbon dioxide trapping

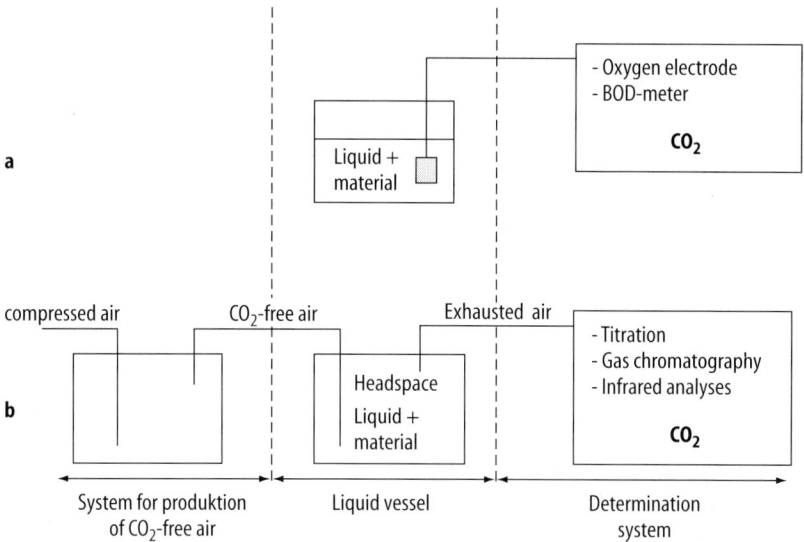

Fig. 1. Principle of test system with (a) measurement of consumed oxygen and (b) measurement of released carbon dioxide

or measurement system (Fig. 1b). According to the standard, the design of the system varies. OECD tests do not state any requirements for the CO_2-free air producing system whereas ASTM standard requires a series of Erlenmeyer flasks. In both tests, stirred reactors are incubated at 23 °C for 28 days. Carbon dioxide produced is trapped in a basic solution ($Ba(OH)_2$ or KOH) and its concentration is determined by titration of the base remaining using hydrochloric acid (OECD, ASTM). Carbon dioxide production can be also determined by using a carbon dioxide analyzer (either infrared or chromatographic) (OECD).

In conclusion, the manual determination of CO_2 production is time-consuming and thus expensive. The automation of the measurement is the only way to make the test easily applicable and to use more widespread. However, comparison between errors in manual and automatic detection has not yet been made.

4.3
Overall System Structure

The assessment of biodegradability cannot be done using measurements on the samples only. It is necessary to simultaneously use several individual lines in parallel, in order to also control the viability of microorganisms, the CO_2 production of the inoculum alone or the absence of inhibitory substances in the polymer. In Table 3, the control tests are listed that are generally used in the various standards, but standard methods do not all include the same set of controls; these sets are listed in Table 4. In all cases, the material is tested twice.

The three controls necessary are the "sample" which shows the biotic and abiotic degradation, the blank (inoculum alone) which checks the endogenous res-

Table 3. Controls used in standard tests: reactor composition and test goals

No	Reactor name	Composition	Quantify and verify the
0	"Sample"	Inoculum + Sample	CO_2 produced by bio- and abiotic degradation
1	"Blank"	Inoculum	CO_2 produced only by the microorganisms
2	"Positive"	Inoculum+Positive control material (sodium acetate)	inoculum activity
3	"Negative"	Inoculum+Negative control material (PE)	interlaboratory comparisons
4	"Sterile"	Sample+Sterilizer ($HgCl_2$)	abiotic sterile degradation
5	"Toxicity"	Inoculum+Sample+Positive control	inhibition by test sample
6	"Adsorption"	Inoculum+Sample+Sterilizer	adsorption

Table 4. Number of controls recommended in the various standard tests

Number name	Sample "0"	Blank "1"	Positive reference "2"	Negative reference "3"	Sterile "4"	Toxicity "5"	Adsorption "6"
ASTM D 5209–92	2	1	1 cell/amide*	no	no	no	no
ASTM D 5247–92	4	4	4 cellophane	4 PE	4	no	no
ASTM D 5271–93	3	3	3 cell/amide*	3 optional	1	1 optional	no
OCDE 301B	≥2	≥2	1 aniline	no	no	no	no
OCDE 301C	3	1	1	no	no	no	no
OCDE 301D	≥10**	≥10**	10**	no	no	#6	no
OCDE 301F	≥2	≥2	1	no	1	1	1

* Cellulose or amidon
** Destructive test: 5 essays·2 bottles=10 bottles

piration and the positive control (inoculum + positive control material such as sodium acetate) which verifies the viability of the microorganisms. These controls are used in all the standard tests. The negative control, which according to ASTM D5247 enables interlaboratory comparisons, may be redundant with the "blank".

"Sterile" tests enable abiotic degradation to be assessed. They are useful only for identifying biotic degradation.

"Toxicity" tests are open to criticism: they are liable to mix the sample, the inoculum and the positive reference at the beginning of the test. As shown in Figure 2, this procedure can lead to confusion between partial degradation and inhibitory substances:

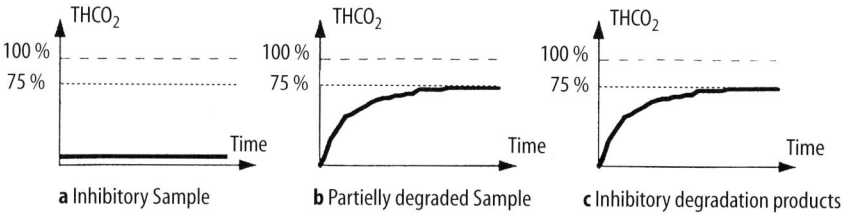

Fig. 2. Toxicity test: Effect of introduction of the positive reference in the beginning of experiments for various scenarios of toxicity

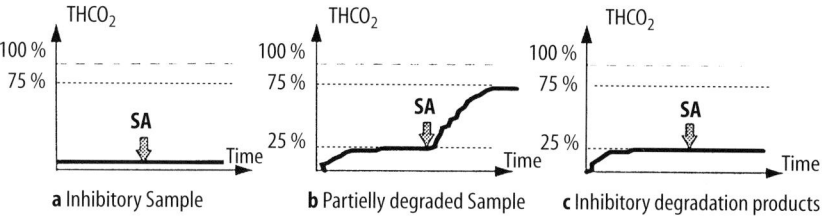

Fig. 3. Toxicity test: Effect of introduction of the positive reference (SA: Sodium acetate) in the course of experiments for various scenarios of toxicity

- if the sample is toxic (Fig. 2a) this test is meaningful because the sample inhibits the degradation of the positive reference.
- if the sample is partially degraded (for instance at 50%, i.e., 25% of the total organic carbon of the reactor), the positive control is totally degraded (50% of the total carbon), this means that 75% of the total organic carbon will be recovered (Fig. 2b).
- if the degradation products of the sample are toxic (Fig. 2c), the positive reference is degraded first (50% of the total organic carbon) then the sample begins to degrade, but as soon as the concentration of degradation products is high enough to be inhibitory (for instance when 50% of the sample is degraded, i.e., 25% of the total carbon), the degradation stops. At this point the amount of recovered carbon is 75%, which is similar to the partial degradation.

Thus, with this procedure, it is not possible to conclude whether the degradation is partial or the degradation products are inhibitory. To overcome this problem, it is suggested that the positive reference be introduced in the course of the experiment, for instance in the middle of the test (14 days). In this case, the system will give the following results:

- if the sample is inhibitory, no degradation is seen, neither at the beginning nor when the positive reference is added (Fig. 3a);
- if the degradation is partial (Fig. 3b) a first degradation (up to 25% of the total carbon) will be seen and addition of the positive reference will lead to another degradation (up to 75% of the total carbon).

- if the degradation products of the sample are inhibitory, the sample begins to degrade until the concentration of the degradation products is high enough; the addition of the positive reference will not induce any further degradation because it will be inhibited (Fig. 3c). In this case, it is easy to recognize whether the sample is partially degraded or its degradation products are inhibitory.

The OCDE 301F standard test also includes an "adsorption" test: inoculum, sample and sterilizer are mixed together. However, the meaning of this test is not clear and there is no explanation about the use of the results.

In conclusion, the standard tests all include different controls. The three most important ones are "sample", "blank" and "positive reference". A toxicity test, with a modified procedure, is also desperately needed. It is more useful to replicate the tests that will give the degradation rate of the sample (i.e., the "sample" and "blank" reactors) than tests which work better. For respiratory tests, a good compromise between the representativeness of the measure and the number of reactors would have to be:

- 2 "sample" reactors
- 2 "blank" reactors
- 1 "positive" reactor
- 1 "toxicity" reactor.

4.4
Sample Preparation

Concentration of material (measured in mg of organic carbon/L of medium) varies from 2 mg/L to 500 mg/L as shown in Table 5. Generally, the material is ground to a powder to increase the specific surface and thus the microbial accessibility. The material can also be studied in the shape of a film to enable the

Table 5. Quantity and form of samples recommended in the various standard tests

Number name	Type of material	Quantity	Form	Parameters to determine
ASTM D 5209-92	Plastics	unknown	Films, fragments, pieces or formed articles	–Weight-% C,H,O,N,P,S. –CPG (optional)
ASTM D 5271-93		<500 mg/L TOC		
ASTM D 5247-92		4·inch.	Strip (ASTM 882)	
OCDE 301B	Insoluble chemical products	10–20 mg/L TOC		
OCDE 301C		100 mg/L BOD	Powder	
OCDE 301D		5–10 mg/L BOD		
OCDE 301F		50–100 mg/L BOD		

mechanical properties to be measured. In all cases, any information available about the size and surface of the sample must be included in the report, as well as the carbon content of the sample, obtained either by stoichiometry or by elementary analysis.

This vagueness is open to criticism: the shape of the sample influences the degradation rate. Thus, the same sample could be assessed as "easily biodegradable", when in powder form and as "inherently biodegradable" when in the form of a film. As it is not easy to grind the film down to a precise particle size, and as the polymers are often used in film shape, the best procedure would be to measure the biodegradability on film strips of a standard size and thickness.

4.5
Inoculum Choice

4.5.1
Inoculum Quality

The inoculum quality is variable and difficult to standardize. First, the inoculum sources vary (even in the same standard method): activated sewage sludge from an aerated municipal wastewater treatment plant, secondary effluent, eluate from compost or soil, or even a mixture of several inocula. Second, the preparations of inoculum differ significantly (Table 6): inoculum may be introduced directly or after pretreatment, such as filtration or decantation, to eliminate the voluminous particles. Aeration, without supplementary substrate, is necessary to purge the system from organic carbon present in the inoculum sample. Moreover, during storage, inoculum sometimes undergoes a preculture in the presence of a positive reference or material sample, precultures that microbiologists name "acclimatation" or "adaptation" respectively [48, 49]. Third, the inoculum extract is variable: tests recommend sampling either the supernatant or a suspension from sewage sludge as well as a direct water sample.

In conclusion, the quality of the inoculum is not precise at all. Moreover, even if the same procedure is used (for instance sewage sludge), the inoculum varies according to the place and the season of sampling. In order to get interlaboratory reproducibility, it is important to try to standardize the quality of inoculum.

4.5.2
Inoculum Quantity

The volume of inoculum sample is not always indicated (Table 6). The quantity of inoculum represents a compromise: on the one hand, a small amount of inoculum (in dry weight) does not bias the measurement because the microbial carbon proportion versus the sample carbon is low; on the other hand, a "large" amount of inoculum accelerates the course of degradation and makes the measurement more repeatable (ASTM D 5271-93). From one test to another, the number of Colony-Forming Units/ml (CFU/ml) varies from 10^4 to 10^8.

The problem with this method is it is time-consuming (24 h to get an answer), requires skilled operators and gives only the viable biomass (not the active one).

Table 6. Preparation, storage and quantity of inoculum for standard tests

Number name	Origin	A[1]	F[2]	D[3]	Storage	Inoculation	Quantity	Cellular density (CFU/ml)
ASTM D 5209–92 5271–93	Well-operated municipal sewage treatment plant	4 h	no	yes	Storage at 4 °C for 2 weeks without nutrients Not specified Possible acclimatization.	Supernatant Sewage sludge suspension	1% (v/v) of the volume of degradation medium 0.1 <VSS <1.0 g/L	10^6 to $20 \cdot 10^6$ Optional Not specified
5247–92	Pure culture	no		no	Acclimatization on gelose at 4 °C for 2 to 8 weeks.	Stock culture	0.5 ml of a spore/cell. suspension	
OCDE 301B and F	- Sewage sludge - Secondary effluent* - Soils* - Mixture of all	yes	yes	yes	Aeration storage. No adaptation.	Supernatant of: - sewage sludge - effluent secondary	TSS <30 mg/L (10 ml/L) 100 ml effluent/L	10^7 to 10^8 (Approx.)
301C	At least 10 fresh samples → mixture				Acclimatization on gelose or peptone for 1–4 months.	Sewage sludge suspension	TSS=30 mg/L	
301D	Sewage sludge or surface water				Aeration storage.	Supernatant or sample	Between 0.05 ml to 5 ml of supernatant/L	10^4 to 10^6 (Approx.)

[1] Aeration, [2] Filtration, [3] Decantation
VSS: Volatile suspended solids, TSS: Total suspended solids
* No indication is given on the inoculum preparation and on inoculation.

It could be replaced by other methods such as I.N.T which is fast (1 h), requires little operator timing and gives active biomass.

4.6
Experimental Conditions

As shown in Table 7, environmental conditions (temperature, pH) can also be very variable. The suggested exposure period is generally 28 days. However, the OECD and ASTM standard tests state that the duration can be extended beyond 28 days if appropriate.

4.7
Conclusions

All standard test methods in liquid conditions are based on respiratory tests with a diversity of methodologies: measurement of oxygen consumption is easily achieved (i.e., closed bottles with DBO meter), whereas determination of the carbon dioxide concentration requires costly equipment. Moreover, inoculum shows huge variability (several sources, treatments and quantities), causing variability in the results. The test sets that combine various types of controls (blank, sterile, toxicity) are not the same from one standard test to another and are sometimes open to criticism in their present form. Finally, as seen in Table 8, there is not common agreement between the standardization committees about the threshold for

Table 7. Environmental conditions of standard methods

Number name	Period test	Temperature	pH	Darkness
ASTM D 5209-92	not specified	23±1 °C	7±1	not specified
ASTM D 5247-92	at least 14 days	Bacteria: 37 °C	7.1	
		Mushroom: 30 °C	4.5	
ASTM D 5271-93	4–6 weeks	23±2 °C	7.3±0.2	
OCDE 301B	28 days	22±2 °C	7.4±0.2	yes
OCDE 301C		25±1 °C	7.0	yes
OCDE 301D		20 °C	7.4±0.2	yes
OCDE 301F		22±2 °C	7.4±0.2	not specified

Table 8. Threshold for validity and assessment of biodegradability

Test	Criterion	Validity of the test "positive" control	Biodegradability of the material "sample" test
OECD (exept 301B)	% BOD	> 40% (7 days)	> 60%
		> 65% (14 days)	
OECD 301B	% CO_2 produced	60–70%	> 60%
ASTM	% CO_2 produced	> 70%	No threshold

controlling the validity of the test (%CO_2 produced from the positive control) or for assessing for the biodegradability of the material (%CO_2 produced from the test material).

5
Standard Test Methods in Solid Media

Few methods have been established in solid media. Two types of test can be distinguished, one dealing with model solid media (Petri disk screen), the other, closer to real conditions, dealing with soil or compost.

5.1
Petri Disk Screen

American (ASTM G21-70, G22-76, G29-75), French (AFNOR X 41-514-81 and X 41-517-69), German (DIN 53739) and International Committees (ISO 846) have set up methods for assessing the resistance of plastics to microbial degradation. The material (e.g., a 2.5 cm×2.5 cm sample) is placed on the surface of a mineral salt agar in a Petri dish containing no additional carbon source. The test material and the agar medium are sprayed with a standardized mixed inoculum of known fungi or bacteria (e.g.: *Aspergillus niger, Penicillium funiculosum*). Petri dishes are incubated at a constant temperature for 21 to 28 days. The test material is then subjected to the following measurements:
- visual assessment: A qualitative measurement shows the amount of growth on the surface of the material or clear zones due to hydrolysis of substrate by the enzymes released (Fig. 4). This estimate is visual, sometimes using microscopy if necessary.
- weight loss measurement which gives a quantitative value of degradability.
- measurement of mechanical properties if samples are tailored in normalized shapes (e.g. dumbbells).

Petri dish tests are more likely to give an estimate of primary biodegradability than ultimate biodegradability: these tests give an estimate of the resistance of plastic to microbiological degradation and the clear zones do not necessarily show the complete consumption of the material but degradation of the polymer bonds.

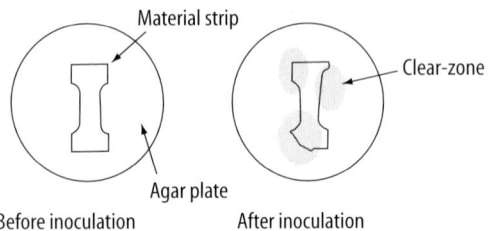

Fig. 4. Aspect of clear-zone formed by bacteria and other microorganisms on material put on agar plate

The advantage of these tests is that, in a single Petri dish, several materials (up to 10) can be studied under the same conditions.

5.2
Compost Chamber Method

Only one standard method is available for assessing the biodegradability of material in compost (ASTM D 5338-92). OECD has published a test (OECD 304A) for the biodegradation of chemical products (not materials) in reconstituted soils. This test uses a specific radioactive substrate; the material must be labeled with ^{14}C and the rate of biodegradation is given in relation to the labeled carbon only. This method is not easily applicable to plastic materials because of the difficulty in obtaining labelled materials and in using radioactivity detectors. In Europe, the E.C.N Technical Committee (TC261) is working on a test draft on the biodegradation and disintegration of packaging materials under composting conditions (E.C.N draft, 1995).

Both standard tests determine the biodegradability under aerobically controlled composting conditions and are based on organic carbon conversion into CO_2. The inoculum consists of mature compost produced from organic fraction municipal solid waste. In order to be acceptable it must have the following characteristics: a pH of between 7 and 8.2, a total solid content of roughly 50% and an ash content of less than 70%.

The set-up is the same in these two standard tests and is very similar to that of liquid tests (Fig. 1b). The samples are intensively composted in a static reactor over a 45-day period. A continuous air flow free of CO_2 is provided to the test vessels. The carbon dioxide produced in the test and blank vessels is either continuously monitored (by gas chromatography or infrared) or measured at regular intervals (titration). The biodegradability of the sample is also reported as a percentage of the biodegradability of a reference positive substance (e.g., cellulose for the thin-layer chromatography). The tests are valid if more than 70 % of the positive reference is degraded.

The single important difference in these tests is the incubation temperature. The American standard test recommends a profile incubation temperature to simulate the temperature profile present in a compostage: one day at 35 °C, 4 days at 58 °C, 28 days at 50 °C and finally a reduction of the temperature to 35 °C until 45 days. The European draft test procedure uses a constant temperature (58 °C).

Results obtained, by both methods, on Kraft paper and polyhydroxybutyrate-hydroxyvalerate (PHB-HV) did not differ [50].

5.3
Conclusions

Standard tests in solid conditions are rare. Petri dishes do not determine mineralization but only the resistance of a material to selective exposure to microorganisms and thus are not relevant for the problem of waste elimination. Controlled composting tests are more representative of the natural environment (i.e. waste management treatment).

6
Discussion and Conclusion

Standard tests for estimating the biodegradation of materials in a given environment have been reviewed in this paper. Several conclusions arise:

Standard test methods are based on the same principle, i.e., respiratory tests. However, it has been shown that a single measurement of consumed O_2 or produced CO_2 is not enough and must be completed by a biomass evaluation and residual organic carbon, in order to make the carbon balance.

Experimental parameters that may influence the mineralization rate differ from one liquid standard test to another:
- Environmental conditions (e.g. temperature, pH, humidity, light, nutriment in the medium)
- Microbial population (e.g. quantity, variety, source, activity...) and its conditioning (extraction, preculture, aeration...)
- Sample concentration and form (e.g. powder, film) that can interfere with the contact between the polymer and the microorganisms (specific surface, availability).

In order to produce a reproducible method it is necessary to keep all these conditions constant from one test to another (and to avoid different conditions in the same standard from one experiment to another). It appears that the major source of variability, and the most difficult to monitor, is the inoculum.

"Control" tests (i.e. other than the "sample" and "blank" ones) are set up in order to check the validity of the test. However, the inhibition test procedure is not satisfactory because it does not take into account the toxicity of degradation products. A modified protocol has been proposed.

These test methods, and especially liquid medium tests, are considered as accelerated tests and do not necessarily correlate with the real exposition conditions of biodegradable plastics. Most representative tests are compost tests but they are more difficult to set up (especially when they use labeled material). It is necessary to assess the representativeness of these tests, especially liquid medium ones, in real-life conditions.

For all these reasons, tests are still far from perfect. There is additional work to be done to make them more reproducible and especially to be sure that the same material will give the same result. If not, standard test results might allow the marketing of dangerous products internationally.

Acknowledgment. We are grateful for the financial support received from the A.D.E.M.E. (France).

References

1. Kirkman A, Kline CH (1991) Recycling plastics today. Chemtech 10: 606-614
2. Filiol B (1991) Les plastiques biodégradables. Rapport du Ministère de l'Industrie et du Commerce Extérieur. France. June 1991. Pp 1-67
3. Cheverry M (1995) Enjeux de la directive européenne sur les emballages. La lettre ADEME 21: 6

4. Ceccaldi P (1993) Le déchet en mutation. Biofutur 12: 30-31
5. Evans JD, Sikdar SK (1990) Biodegradable plastics: an idea whose time has come. Chemtech 20: 38-42
6. Cimmino A, Conte C, Incitte S (1991) Biodegradability of plastic bag: chemical and regulatory aspects. Rass Chem 43: 109-116
7. Musmeci L, Gucci PM, Voltera L (1994) Paper as reference material in „Sturm test" applied to insoluble substances. Env. Toxicol. Water Quality. 9: 83-86
8. Kaplan DL, Mayer JM, Ball D, McCassie J, Allan AL, Stenhouse P (1993) Fundamentals of biodegradable polymers. Biodegradable polymers and packaging. pp 1-43
9. Thayer A (1990) Degradable plastics generate contreversy in solid waste issues. Chemical and Engineering News 68: 7-24
10. Thiebaud S, Borredon ME (1995) Solvent-free esterification of wood with fatty acid chloride. Bioresource Technology 52
11. Brandl H, Gross RA, Lenz RW, Fuller RC (1988) *Pseudomonas oleovorans* as source of poly(β-hydroxyalkanoates) for potential applications as biodegradable polyesters. Appl Env Microbiol 54: 1977-1982
12. Doi Y, Kumagai Y, Tanahashi N, Mukai K (1992) Structural effects on the biodegradation of microbial and synthetic Poly(hydroxyalkanoates). In: Vert M, Feijen J, Albertsson A, Scott G, Chiellini E (eds) Biodegradable Polymers and Plastics. The Royal Society of Chemistry, Cambridge, pp 139-148
13. Bastioli C, Belloti V, Rallis A (1994) Microstructure and melt flow behavior of a starch-based polymer. Rheol Acta 33: 307-316
14. Mayer J, Allen AL, Dell PA, Kaplan DL (1994) Development of biodegradable materials: balancing degradability and performance. Polym Prep 34: 910-911
15. Battersby NS, Pack SE, Watkinson RJ (1992) A correlation between the biodegradability of oil products in the CEC L-33-T82 and modified sturm tests. Chemosphere 24: 1989-2000
16. Krupp LR, Jewell WJ (1992) Biodegradability of modified plastics films in controlled biological environments. Environ Sci. Technol 26: 193-198
17. Nyholm N (1991) The european system of standardized legal tests for assesing the biodegradability of chemicals. Env Technol & Chem 10: 1237-1246
18. Seal K (1991) A review of biodegradability test for new chemical notifications scheme. Chimica Oggi 9: 30-32
19. Weytjens D, Van Ginneken I, Painter HA (1994) The recovery of carbon dioxide in the sturm test for ready biodegradability. Chemosphere 28: 801-812
20. Cha Y, Pitt CG (1990) The biodegradability of polyester blends. Biomaterials 11: 108-112
21. David C, De Kesel C, Lefebvre F, Weiland M (1994) The biodegradation of polymers: recent results. Die Ang. Makro. Chem. 216: 21-35.
22. Ottenbrite RM, Albertsson AC (1992) Discussion on degradation terminology. In: Vert M, Feijen J, Albertsson A, Scott G, Chiellini E (eds) Biodegradable Polymers and Plastics. The Royal Society of Chemistry, Cambridge, pp 73-92
23. Swift G (1992) Biodegradability of polymers in the environment: complexities and signifiance of definitions and measurements. FEMS Microbiol Rev 103: 339-346
24. Van Volkenburgh B, White M (1994) Overview of biodegradable polymers and solid waste issues. Personal communication
25. Barenberg SA, Brash JL, Narayan R, Redpath AE (1990) Degradable materials: perspectives, issues and opportunities. CRC, Boca Raton, FL
26. Vert M, Feijen J, Albertsson A, Scott G, Chiellini E (1992) Biodegradable polymers and plastics. The Royal Society of Chemistry, Cambridge
27. Doi Y, Fukuda K (1994) Biodegradable plastics and polymers. Elsevier, Amsterdam
28. Andrady AL (1994) Assessment of environmental biodegradation of synthetic polymers. JMS – Rev Macromol Chem Phys C34: 25-76
29. Kimura M, Toyota K, Iwatsuki M, Sawada H (1994) Effects of soil conditions on biodegradation of plastics and responsible microorganisms. In: Doi Y, Fukuda K (eds) Biodegradable Plastics and Polymers. Elsevier, pp 92-108

30. Battersby NS, Fieldwick PA, Ablitt T, Lee SA, Moys GR (1994) The interpretation of CEC L-33-T-82 biodegradability data. Chemosphere 28: 787-800
31. Buchanan CM, Gardner RM, Komarek RJ (1993) Aerobic biodegradation of cellulose acetate. J Appl Polym Sci 47: 1709-1719
32. Seal K (1991) A review of biodegradability test for new chemical notifications scheme. Chimica Oggi 9: 30-32
33. Swift G (1994) Expectations for biodegradation testing methods. In: Doi Y, Fukuda K (eds) Biodegradable Plastics and Polymers. Elsevier, pp 228-236
34. Carrick DT (1994) Composting of PHBV co-polymers: the relationship between laboratory biodegradation testing and practical composting. Symposium on Polymers from Renewable Resources and their degradation. Stockholm November 1994, p 26
35. Lenz RW (1993) Biodegradable polymers. Adv Polym Sci 107: 1-40
36. Sawada H (1994) Field testing of biodegradable plastics. In: Doi Y, Fukuda K (eds) Biodegradable Plastics and Polymers. Elsevier, pp 298-312
37. ASTM (1992) American Standardization of Testing Material. In: Annual Book of ASTM Standards. vol 08-02
38. McCassie JE, Mayer JM, Stoto RZ, Shupe AE (1992) Current methods for determining biodegradation of polymeric materials. Polym Mate Sci Eng 67: 353-355
39. Narayan R (1992) Development of standards for degradable plastics by ASTM Subcommittee D-20.96 on environmentally degradable plastics. In: Vert M, Feijen J, Albertsson A, Scott G, Chiellini E (eds) Biodegradable Polymers and Plastics. The Royal Society of Chemistry, Cambridge, pp 176-190
40. JETRO (1992) R & D on biodegradable plastics in Japan. Report of Japan External Trade Organization, pp 1-35
41. Breant P, Aitken Y (1992) Regulations and standards in Europe. In: Vert M, Feijen J, Albertsson A, Scott G, Chiellini E (eds) Biodegradable Polymers and Plastics. The Royal Society of Chemistry, Cambridge, pp 165-168
42. Nyholm N (1991) The European system of standardized legal tests for assessing the biodegradability of chemicals. Env Toxicol & Chem 10: 1237-1246
43. Ziegahm KF (1994) Degradation of packaging and packaging materials – Requirements and recent legislation in Europe. Symposium on Polymers from Renewable Resources and their degradation. Stockholm, November 1994, p 18
44. OECD. Organisation for Economic Cooperation and Development (1992) Guidelines for Testing of Chemicals. Paris France
45. Jones PH, Prasad D, Heskins M, Morgan MH, Guillet JE (1975) Bidegradability of photodegraded polymers. I-development of experimental procedure. Environmental Science and Technology 8: 919-923
46. Van Der Zee M, Sitsma L, Tournois H, DeWit D (1994) Assessment of biodegradation of water insoluble polymeric materials in aerobic and anaerobic aquatic environments. Chemosphere. 28: 1757-1771
47. Thouand G, Block JC (1993) The use of precultured inocula for biodegradability tests. Env Technol 14: 601-614
48. Sturm RN (1973) Biodegradability of nonionic surfactants: screebning test for predicting rate and ultimate biodegradation. J Oil Chem Soc 50: 159-167
49. Raghavan D, Wagner GC, Wool RP (1993) Aerobic biometer analysis of glucose and starch biodegradation. J Environ Polym Degrad 1: 203-211
50. Bloenberger S, David J, Geyer D, Gusterfson A, Snook J, Narayan R (1994) In: Doi Y, Fukuda K (eds) Biodegradable Plastics and Polymers. Elsevier, pp 601-609

Editor: Prof. T. Saegusa
Received: January 1997

Author Index Volumes 101–135

Author Index Volumes 1–100 see Volume 100

Adolf, D. B. see *Ediger, M. D.*: Vol. 116, pp. 73-110.
Aharoni, S. M. and *Edwards, S. F.*: Rigid Polymer Networks. Vol. 118, pp. 1-231.
Améduri, B., Boutevin, B. and *Gramain, P.*: Synthesis of Block Copolymers by Radical Polymerization and Telomerization. Vol. 127, pp. 87-142.
Améduri, B. and *Boutevin, B.*: Synthesis and Properties of Fluorinated Telechelic Monodispersed Compounds. Vol. 102, pp. 133-170.
Amselem, S. see *Domb, A. J.*: Vol. 107, pp. 93-142.
Andrady, A. L.: Wavelenght Sensitivity in Polymer Photodegradation. Vol. 128, pp. 47-94.
Andreis, M. and *Koenig, J. L.*: Application of Nitrogen-15 NMR to Polymers. Vol. 124, pp. 191-238.
Angiolini, L. see *Carlini, C.*: Vol. 123, pp. 127-214.
Anseth, K. S., Newman, S. M. and *Bowman, C. N.*: Polymeric Dental Composites: Properties and Reaction Behavior of Multimethacrylate Dental Restorations. Vol. 122, pp. 177-218.
Armitage, B. A. see *O'Brien, D. F.*: Vol. 126, pp. 53-58.
Arndt, M. see *Kaminski, W.*: Vol. 127, pp. 143-187.
Arnold Jr., F. E. and *Arnold, F. E.*: Rigid-Rod Polymers and Molecular Composites. Vol. 117, pp. 257-296.
Arshady, R.: Polymer Synthesis via Activated Esters: A New Dimension of Creativity in Macromolecular Chemistry. Vol. 111, pp. 1-42.

Bahar, I., Erman, B. and *Monnerie, L.*: Effect of Molecular Structure on Local Chain Dynamics: Analytical Approaches and Computational Methods. Vol. 116, pp. 145-206.
Baltá-Calleja, F. J., González Arche, A., Ezquerra, T. A., Santa Cruz, C., Batallón, F., Frick, B. and *López Cabarcos, E.*: Structure and Properties of Ferroelectric Copolymers of Poly(vinylidene) Fluoride. Vol. 108, pp. 1-48.
Barshtein, G. R. and *Sabsai, O. Y.*: Compositions with Mineralorganic Fillers. Vol. 101, pp.1-28.
Batallán, F. see *Baltá-Calleja, F. J.*: Vol. 108, pp. 1-48.
Barton, J. see *Hunkeler, D.*: Vol. 112, pp. 115-134.
Bell, C. L. and *Peppas, N. A.*: Biomedical Membranes from Hydrogels and Interpolymer Complexes. Vol. 122, pp. 125-176.
Bellon-Maurel, A. see *Calmon-Decriaud, A.*: Vol. 135, pp. 207-226
Bennett, D. E. see *O'Brien, D. F.*: Vol. 126, pp. 53-84.
Berry, G.C.: Static and Dynamic Light Scattering on Moderately Concentraded Solutions: Isotropic Solutions of Flexible and Rodlike Chains and Nematic Solutions of Rodlike Chains. Vol. 114, pp. 233-290.
Bershtein, V. A. and *Ryzhov, V. A.*: Far Infrared Spectroscopy of Polymers. Vol. 114, pp. 43-122.
Bigg, D. M.: Thermal Conductivity of Heterophase Polymer Compositions. Vol. 119, pp. 1-30.
Binder, K.: Phase Transitions in Polymer Blends and Block Copolymer Melts: Some Recent Developments. Vol. 112, pp. 115-134.
Bird, R. B. see *Curtiss, C. F.*: Vol. 125, pp. 1-102.

Biswas, M. and *Mukherjee, A.*: Synthesis and Evaluation of Metal-Containing Polymers. Vol. 115, pp. 89-124.
Boutevin, B. and *Robin, J. J.*: Synthesis and Properties of Fluorinated Diols. Vol. 102. pp. 105-132.
Boutevin, B. see Amédouri, B.: Vol. 102, pp. 133-170.
Boutevin, B. see Améduri, B.: Vol. 127, pp. 87-142.
Bowman, C. N. see Anseth, K. S.: Vol. 122, pp. 177-218.
Boyd, R. H.: Prediction of Polymer Crystal Structures and Properties. Vol. 116, pp. 1-26.
Bronnikov, S. V., Vettegren, V. I. and *Frenkel, S. Y.*: Kinetics of Deformation and Relaxation in Highly Oriented Polymers. Vol. 125, pp. 103-146.
Bruza, K. J. see Kirchhoff, R. A.: Vol. 117, pp. 1-66.
Burban, J. H. see Cussler, E. L.: Vol. 110, pp. 67-80.

Calmon-Decriaud, A. Bellon-Maurel, V., Silvestre, F.: Standard Methods for Testing the Aerobic Biodegradation of Polymeric Materials. Vol 135, pp. 207-226
Cameron, N. R. and *Sherrington, D. C.*: High Internal Phase Emulsions (HIPEs)-Structure, Properties and Use in Polymer Preparation. Vol. 126, pp. 163-214.
Candau, F. see Hunkeler, D.: Vol. 112, pp. 115-134.
Canelas, D. A. and *DeSimone, J. M.*: Polymerizations in Liquid and Supercritical Carbon Dioxide. Vol. 133, pp. 103-140.
Capek, I.: Kinetics of the Free-Radical Emulsion Polymerization of Vinyl Chloride. Vol. 120, pp. 135-206.
Carlini, C. and *Angiolini, L.*: Polymers as Free Radical Photoinitiators. Vol. 123, pp. 127-214.
Casas-Vazquez, J. see Jou, D.: Vol. 120, pp. 207-266.
Chandrasekhar, V.: Polymer Solid Electrolytes: Synthesis and Structure. Vol 135, pp. 139-206
Chen, P. see Jaffe, M.: Vol. 117, pp. 297-328.
Choe, E.-W. see Jaffe, M.: Vol. 117, pp. 297-328.
Chow, T. S.: Glassy State Relaxation and Deformation in Polymers. Vol. 103, pp. 149-190.
Chung, T.-S. see Jaffe, M.: Vol. 117, pp. 297-328.
Connell, J. W. see Hergenrother, P. M.: Vol. 117, pp. 67-110.
Criado-Sancho, M. see Jou, D.: Vol. 120, pp. 207-266.
Curro, J.G. see Schweizer, K.S.: Vol. 116, pp. 319-378.
Curtiss, C. F. and *Bird, R. B.*: Statistical Mechanics of Transport Phenomena: Polymeric Liquid Mixtures. Vol. 125, pp. 1-102.
Cussler, E. L., Wang, K. L. and *Burban, J. H.*: Hydrogels as Separation Agents. Vol. 110, pp. 67-80.

DeSimone, J. M. see Canelas D. A.: Vol. 133, pp. 103-140.
Dimonie, M. V. see Hunkeler, D.: Vol. 112, pp. 115-134.
Dodd, L. R. and *Theodorou, D. N.*: Atomistic Monte Carlo Simulation and Continuum Mean Field Theory of the Structure and Equation of State Properties of Alkane and Polymer Melts. Vol. 116, pp. 249-282.
Doelker, E.: Cellulose Derivatives. Vol. 107, pp. 199-266.
Domb, A. J., Amselem, S., Shah, J. and *Maniar, M.*: Polyanhydrides: Synthesis and Characterization. Vol.107, pp. 93-142.
Dubrovskii, S. A. see Kazanskii, K. S.: Vol. 104, pp. 97-134.
Dunkin, I. R. see Steinke, J.: Vol. 123, pp. 81-126.

Economy, J. and *Goranov, K.*: Thermotropic Liquid Crystalline Polymers for High Performance Applications. Vol. 117, pp. 221-256.
Ediger, M. D. and *Adolf, D. B.*: Brownian Dynamics Simulations of Local Polymer Dynamics. Vol. 116, pp. 73-110.
Edwards, S. F. see Aharoni, S. M.: Vol. 118, pp. 1-231.
Endo, T. see Yagci, Y.: Vol. 127, pp. 59-86.
Erman, B. see Bahar, I.: Vol. 116, pp. 145-206.

Ewen, B, Richter, D.: Neutron Spin Echo Investigations on the Segmental Dynamics of Polymers in Melts, Networks and Solutions. Vol. 134, pp. 1-130.
Ezquerra, T. A. see Baltá-Calleja, F. J.: Vol. 108, pp. 1-48.

Fendler, J.H.: Membrane-Mimetic Approach to Advanced Materials. Vol. 113, pp. 1-209.
Fetters, L. J. see Xu, Z.: Vol. 120, pp. 1-50.
Förster, S. and *Schmidt, M.*: Polyelectrolytes in Solution. Vol. 120, pp. 51-134.
Frenkel, S. Y. see Bronnikov, S. V.: Vol. 125, pp. 103-146.
Frick, B. see Baltá-Calleja, F. J.: Vol. 108, pp. 1-48.
Fridman, M. L.: see Terent´eva, J. P.: Vol. 101, pp. 29-64.

Ganesh, K. see Kishore, K.: Vol. 121, pp. 81-122.
Geckeler, K. E. see Rivas, B.: Vol. 102, pp. 171-188.
Geckeler, K. E.: Soluble Polymer Supports for Liquid-Phase Synthesis. Vol. 121, pp. 31-80.
Gehrke, S. H.: Synthesis, Equilibrium Swelling, Kinetics Permeability and Applications of Environmentally Responsive Gels. Vol. 110, pp. 81-144.
Godovsky, D. Y.: Electron Behavior and Magnetic Properties Polymer-Nanocomposites. Vol. 119, pp. 79-122.
González Arche, A. see Baltá-Calleja, F. J.: Vol. 108, pp. 1-48.
Goranov, K. see Economy, J.: Vol. 117, pp. 221-256.
Gramain, P. see Améduri, B.: Vol. 127, pp. 87-142.
Grosberg, A. and *Nechaev, S.*: Polymer Topology. Vol. 106, pp. 1-30.
Grubbs, R., Risse, W. and *Novac, B.*: The Development of Well-defined Catalysts for Ring-Opening Olefin Metathesis. Vol. 102, pp. 47-72.
van Gunsteren, W. F. see Gusev, A. A.: Vol. 116, pp. 207-248.
Gusev, A. A., Müller-Plathe, F., van Gunsteren, W. F. and *Suter, U. W.*: Dynamics of Small Molecules in Bulk Polymers. Vol. 116, pp. 207-248.
Guillot, J. see Hunkeler, D.: Vol. 112, pp. 115-134.
Guyot, A. and *Tauer, K.*: Reactive Surfactants in Emulsion Polymerization. Vol. 111, pp. 43-66.

Hadjichristidis, N. see Xu, Z.: Vol. 120, pp. 1-50.
Hadjichristidis, N. see Pitsikalis, M.: Vol. 135, pp. 1-138
Hall, H. K. see Penelle, J.: Vol. 102, pp. 73-104.
Hammouda, B.: SANS from Homogeneous Polymer Mixtures: A Unified Overview. Vol. 106, pp. 87-134.
Harada, A.: Design and Construction of Supramolecular Architectures Consisting of Cyclodextrins and Polymers. Vol. 133, pp. 141-192.
Hedrick, J. L. see Hergenrother, P. M.: Vol. 117, pp. 67-110.
Heller, J.: Poly (Ortho Esters). Vol. 107, pp. 41-92.
Hemielec, A. A. see Hunkeler, D.: Vol. 112, pp. 115-134.
Hergenrother, P. M., Connell, J. W., Labadie, J. W. and *Hedrick, J. L.*: Poly(arylene ether)s Containing Heterocyclic Units. Vol. 117, pp. 67-110.
Hiramatsu, N. see Matsushige, M.: Vol. 125, pp. 147-186.
Hirasa, O. see Suzuki, M.: Vol. 110, pp. 241-262.
Hirotsu, S.: Coexistence of Phases and the Nature of First-Order Transition in Poly-N-isopropylacrylamide Gels. Vol. 110, pp. 1-26.
Hunkeler, D., Candau, F., Pichot, C., Hemielec, A. E., Xie, T. Y., Barton, J., Vaskova, V., Guillot, J., Dimonie, M. V., Reichert, K. H.: Heterophase Polymerization: A Physical and Kinetic Comparision and Categorization. Vol. 112, pp. 115-134.

Ichikawa, T. see Yoshida, H.: Vol. 105, pp. 3-36.
Ihara, E. see Yasuda, H.: Vol. 133, pp. 53-102.
Ilavsky, M.: Effect on Phase Transition on Swelling and Mechanical Behavior of Synthetic Hydrogels. Vol. 109, pp. 173-206.

Inomata, H. see Saito, S.: Vol. 106, pp. 207-232.
Irie, M.: Stimuli-Responsive Poly(N-isopropylacrylamide), Photo- and Chemical-Induced Phase Transitions. Vol. 110, pp. 49-66.
Ise, N. see Matsuoka, H.: Vol. 114, pp. 187-232.
Ivanov, A. E. see Zubov, V. P.: Vol. 104, pp. 135-176.

Jaffe, M., Chen, P., Choe, E.-W., Chung, T.-S. and *Makhija, S.*: High Performance Polymer Blends. Vol. 117, pp. 297-328.
Jou, D., Casas-Vazquez, J. and *Criado-Sancho, M.*: Thermodynamics of Polymer Solutions under Flow: Phase Separation and Polymer Degradation. Vol. 120, pp. 207-266.

Kaetsu, I.: Radiation Synthesis of Polymeric Materials for Biomedical and Biochemical Applications. Vol. 105, pp. 81-98.
Kaminski, W. and *Arndt, M.*: Metallocenes for Polymer Catalysis. Vol. 127, pp. 143-187.
Kammer, H. W., Kressler, H. and *Kummerloewe, C.*: Phase Behavior of Polymer Blends - Effects of Thermodynamics and Rheology. Vol. 106, pp. 31-86.
Kandyrin, L. B. and *Kuleznev, V. N.*: The Dependence of Viscosity on the Composition of Concentrated Dispersions and the Free Volume Concept of Disperse Systems. Vol. 103, pp. 103-148.
Kaneko, M. see Ramaraj, R.: Vol. 123, pp. 215-242.
Kang, E. T., Neoh, K. G. and *Tan, K. L.*: X-Ray Photoelectron Spectroscopic Studies of Electroactive Polymers. Vol. 106, pp. 135-190.
Kazanskii, K. S. and *Dubrovskii, S. A.*: Chemistry and Physics of „Agricultural" Hydrogels. Vol. 104, pp. 97-134.
Kennedy, J. P. see Majoros, I.: Vol. 112, pp. 1-113.
Khokhlov, A., Starodybtzev, S. and *Vasilevskaya, V.*: Conformational Transitions of Polymer Gels: Theory and Experiment. Vol. 109, pp. 121-172.
Kilian, H. G. and *Pieper, T.*: Packing of Chain Segments. A Method for Describing X-Ray Patterns of Crystalline, Liquid Crystalline and Non-Crystalline Polymers. Vol. 108, pp. 49-90.
Kishore, K. and *Ganesh, K.*: Polymers Containing Disulfide, Tetrasulfide, Diselenide and Ditelluride Linkages in the Main Chain. Vol. 121, pp. 81-122.
Klier, J. see Scranton, A. B.: Vol. 122, pp. 1-54.
Kobayashi, S., Shoda, S. and *Uyama, H.*: Enzymatic Polymerization and Oligomerization. Vol. 121, pp. 1-30.
Koenig, J. L. see Andreis, M.: Vol. 124, pp. 191-238.
Kokufuta, E.: Novel Applications for Stimulus-Sensitive Polymer Gels in the Preparation of Functional Immobilized Biocatalysts. Vol. 110, pp. 157-178.
Konno, M. see Saito, S.: Vol. 109, pp. 207-232.
Kopecek, J. see Putnam, D.: Vol. 122, pp. 55-124.
Koßmehl, G. see Schopf, G.: Vol. 129, pp. 1-145.
Kressler, J. see Kammer, H. W.: Vol. 106, pp. 31-86.
Kirchhoff, R. A. and *Bruza, K. J.*: Polymers from Benzocyclobutenes. Vol. 117, pp. 1-66.
Kuchanov, S. I.: Modern Aspects of Quantitative Theory of Free-Radical Copolymerization. Vol. 103, pp. 1-102.
Kuleznev, V. N. see Kandyrin, L. B.: Vol. 103, pp. 103-148.
Kulichkhin, S. G. see Malkin, A. Y.: Vol. 101, pp. 217-258.
Kummerloewe, C. see Kammer, H. W.: Vol. 106, pp. 31-86.
Kuznetsova, N. P. see Samsonov, G. V.: Vol. 104, pp. 1-50.Labadie, J. W. see Hergenrother, P. M.: Vol. 117, pp. 67-110.

Lamparski, H. G. see O´Brien, D. F.: Vol. 126, pp. 53-84.
Laschewsky, A.: Molecular Concepts, Self-Organisation and Properties of Polysoaps. Vol. 124, pp. 1-86.
Laso, M. see Leontidis, E.: Vol. 116, pp. 283-318.

Lazár, M. and *Rychlǫ, R.*: Oxidation of Hydrocarbon Polymers. Vol. 102, pp. 189-222.
Lenz, R. W.: Biodegradable Polymers. Vol. 107, pp. 1-40.
Leontidis, E., de Pablo, J. J., Laso, M. and *Suter, U. W.*: A Critical Evaluation of Novel Algorithms for the Off-Lattice Monte Carlo Simulation of Condensed Polymer Phases. Vol. 116, pp. 283-318.
Lesec, J. see *Viovy, J.-L.*: Vol. 114, pp. 1-42.
Liang, G. L. see Sumpter, B. G.: Vol. 116, pp. 27-72.
Lin, J. and *Sherrington, D. C.*: Recent Developments in the Synthesis, Thermostability and Liquid Crystal Properties of Aromatic Polyamides. Vol. 111, pp. 177-220.
López Cabarcos, E. see Baltá-Calleja, F. J.: Vol. 108, pp. 1-48.

Majoros, I., Nagy, A. and *Kennedy, J. P.*: Conventional and Living Carbocationic Polymerizations United. I. A Comprehensive Model and New Diagnostic Method to Probe the Mechanism of Homopolymerizations. Vol. 112, pp. 1-113.
Makhija, S. see Jaffe, M.: Vol. 117, pp. 297-328.
Malkin, A. Y. and *Kulichkhin, S. G.*: Rheokinetics of Curing. Vol. 101, pp. 217-258.
Maniar, M. see Domb, A. J.: Vol. 107, pp. 93-142.
Mashima, K., Nakayama, Y. and *Nakamura, A.*: Recent Trends in Polymerization of a-Olefins Catalyzed by Organometallic Complexes of Early Transition Metals. Vol. 133, pp. 1-52.
Matsumoto, A.: Free-Radical Crosslinking Polymerization and Copolymerization of Multivinyl Compounds. Vol. 123, pp. 41-80.
Matsuoka, H. and *Ise, N.*: Small-Angle and Ultra-Small Angle Scattering Study of the Ordered Structure in Polyelectrolyte Solutions and Colloidal Dispersions. Vol. 114, pp. 187-232.
Matsushige, K., Hiramatsu, N. and *Okabe, H.*: Ultrasonic Spectroscopy for Polymeric Materials. Vol. 125, pp. 147-186.
Mattice, W. L. see Rehahn, M.: Vol. 131/132, pp. 1-475
Mays, W. see Xu, Z.: Vol. 120, pp. 1-50.
Mays, J.W. see Pitsikalis, M.: Vol.135, pp. 1-138
Mikos, A. G. see Thomson, R. C.: Vol. 122, pp. 245-274.
Miyasaka, K.: PVA-Iodine Complexes: Formation, Structure and Properties. Vol. 108. pp. 91-130.
Monnerie, L. see Bahar, I.: Vol. 116, pp. 145-206.
Morishima, Y.: Photoinduced Electron Transfer in Amphiphilic Polyelectrolyte Systems. Vol. 104, pp. 51-96.
Mours, M. see Winter, H. H.: Vol. 134, pp. 165-234.
Müllen, K. see Scherf, U.: Vol. 123, pp. 1-40.
Müller-Plathe, F. see Gusev, A. A.: Vol. 116, pp. 207-248.
Mukerherjee, A. see Biswas, M.: Vol. 115, pp. 89-124.
Mylnikov, V.: Photoconducting Polymers. Vol. 115, pp. 1-88.

Nagy, A. see Majoros, I.: Vol. 112, pp. 1-11.
Nakamura, A. see Mashima, K.: Vol. 133, pp. 1-52.
Nakayama, Y. see Mashima, K.: Vol. 133, pp. 1-52.
Narasinham, B., Peppas, N. A.: The Physics of Polymer Dissolution: Modeling Approaches and Experimental Behavior. Vol. 128, pp. 157-208.
Nechaev, S. see Grosberg, A.: Vol. 106, pp. 1-30.
Neoh, K. G. see Kang, E. T.: Vol. 106, pp. 135-190.
Newman, S. M. see Anseth, K. S.: Vol. 122, pp. 177-218.
Nijenhuis, K. te: Thermoreversible Networks. Vol. 130, pp. 1-252.
Noid, D. W. see Sumpter, B. G.: Vol. 116, pp. 27-72.
Novac, B. see Grubbs, R.: Vol. 102, pp. 47-72.
Novikov, V. V. see Privalko, V. P.: Vol. 119, pp. 31-78.

O'Brien, D. F., Armitage, B. A., Bennett, D. E. and *Lamparski, H. G.*: Polymerization and Domain Formation in Lipid Assemblies. Vol. 126, pp. 53-84.
Ogasawara, M.: Application of Pulse Radiolysis to the Study of Polymers and Polymerizations. Vol.105, pp.37-80.

Okabe, H. see Matsushige, K.: Vol. 125, pp. 147-186.
Okada, M.: Ring-Opening Polymerization of Bicyclic and Spiro Compounds. Reactivities and Polymerization Mechanisms. Vol. 102, pp. 1-46.
Okano, T.: Molecular Design of Temperature-Responsive Polymers as Intelligent Materials. Vol. 110, pp. 179-198.
Onuki, A.: Theory of Phase Transition in Polymer Gels. Vol. 109, pp. 63-120.
Osad'ko, I.S.: Selective Spectroscopy of Chromophore Doped Polymers and Glasses. Vol. 114, pp. 123-186.

de Pablo, J. J. see Leontidis, E.: Vol. 116, pp. 283-318.
Padias, A. B. see Penelle, J.: Vol. 102, pp. 73-104.
Pascault, J.-P. see Williams, R. J. J.: Vol. 128, pp. 95-156.
Pasch, H.: Analysis of Complex Polymers by Interaction Chromatography. Vol. 128, pp. 1-46.
Penelle, J., Hall, H. K., Padias, A. B. and *Tanaka, H.*: Captodative Olefins in Polymer Chemistry. Vol. 102, pp. 73-104.
Peppas, N. A. see Bell, C. L.: Vol. 122, pp. 125-176.
Peppas, N. A. see Narasimhan, B.: Vol. 128, pp. 157-208.
Pichot, C. see Hunkeler, D.: Vol. 112, pp. 115-134.
Pieper, T. see Kilian, H. G.: Vol. 108, pp. 49-90.
Pispas, S. see Pitsikalis, M.: Vol. 135, pp. 1-138
Pitsikalis, M., Pispas, S., Mays, J. W., Hadjichristidis, N.: Nonlinear Block Copolymer Architectures. Vol. 135, pp. 1-138
Pospíšil, J.: Functionalized Oligomers and Polymers as Stabilizers for Conventional Polymers. Vol. 101, pp. 65-168.
Pospíšil, J.: Aromatic and Heterocyclic Amines in Polymer Stabilization. Vol. 124, pp. 87-190.
Priddy, D. B.: Recent Advances in Styrene Polymerization. Vol. 111, pp. 67-114.
Priddy, D. B.: Thermal Discoloration Chemistry of Styrene-co-Acrylonitrile. Vol. 121, pp. 123-154.
Privalko, V. P. and *Novikov, V. V.*: Model Treatments of the Heat Conductivity of Heterogeneous Polymers. Vol. 119, pp 31-78.
Putnam, D. and *Kopecek, J.*: Polymer Conjugates with Anticancer Acitivity. Vol. 122, pp. 55-124.

Ramaraj, R. and *Kaneko, M.*: Metal Complex in Polymer Membrane as a Model for Photosynthetic Oxygen Evolving Center. Vol. 123, pp. 215-242.
Rangarajan, B. see Scranton, A. B.: Vol. 122, pp. 1-54.
Reichert, K. H. see Hunkeler, D.: Vol. 112, pp. 115-134.
Rehahn, M., Mattice, W. L., Suter, U. W.: Rotational Isomeric State Models in Macromolecular Systems. Vol. 131/132, pp. 1-475.
Richter, D. see Ewen, B.: Vol. 134, pp.1-130.
Risse, W. see Grubbs, R.: Vol. 102, pp. 47-72.
Rivas, B. L. and *Geckeler, K. E.*: Synthesis and Metal Complexation of Poly(ethyleneimine) and Derivatives. Vol. 102, pp. 171-188.
Robin, J. J. see Boutevin, B.: Vol. 102, pp. 105-132.
Roe, R.-J.: MD Simulation Study of Glass Transition and Short Time Dynamics in Polymer Liquids. Vol. 116, pp. 111-114.
Rozenberg, B. A. see Williams, R. J. J.: Vol. 128, pp. 95-156.
Ruckenstein, E.: Concentrated Emulsion Polymerization. Vol. 127, pp. 1-58.
Rusanov, A. L.: Novel Bis (Naphtalic Anhydrides) and Their Polyheteroarylenes with Improved Processability. Vol. 111, pp. 115-176.
Rychlý, J. see Lazár, M.: Vol. 102, pp. 189-222.
Ryzhov, V. A. see Bershtein, V. A.: Vol. 114, pp. 43-122.

Sabsai, O. Y. see Barshtein, G. R.: Vol. 101, pp. 1-28.
Saburov, V. V. see Zubov, V. P.: Vol. 104, pp. 135-176.
Saito, S., Konno, M. and *Inomata, H.*: Volume Phase Transition of N-Alkylacrylamide Gels. Vol. 109, pp. 207-232.
Samsonov, G. V. and *Kuznetsova, N. P.*: Crosslinked Polyelectrolytes in Biology. Vol. 104, pp. 1-50.
Santa Cruz, C. see Baltá-Calleja, F. J.: Vol. 108, pp. 1-48.
Sato, T. and *Teramoto, A.*: Concentrated Solutions of Liquid-Christalline Polymers. Vol. 126, pp. 85-162.
Scherf, U. and *Müllen, K.*: The Synthesis of Ladder Polymers. Vol. 123, pp. 1-40.
Schmidt, M. see Förster, S.: Vol. 120, pp. 51-134.
Schopf, G. and *Koßmehl, G.*: Polythiophenes - Electrically Conductive Polymers. Vol. 129, pp. 1-145.
Schweizer, K. S.: Prism Theory of the Structure, Thermodynamics, and Phase Transitions of Polymer Liquids and Alloys. Vol. 116, pp. 319-378.
Scranton, A. B., Rangarajan, B. and *Klier, J.*: Biomedical Applications of Polyelectrolytes. Vol. 122, pp. 1-54.
Sefton, M. V. and *Stevenson, W. T. K.*: Microencapsulation of Live Animal Cells Using Polycrylates. Vol. 107, pp. 143-198.
Shamanin, V. V.: Bases of the Axiomatic Theory of Addition Polymerization. Vol. 112, pp. 135-180.
Sherrington, D. C. see Cameron, N. R. , Vol. 126, pp. 163-214.
Sherrington, D. C. see Lin, J.: Vol. 111, pp. 177-220.
Sherrington, D. C. see Steinke, J.: Vol. 123, pp. 81-126.
Shibayama, M. see Tanaka, T.: Vol. 109, pp. 1-62.
Shiga, T.: Deformation and Viscoelastic Behavior of Polymer Gels in Electric Fields. Vol. 134, pp. 131-164.
Shoda, S. see Kobayashi, S.: Vol. 121, pp. 1-30.
Siegel, R. A.: Hydrophobic Weak Polyelectrolyte Gels: Studies of Swelling Equilibria and Kinetics. Vol. 109, pp. 233-268.
Silvestre, F. see Calmon-Decriaud, A.: Vol. 207, pp. 207-226
Singh, R. P. see Sivaram, S.: Vol. 101, pp. 169-216.
Sivaram, S. and *Singh, R. P.*: Degradation and Stabilization of Ethylene-Propylene Copolymers and Their Blends: A Critical Review. Vol. 101, pp. 169-216.
Starodybtzev, S. see Khokhlov, A.: Vol. 109, pp. 121-172.
Steinke, J., Sherrington, D. C. and *Dunkin, I. R.*: Imprinting of Synthetic Polymers Using Molecular Templates. Vol. 123, pp. 81-126.
Stenzenberger, H. D.: Addition Polyimides. Vol. 117, pp. 165-220.
Stevenson, W. T. K. see Sefton, M. V.: Vol. 107, pp. 143-198.
Sumpter, B. G., Noid, D. W., Liang, G. L. and *Wunderlich, B.*: Atomistic Dynamics of Macromolecular Crystals. Vol. 116, pp. 27-72.
Suter, U. W. see Gusev, A. A.: Vol. 116, pp. 207-248.
Suter, U. W. see Leontidis, E.: Vol. 116, pp. 283-318.
Suter, U. W. see Rehahn, M.: Vol. 131/132, pp. 1-475
Suzuki, A.: Phase Transition in Gels of Sub-Millimeter Size Induced by Interaction with Stimuli. Vol. 110, pp. 199-240.
Suzuki, A. and *Hirasa, O.*: An Approach to Artifical Muscle by Polymer Gels due to Micro-Phase Separation. Vol. 110, pp. 241-262.

Tagawa, S.: Radiation Effects on Ion Beams on Polymers. Vol. 105, pp. 99-116.
Tan, K. L. see Kang, E. T.: Vol. 106, pp. 135-190.
Tanaka, T. see Penelle, J.: Vol. 102, pp. 73-104.
Tanaka, H. and *Shibayama, M.*: Phase Transition and Related Phenomena of Polymer Gels. Vol. 109, pp. 1-62.

Tauer, K. see Guyot, A.: Vol. 111, pp. 43-66.
Teramoto, A. see Sato, T.: Vol. 126, pp. 85-162.
Terent´eva, J. P. and *Fridman, M. L.*: Compositions Based on Aminoresins. Vol. 101, pp. 29-64.
Theodorou, D. N. see *Dodd, L. R.*: Vol. 116, pp. 249-282.
Thomson, R. C., Wake, M. C., Yaszemski, M. J. and *Mikos, A. G.*: Biodegradable Polymer Scaffolds to Regenerate Organs. Vol. 122, pp. 245-274.
Tokita, M.: Friction Between Polymer Networks of Gels and Solvent. Vol. 110, pp. 27-48.
Tsuruta, T.: Contemporary Topics in Polymeric Materials for Biomedical Applications. Vol. 126, pp. 1-52.

Uyama, H. see Kobayashi, S.: Vol. 121, pp. 1-30.

Vasilevskaya, V. see Khokhlov, A.: Vol. 109, pp. 121-172.
Vaskova, V. see Hunkeler, D.: Vol.:112, pp. 115-134.
Verdugo, P.: Polymer Gel Phase Transition in Condensation-Decondensation of Secretory Products. Vol. 110, pp. 145-156.
Vettegren, V. I.: see Bronnikov, S. V.: Vol. 125, pp. 103-146.
Viovy, J.-L. and *Lesec, J.*: Separation of Macromolecules in Gels: Permeation Chromatography and Electrophoresis. Vol. 114, pp. 1-42.
Volksen, W.: Condensation Polyimides: Synthesis, Solution Behavior, and Imidization Characteristics. Vol. 117, pp. 111-164.

Wake, M. C. see Thomson, R. C.: Vol. 122, pp. 245-274.
Wang, K. L. see Cussler, E. L.: Vol. 110, pp. 67-80.
Williams, R. J. J., Rozenberg, B. A., Pascault, J.-P.: Reaction Induced Phase Separation in Modified Thermosetting Polymers. Vol. 128, pp. 95-156.
Winter, H. H., Mours, M.: Rheology of Polymers Near Liquid-Solid Transitions. Vol. 134, pp. 165-234.
Wunderlich, B. see Sumpter, B. G.: Vol. 116, pp. 27-72.

Xie, T. Y. see Hunkeler, D.: Vol. 112, pp. 115-134.
Xu, Z., Hadjichristidis, N., Fetters, L. J. and *Mays, J. W.*: Structure/Chain-Flexibility Relationships of Polymers. Vol. 120, pp. 1-50.

Yagci, Y. and *Endo, T.*: N-Benzyl and N-Alkoxy Pyridium Salts as Thermal and Photochemical Initiators for Cationic Polymerization. Vol. 127, pp. 59-86.
Yannas, I. V.: Tissue Regeneration Templates Based on Collagen-Glycosaminoglycan Copolymers. Vol. 122, pp. 219-244.
Yamaoka, H.: Polymer Materials for Fusion Reactors. Vol. 105, pp. 117-144.
Yasuda, H. and *Ihara, E.*: Rare Earth Metal-Initiated Living Polymerizations of Polar and Nonpolar Monomers. Vol. 133, pp. 53-102.
Yaszemski, M. J. see Thomson, R. C.: Vol. 122, pp. 245-274.
Yoshida, H. and *Ichikawa, T.*: Electron Spin Studies of Free Radicals in Irradiated Polymers. Vol. 105, pp. 3-36.

Zubov, V. P., Ivanov, A. E. and *Saburov, V. V.*: Polymer-Coated Adsorbents for the Separation of Biopolymers and Particles. Vol. 104, pp. 135-176.

Subject Index

Aerobic 212
AFNOR 222
Agar 222
Aggregation number 115-117
Aminotelechelic star polymers 6
Amphiphilic copolymers 93-94, 115, 116, 118
Amphiphilic star block copolymers 8, 11, 115
Anionic ring-opening polymerization 39-41, 43
Arborescent graft copolymers 103, 107
Architectural asymmetry 121, 125-126
Arrhenius plot 146
ASTM 210-211, 222, 223

Bicontinuous morphology 121, 123-126
Biodegradability, definition 209-210
-, inherent 211, 212
-, primary 210, 212
-, ready 211, 212
-, threshold 221, 223
-, ultimate 210, 212
Biodegradable polymers, environmentally acceptable 211
Block-graft copolymers 22-24, 32-33
Bonding, $d\pi$-$p\pi$ 171

Carbon balance 224
Carbon dioxide measurement 214-215, 223
Catenated block copolymers 93, 108, 109
Cellulose 21
Chain transfer agents 25, 47-48
- - -, macromolecular 25
Chloromethylation of PS 20, 103
- -, PVC 20, 21
Clear-zone 222
Compatibilizers 128, 129
Compost 213, 223
Compostability 211

Conductivity, ionic 145
Conductivity plots 150
Conformation characteristics 111-119
Conformational asymmetry 121, 127
Control tests 215
Critical micelle concentration 123
Cyclic block copolymers 93, 108, 109, 113, 117, 118, 127
Cyclophosphazene 181

Decomposition 209
Degradation 209
Dendrimers 93, 96, 100-104, 114, 118
Deterioration 209
Dielectric spectroscopy 126
Diffusion coefficient 111, 118, 127, 128
DIN 210, 222
DOC 214
Dumbbell copolymers 117
Dynamic structure factor 111

ECN 210, 213, 223
Electrolytes 163
-, solid 141
EXAFS 186
Expansion factors 111-113
Experimental conditions 221, 223

Flexible-rigid rod star block copolymers 6
Flory hydrodynamic parameter 113
Flory interaction parameter 114, 119-124, 126, 127
Friedel-Crafts acetylation 27-28

Glass transition temperature 128
Graft copolymers, arborescent 103, 107
Grafting onto, in-situ 25

H copolymers   93-95
H polymers   93
Heterointeractions   111
Hydroboration-oxidation reaction   98
Hydrodynamic radius   111, 112, 115
Hydrosilylation   16, 17, 29, 39, 64, 67, 83, 96, 97
Hydroxyl groups, metallation   32

Impedance spectroscopy   145
Initiators, functional   53
Inoculum   219
Interfacial properties   115-118, 124, 128
Intrinsic viscosity   113, 115, 116, 118
Inverse star block copolymer   7
Ionic conductivity   145
Ionic star block copolymers   6
IR spectroscopy   193
ISO   210, 222

Japanese Biodegradable Plastics Society   210
JIS   210-211

Lamellar domain spacing   124
LC phases   115, 127
Lignin   35
Liquid crystalline phases   115, 127
Liquid test   213

Macrocylinders, molecular   103
Macromolecular chain transfer agent   25
Macrophase separation   122
Mean field theory   114, 119, 121
Mechanical properties   127-129, 222
Media, solid   222
MEEMA   165
MEEP   172
-, crosslinked   174
Membranes   117
Metallation, hydroxyl groups   32
-, polydiene backbones   16, 29-30
Metallocene catalysts   73
Methoxy ethoxy ethoxy polyphosphazene (MEEP)   172
Micelle concentration, critical   123
Micellization   114, 116-118
Microdomain structure/formation   115
Microphase separation   114, 119-128
Mineralization   223, 224
Monte Carlo calculations   111, 113, 114, 122

Morphology   121-127
-, bicontinuous   121, 123-126
NMR   196
Nucleophilic polyaddition   64

OBDD morphology   121, 123, 124
OECD   213, 216, 223
Order-disorder transition   119-122, 124-126
Oxygen measurement   214-215

PEO, addition of plasticizers   151
-, crosslinking   158
-, grafting   158
-, modification   151
-, related polymers   160
PEO-metal salts, phase diagram   148
Petri disk screen   222
Phase diagram   119-122, 125
- -, PEO-metal salts   148
Photodegradation   209
Plasticizing salts   155
Poly crown ether   168
Polyacrylates   163
Polydichlorophosphazene   169
Polydiene backbones, metallation   16, 29-30
Polydienes, microstructure   30, 87, 94-96
Polyethylene oxide   147
Polyitaconates   163
Polymer brushes   121, 122
Polymer electrolytes   163
Polymer gel electrolytes   162
Polymer-in-salt electrolytes   163
Polymethoxyethoxyethyl methacrylate (poly MEEMA)   165
Polyphosphazenes   168
-, etheroxy   179
-, ionic   183
-, mixed substituent   179
-, surfactant linked   176
Polysiloxanes   183
Post polymerization reactions   64-70
Pressure sensitive adhesives   128-129
Propellanes   103
PS, chloromethylation   20, 103
Pseudo-living polymerization   34
PVC, chloromethylation   20, 21

Radius of gyration   111-116
Raman spectroscopy   193
Randomly oriented wormlike micelles   126
Refractive index   111-116

Subject Index

Renormalization group theory   111, 113
Resistance of plastic   222
Respiration   214
Rheology   115, 125-128
Rigid rod star block copolymers   115
Ring-opening metathesis polymerization (ROMP)   13
ROMP   13

Scattering functions   120
Second virial coefficients   118
Segmental dynamics   126
Segregation   113-117
- limit   121
Sewage sludge   220
Solid media   222
Spinodal   119, 122, 124
Star block copolymers, ionic   6
- - -, amphiphilic   8, 11, 115
- - -, flexible-rigid rod   6
- - -, inverse   7
- - -, rigid rod   115
Static structure factor   111, 119-120
Stoichiometric titration   80-82
Super-H copolymers   93, 95, 96, 117
Surface properties   115-118

Thermodynamic cross interactions   111, 113
Titration, stoichiometric   80-82
Toxicity   216-217

Umbrella copolymers   93, 96, 97
Umbrella star copolymers   96, 97
UV irradiation   34-35

Virial coefficients, second   118
Viscoelastic properties   115, 127, 128, 129
Viscosity, intrinsic   113, 115, 116, 118
Visual assessment   222
VTF equation   146

Weight loss   222
Wittig reaction   73-74

X-ray structure   188

Ziegler-Natta polymerization   73

Springer and the environment

At Springer we firmly believe that an international science publisher has a special obligation to the environment, and our corporate policies consistently reflect this conviction.

We also expect our business partners – paper mills, printers, packaging manufacturers, etc. – to commit themselves to using materials and production processes that do not harm the environment. The paper in this book is made from low- or no-chlorine pulp and is acid free, in conformance with international standards for paper permanency.

Printing: Saladruck, Berlin
Binding: Buchbinderei Lüderitz & Bauer, Berlin